Design of Transistor Circuits, With Experiments

by

Dr. Keats A. Pullen, Jr.

Howard W. Sams & Co., Inc.

4300 WEST 62ND ST. INDIANAPOLIS, INDIANA 46268 USA

International Standard Book Number: 0-672-21626-4
Library of Congress Catalog Card Number: 79-65746

Printed in the United States of America.

Preface

The purpose of this book is to provide the reader with a self-teaching textbook to increase his or her understanding of the operation of active semiconductors. The text is experiment oriented to show how to verify the described characteristics and how to take advantage of them in useful ways. It further shows what should be strictly regulated in a circuit, and how to achieve that regulation.

The reader will find it helpful to perform at least some of the described experiments, as the principles demonstrated are fundamental even though they may be misunderstood by many workers in the field. Part of the reason for this misunderstanding is that measuring equipment with special sensitivity characteristics is required. This is to assure that the measurements being made do not distort the operation of the device. (Current measurements that result in as little as a 10 mV change in the base-to-emitter voltage can change the effective transconductance of a bipolar transistor by nearly 30%.) This writer knows on only one digital meter that is capable of approaching this kind of sensitivity, and knows of no analog meter that is capable of comparable sensitivity.

An attempt has been made to grade the material on a chapter-by-chapter basis so that as the reader gains an understanding of the operating principles and procedures, he will be able to select the component values in those parts of the circuits where the designs parallel the designs already studied and analyzed. It is particularly fortunate that this approach is possible, as it means that an ordered series of steps will usually lead to an effective design. These steps are typically:

1. Select a trial device.
2. Select a device output current.
3. Determine the resulting approximate device transconductance.
4. Select a value of load resistance (R_L) suitable for use, keeping in mind the importance of limiting voltage gain.
5. Select an output supply voltage as low as will give the full required range of operation.
6. Remember that the amount of voltage change in the input circuit required to show significant nonlinearity is given by $0.026/\kappa$ volts, where kappa is the transconductance efficiency.
7. Set the bias voltage or current required to provide the selected output current, and adjust as needed.
8. Be sure that interstage loading does not degrade operation— if it does, lower the output load resistance.

Based on these rules and the way that the chapters are organized, the reader should find that he or she can easily assume the responsibility for making circuit-design decisions bearing on the use of devices as amplifiers. The reader is encouraged to do this based on the above rules.

The series of steps listed above can prove useful as initial steps for design or circuit adjustment with both bipolar and field-effect transistors (as well as with electron tubes). Once these steps have been mastered and accomplished, distortion can be determined and reduced by emitter or source degeneration as applicable, if such is required.

Since this book is intended for the use of individuals who have at least a modest basic knowledge of elementary electricity but little in electronics per se, a variety of material that is correlary to helping develop the necessary background has had to be included. Some of this material is included in the Appendices, but at least a superficial review is included in Chapters 1 and 7. Chapter 1 indicates the basic physics involved and gives a general outline of the contents of the book, while Chapter 7 is devoted to a variety of special problems that may otherwise trouble a reader. The latter is designed to require essentially no additional background in electronics, as it treats rectifiers as the switches they almost are, and amplifiers as if they were perfectly linear devices. The more experienced reader will not need to study Chapter 7 in detail, although even he or she may find many points of interest there. On the other hand, the beginner in semiconductor circuits may well find a study of Chapter 7 first extremely helpful. The material in the Appendices is directed toward rounding out the reader's information as well as to help him or her prepare the special devices required to take the best advantage of the balance of the book. Most of our readers will want to get their hands "dirty" to some extent by at least building some of the instruments, as the experience can prove to be very helpful. (Construction of a Heathkit instrument is particularly productive, as they contain very complete instructions that will help people who have never used a soldering iron before.) The author hopes that this book will stimulate your interest in transistors and other semiconductors so that you will search further for more information on solid-state circuit designing.

KEATS A. PULLEN, JR.

Contents

for Statistical Differences–The Use of Emitter Degeneration—What Have You Learned?

CHAPTER 7

APPENDIX A

APPENDIX B

APPENDIX C

APPENDIX D

APPENDIX E

Cascode Transistor Amplifier–The Use of Darlington Compound
Circuits

APPENDIX F

CHAPTER 1

Introduction

The purpose of this book is to help hobbyist, amateurs, experimenters, scientists, and engineers whose principal areas of activity are in fields other than electronics to develop an understanding of electronic circuits. This will enable them to achieve a sense of confidence when they are working with transistors and related devices, thereby assuring desired results when working with electronic circuits. It will also provide a sound basic understanding that will make the study of other books on electronics much easier. The goal of this book is to provide simple, yet valid, explanations of the way solid-state devices work and how they should be used, and to back up these explanations with experiments which can be performed by you, the reader, so that you can verify that the statements are in fact correct. With this in mind, spaces for your calculations and comments will be found throughout the book. Tabular and graph forms will also be found. These can be used for recording and/or plotting the data you obtained while making the tests and experiments.

A brief sumary of the objectives of each of the chapters and of the appendices is given in this chapter. In this respect, the contents of this chapter and Chapter 7 should be noted particularly. Some readers may wish to study these chapters before going into the intricacies of nonlinear devices and their applications. The first few sections of this chapter are devoted to helping the reader, whose sole knowledge of electricity is limited to Ohm's law, to extend that knowledge to simple networks and to the inverse form of Ohm's law. The simple networks have separate input and output terminals, or separate "ports" as they are presently called. The inverse form of Ohm's law states that the current flow in a circuit is the product of a conductance (reciprocal of resistance) and the applied voltage. This latter form of Ohm's law is of vital importance in electronics since the controlling variable (independent variable) is voltage, and the controlled variable (dependent variable) is current.

Then, many readers may choose to study Chapter 7 before reading the balance of the book in any detail. Chapter 7 considers some important measurement problems that are encountered with solid-state devices and explains simple methods for taking care of them in terms of practical circuits. Where diodes are utilized, they are treated as switches. Transistors are similarly treated as ideal amplifiers to make consideration practical for those readers helped by such an approach. (Some further constructional details for the special circuits used are included in Appendix B.)

The rest of the chapters of the book go from comparing the differences of resistors, both linear and highly nonlinear, through the various kinds of transistors (based on their nonlinear resistance properties). Some additional complex combinations of nonlinear resistance elements complete the chapters dealing with semiconductor devices. These nonlinear resistance elements generate some particularly interesting features for the student who is interested in computers and the control systems based on them.

You will find concepts explained herein that are rarely encountered in standard textbooks. Such concepts are supported by experiments that show, in fact, that the concepts are valid. It will also be found that these concepts prove important in the practical application of active devices. One of the important consequences of the approach used herein is that it leads to a unified explanation for the operation of solid-state devices. This can result in an improved understanding and can lead to a greater ease in the handling of the devices and their circuits.

In this chapter, you will find a very brief review of the basic fundamentals that are needed in the balance of the book. This review is followed by an outline of the rest of the chapters and the appendices. In this chapter, it has to be assumed that you understand, at least in a general way, most of the basics. Thus, the explanation will be brief while, at the same time, clarifying some points that are seldom explained very well. (Some of the points are usually totally ignored.) For example, a basic acquaintance with Ohm's law must be assumed, but its reciprocal form, where the current is given in terms of conductance and voltage, can require some clarification.

TWO-PORT EQUATIONS

The expansion of Ohm's law to a circuit having two voltages and two currents, and the converse conductance form, are both very important. They take the forms (see Fig. 1-1):

$$E_1 = Z_{11}I_1 + Z_{12}I_2$$
$$E_2 = Z_{21}I_1 + Z_{22}I_2 \qquad \text{(Eq. 1-1)}$$

$$I_1 = Y_{11}E_1 + Y_{12}E_2$$
$$I_2 = Y_{21}E_1 + Y_{22}E_2 \qquad \text{(Eq. 1-2)}$$

Where Equations 1 and 2 represent circuits having only resistances, conductances, inductances, capacitances, or other nonactive devices, $Z_{12} = Z_{21}$ and $Y_{12} = Y_{21}$. Where active devices are included, these terms need not be equal. These equations are said to represent ideal "two-port" networks, or networks having separate input and output terminals. (Typically, they have at least three terminals.) The solu-

Fig. 1-1. A 2-Port network representation.

tion of these equations for circuits that include transistors and other admittance-type devices is the goal of the following chapters, but mathematics will be avoided to the greatest extent possible. The symbol R will often be used for Z where reactance is comparatively unimportant, and the symbol G will be used for Y where susceptance is unimportant. When talking of both impedance and admittance concepts without wishing to distinguish between them, the term *immittance* will be used.

LAWS AND THEOREMS

Kirchhoff's Laws

By using the 2-port network configuration, it is possible to avoid using Kirchhoff's laws. This enables us to solve our problems with the help of Ohm's law in one of its two forms, and it also enables us to avoid the use of Thevenin's and Norton's theorems. These theorems can be very useful if properly applied, but they are too easy to apply incorrectly. The approach we shall take can be referred to as being based on the *Indefinite Admittance Matrix,* which is a specialized form of Kirchhoff's laws. Actually, what is needed most is the basic concept, and nearly all of the algebraic manipulations can be avoided. Those interested in further details about the mathematics may find other books by the author useful. They can also refer to any standard text on network theory.[1,2]

Kirchhoff's Voltage Law

Where circuit elements are connected in series (in a ring form), and there are no other points of entry along the ring, the current in each element will be the same as in every other element. When this is true, the total voltage across the group will be the product of the common current multiplied by the sum of the resistance or impedances. Where there are taps between the elements and the currents need not be the same, the algebraic sum of the individual voltages will give the total voltage. This relationship is known as Kirchhoff's voltage law. It has a special form for ac circuits known as the *phasor* form. Phasors will be explained later.

Kirchhoff's Current Law and Small Differences

Kirchhoff's current law states that the algebraic sum of the currents flowing into a point of interconnection, called a junction, must be identically zero. A transistor represents such a junction. It is a junction in which most of the current flowing in one lead goes out the other, with just a small amount of current flowing in the third lead. Kirchhoff's current law can be written in equation form as:

$$I_1 + I_2 + I_3 + \ldots = \Sigma I_j = 0 \qquad \text{(Eq. 1-3)}$$

in accordance with the above statement of the law.

The application of this law and the fact that one current is much smaller than the other two has an important bearing on the following discussions presented in this book, as transistor beta, or current gain, is a function of this difference. An attempt will be made to avoid as much as possible the use of parameters whose values are based on small differences of large numbers. Failure to do this can cause complications. Fortunately, this is relatively easily avoided with transistors.

Thevenin's Theorem

This theorem states that an equivalent voltage source that has an internal impedance in series with it can be used to represent one network as a source of energy for some other network. In this form, Thevenin's theorem is much more restrictive than is normally realized, as one often obtains physically nonrealizable series network elements

1. Pullen, K. A., *Theory and Application of Topological and Matrix Methods,* (Rochelle Park: Hayden Book Co., Inc., 1962).
2. Van Valkenburg, M. E., *Network Analysis,* (Englewood Cliffs: Prentice-Hall, Inc., 1955).

for use with the voltage source. In addition, the voltage source itself may be physically unrealizable as well. (In a physically realizable network, one will find that the values of the resistive and reactive elements are not found mathematically to be functions of the applied frequency.) Even a configuration as simple as the base input circuit of a transistor, which in its simplest form consists of a resistance, a conductance, and a capacitance, is unrealizable in this sense. For this reason, you are not asked to become better acquainted with it or to try to use it.

Norton's Theorem

This theorem is really the converse of Thevenin's theorem, in that it states that a current source in parallel with an admittance can be used to represent a network in relation to another network. This theorem is also much more restrictive than is normally realized, since it again is common to encounter equivalent networks which are physically nonrealizable because of elements whose values are dependent on frequency. Both representations are mathematical fictions.

Stockman's Theorem

There is a generalized theorem based on the previous two theorems which can at least partially resolve the equivalency problem. Your attention is called to its existence simply so that you will know about it. It overcomes some of the more severe objections to the use of Thevenin's and Norton's theorems.

PHASORS

You will find that the term *phasor* is used quite often when talking about circuits involving alternating currents and using inductance-resistance and/or capacitance-resistance combinations. A phasor becomes involved when a circuit has an element that is capable of dissipating energy and also has an element that is capable of storing energy and returning it to the circuit. (The same element may be capable of doing both.) This leads to the voltage and the current reaching their respective maximums at a different time. Thus, a phase difference exists between the voltage and current. This difference leads to the use of a phasor representation and a phasor diagram.

It should be remembered that where resistances in series are added with resistances in parallel, the conductances should be added. This leads to the formula:

$$\frac{1}{R_t} = \frac{1}{R_1} + \frac{1}{R_2} + \frac{1}{R_3} + \ldots = \Sigma \left(\frac{1}{R_j}\right) \qquad \text{(Eq. 1-4)}$$

Proper application of Kirchhoff's laws will automatically lead to this result. Just as the series resistance equation can be extended to apply to series impedances, the parallel resistance equation can be extended to apply to parallel impedances. This results in the total admittance being the phasor sum of the parallel admittances, where an admittance is a phasor sum of a conductance and a susceptance.

PRECISION WITH SMALL DIFFERENCES

With a *good* transistor, it is essential that the current in the base lead be very small, with the result that most of the current flow is between the emitter and the collector. If the precision with which I_c and I_e must be measured is "y" percent (to get "x"-percent precision with I_b), it is necessary that the precision for measuring "y" satisfy the equation:

$$y = \frac{x}{(1 + 2\beta)} \qquad \text{(Eq. 1-5)}$$

where,

β (beta) is the current gain of the transistor.

Therefore, to determine the base current (with 10% precision) for a transistor having a current gain of about 100, it is necessary to measure both the collector and emitter currents with a precision of about 10/201, or 0.0498%. Needless to say, that is just a bit impractical.

Unfortunately, beta is extremely dependent on the value of I_b, since it is a derivative with respect to it. Since the minority carrier lifetime in the transistor largely controls the value of I_b, and it cannot be a precisely defined value, the ranges of betas given on transistor data sheets are as high as four to one. In short, beta is a parameter belonging to the class of small differences of large numbers. This should be kept in mind whenever transistor beta is used for selection or design purposes.

SUMMARY OF THIS BOOK

Chapter 2 is directed toward developing the basic ideas of the differences between the linear resistances which are used extensively in electronics and the highly nonlinear resistances that are the bases for solid-state devices. As will be shown later, these latter devices have an exponential relation between voltage and current. This relationship takes the general form[3]:

$$I = I_s \exp \left[(q/kT) \, V \right] \qquad \text{(Eq. 1-6)}$$
$$= I_s \exp (\Lambda V)$$

---

3. Exp (x) means raising 2.71828 to the "x" power.

where,
   V is the independent voltage variable,
   I is the dependent current variable.[4]

This kind of a relationship calls for the use of the conductance form of Ohm's law:

$$I = GV \qquad \text{(Eq. 1-7)}$$

where the conductance G (in simple circuits) is the reciprocal of the resistance R (used in the standard form of Ohm's law). The validity of the above equations is the problem that must be resolved in Chapter 2. For this reason, experiments with various kinds of diodes and rectifiers are of vital importance in preparing you for experiments with transistor devices. The relationships discovered are important when considering field-effect transistors also, although some modification is required there.

Then, Chapter 3 is the first of two chapters that examine the characteristics of bipolar transistors and how to apply them. Here, interest is limited primarily to the properties of one kind of device, the npn silicon transistor. Some basic techniques are developed for fitting the devices into circuits which will utilize them effectively. You will be concerned primarily with showing that the input conductance and the forward conductance (or transconductance) of your devices inherently obey Equation 1-6, above. Current gain has little bearing on this equation beyond the fact that it gives an indication of the relative magnitudes of I_b and I_c. Because of the fact that the ideal transistor obeys Equation 1-6, any shunt conductance or capacitance can be combined directly into the appropriate transistor admittance as an additive term, making the relation appear to be somewhat nonlinear. (See the additive terms of this type in Appendix A—I_{bo} and I_{co}.) For this reason, the next important elements in changing the properties of the ideal transistor have to be series impedances. These appear as the various spreading resistances.

The purposes of Chapter 4 are two-fold—to establish the relationships between npn and pnp transistors, both germanium and silicon, and to develop in considerably more detail, of the characteristics of these devices as they interact with typical circuits. You are asked to assemble many of these circuits on your solderless breadboard and to measure their properties so that you can verify that what you have been told is correct. You will find that this chapter gives explanations,

---

4. The significant feature of an exponential relation is that the value of one variable can be repeatedly doubled or halved by a fixed change in the other variable. The device current can be doubled or halved by about 0.018 volt change.

in simple terms, of many of the phenomena that are encountered with semiconductor devices. Of particular importance is the effect of voltage gain on the properties of a circuit, and why it is much more significant in the determination of the properties of a circuit than beta is. The importance that a limitation on voltage gain can have on the reduction of wasted power and on improved reliability is brought out. The relationship of transistors to transformer-coupled and tuned circuits are explained.

In Chapter 5 the discussion is extended to field-effect transistors. These devices have significant similarities to bipolar devices, and significant differences as well. They are *majority* carrier devices, not *minority* carrier devices. Also, they do not draw current on the gate lead in the same sense that bipolar devices do on their base leads. There may be some charging current and leakage current with some of these devices, however. This chapter develops a picture of the similarities and the differences of FETs with respect to bipolar transistors, and then shows how the principles developed for bipolar devices can be carried over to field-effect transistor application. This is made possible primarily by the acceptance of the fact that both devices are transconductance controlled. The concept of transconductance efficiency is of importance with these devices, and so is developed in great detail.

There are a greater variety of FET devices than there are bipolar devices. In addition to diode types (which can have either P-type channels or N-type channels), there are also insulated-gate devices. There are enhancement-mode devices (which require a forward bias to make them function) and there are depletion-mode devices (which must be reversed biased to stop them from working). In addition, there are lateral FET devices and vertical FET devices. There are several ways of achieving the latter device.

Setting up a circuit configuration for FET devices is based on two different sets of conditions—whether the device selected has a P-type channel or an N-type channel, and whether it operates in the enhancement mode or the depletion mode. It is important to note what the relative input admittance level must be so that a selection can be made between diode FETs and the various classes of IGFETs. These are explained in this chapter, and a variety of circuits are tested to show what the properties of the various devices are in typical circuits.

The discussion in Chapter 6 is devoted to the evaluation of the properties of some of the special devices that can be made (based on transistor-like structures). These are mostly switching-type devices, although some special limiting devices are also considered. The discussion is directed primarily at the characteristics of these devices, with only a limited discussion on the principles of operation. Again, a series of experiments is included to help develop an understanding of the behavior of the devices in a circuit.

Finally, Chapter 7 is directed to the variety of special measurement problems that are encountered when making an evaluation of the various devices studied. You may find it useful to first read all of the chapters through without doing any of the experiments and, then, prepare yourself for making tests and measurements. On the other hand, if you wish to start by reading Chapter 7, it has been written so it can largely stand alone.

Measurement problems that are encountered with bipolar transistors, in particular, are really quite unique, and they would be hard to handle effectively without many of the instruments discussed in Chapter 7. For example, a resistance in the emitter circuit of a bipolar transistor, which develops only 18 mV across it, can reduce the effective transconductance of a transistor to half its original value. This means that you need to be able to measure, with reasonable precision, voltage changes that are as small as a few millivolts. In fact, an 18 mV change in the base voltage of a bipolar transistor can double or halve the current through the device. In addition, because of the significant amount of current typically drawn by the base of a transistor and by its variability, the operation of a transformer-coupled amplifier can be affected by the polarity that is selected for the load connected to the transformer secondary. These unique characteristics also place some special restrictions on the use of a cathode-ray oscilloscope. Thus, some discussion on this topic necessary.

Chapter 7 discusses the characteristics required for many of the special instruments that are useful in experimental work. It also includes some guidance on how to construct some of the instruments. Where existing instruments or kits can fulfill your purposes, they are noted. Probably the most serious problems occur in economically providing the right kinds of power supplies and proper meters. In both of these areas, sample units have been built and tested to verify their utility.

THE APPENDICES

The various appendices provide useful information that ranges from a simplified derivation of some of the Ebers-Moll equations for the bipolar transistor through instruments and other components that you will find useful. Some of this material will be redundant, but it may be handy to have it in a specific location for reference purposes.

In Appendix A, a derivation of equations showing the importance of the transconductance relation is developed from a simplified form of the Ebers-Moll equations. From this, the equations for voltage gain, output supply voltage, and similar relations are generated. These basic equations can be developed from the indefinite admittance matrix equations for a three-terminal active device as well. It is clear from the results that both the input current and the output current for solid-state active devices are dependent on the input and output signal voltages, with the prime dependence being on the input signal voltage. With a good transistor, it is almost independent of the output supply voltage.

Appendix B presents some additional information on the units discussed in Chapter 7; in this case, concentrating on circuits and circuit-board layouts and construction tips. You may wish to build some of these in permanent form, and others, you can set up on your solderless breadboard. These units have been built and tested.

Appendix C discusses some of the instruments you will most likely want to buy, possibly in kit form, rather than to build from scratch. In some instances, you will have several options. In addition, an extensive list of components that you may possibly need is included. You can get them a few at a time or all at once.

Appendix D lists a few possible sources for components that can supplement the electronics jobbers who may be in your neighborhood. It includes both mail-order concerns and chain stores. Other sources are listed in radio hobby magazines, amateur radio magazines, computer magazines, and a variety of similar publications.

Appendix E includes a few additional interesting experiments which you may wish to try after you have worked most of the ones given in the book. The author can assure you that all of the circuits are useful, as he has used them all at one time or another.

Appendix F discusses further the characterization of electron devices. It is unfortunate that much of the application material given in this book has not been available in either technical literature or in hobbyist magazines over the years, as it is a definite help.

Linear and Nonlinear Elements

The purpose of this chapter is to develop an understanding of the underlying properties that are vital to presently known kinds of active devices, which are inherently nonlinear in nature. You will find it helpful to examine the properties of devices that are relatively linear, and those that are not linear. You will also want to find out what properties are stable and dependable, and how you can manage nonlinearities in ways that will minimize their deleterious impact on your tasks, whatever they might be. We will assume that you understand that nonlinear elements have considerable similarity to switches, and can often be represented as switches. In this chapter, you will find out why, and you will also learn how you can represent these properties in a more precise and meaningful way.

OBJECTIVES

After you have studied this chapter and performed the experiments, you will understand the differences between linear and nonlinear resistors (particularly semiconductor devices), and some of the limitations of each. You will be able to describe and explain the following:

1. Properties of linear resistors.
2. How resistance values are established.
3. Factors which can cause resistance values to change.
4. Characteristics of nonlinear semiconductor devices.
5. A little history of nonlinear devices.

6. The importance of (q/kT) for nonlinear and semiconductor devices.
7. The establishment of optimum diode loads.
8. The nature of piecewise linearization.
9. The nature of silicon-controlled rectifiers.
10. The nature of negative immittance devices.
11. Typical rectifier circuits.

You will perform many experiments designed to demonstrate the significance of these points.

DEFINITIONS

You will need to understand the following terms in your study of this chapter:

linear—A relation is linear when a response to a force of some kind is directly proportional to the initiating force.

superposition theorem—A basic theorem of linear systems states that if a series of input actions are applied to the system, the total response of the system can be obtained by taking the summation of the individual responses to each of the individual actions.

linear-time-variant—A linear-time-variant system is one which at any given instant behaves in a linear fashion, but in which the behavior at different instants in time will respond differently to the same input depending on some outside influence.

nonlinear—A nonlinear element is one whose response changes dramatically and nonproportionally to the activating force. Typically, a given small change in one parameter may repeatedly double or halve the value of another parameter in a nonlinear device.

piecewise linearization—Piecewise linearization is the process of representing a nonlinear relationship in terms of a series of linear relations spaced along the nonlinear contour. Usually this is done in terms of possibly three or four values, but it is better to do it in terms of the operating characteristics at a series of distinct points. At these points, the variation of the nonlinear parameter from point to point is on the order of 10 to 20%. This enables one to handle the device as being weakly nonlinear, rather than grossly nonlinear.

immittance—A term used to include both the concept of impedance and the concept of admittance. Because of the reciprocal relation between the two concepts, such a collective term is very useful.

resistor, resistance—A resistor is a device capable of impeding the flow of electrical current, thereby generating a voltage and dissipating enegy. Resistance is the measure of its ability to impede that current flow.

capacitor, capacitance—A capacitor is a device capable of storing electrical energy in a static, or potential energy, form. Capacitance is the measure of its ability to store electrical energy.

inductor, inductance—An inductor is a device capable of storing kinetic energy in the form of a magnetic field. Inductance is the measure of its ability to store energy in this form.

cathode-ray oscilloscope—An oscilloscope (or scope) is a device which can present a visual representation of the behavior of an electrical signal, either in terms of another signal or in terms of time. It is one of the most useful tools available to persons working with electronics and electronic computers.

power supply—The source of electrical energy needed to operate a circuit. A power supply may consist of batteries, or it may consist of a group of components that are capable of changing commercial electrical power into a form that can be used by a circuit.

alternating current—A current which periodically changes polarity. Abbreviated ac.

transformer—A device for use with alternating current that makes the change of magnitude of voltage level possible. It consists of two or more windings coupled by a magnetic core. There often is no direct electrical connection between the windings.

variable transformer—Sometimes called a Variac®. A transformer which has only one winding and a tap, whose position can be varied across the winding, from one end to the other. It can provide a continuously variable alternating voltage from, roughly, zero to the maximum supplied by the power line.

center tap—An additional wire installed at the center of a transformer winding to permit the use of either half of the winding, in addition to the full winding, as a voltage source.

signal generator—A device capable of producing some specialized kind of electrical signal required for test purposes with electronic equipment. Most signal generators generate sine-wave signals, square-wave signals, sawtooth signals, or some combination of these. The amplitude or the frequency of the signal may be varied.

oscillator—Usually, an oscillator is a simplified form of a signal generator. It, usually, is somewhat less precise in its operating characteristics and it may not have as great a flexibility.

passive device—A device which is incapable of processing energy in any way which will, in effect, increase the power level in a signal.

active device—A device which can accept a control signal and increase its amplitude and total available power.

diode—A diode is a device that conducts electricity more strongly in one direction than in the other (a nonlinear relation). Typical diodes have an inherent characteristic which is exponential in nature, in that the output current is an exponential function of the applied voltage:

$$I = I_s \exp (qV/kT) \qquad \text{(Eq. 2-1)}$$

where,

I_s is a reference current called the saturation current,
exp indicates that 2.71828 is to be raised to the power indicated in the parentheses,
q is the charge on the electron,
V is the applied voltage,
k is Boltzman's constant,
T is the absolute temperature.

rectifier—A rectifier is a diode designed to handle power rather than performing small-signal processing. It obeys the same equation as an ordinary diode, but it has limitations for applications in signal processing. It can usually withstand a higher voltage than a diode can.

switching diode—This is a diode designed to switch rapidly, yet have an extremely small leakage current in the reverse direction. It also has a very small capacitance and a large forward conductance, giving it a very good high-frequency response.

Schottky diode—These diodes are similar to switching diodes. They require only an abnormally small bias voltage to achieve conduction. They also have an extremely small charge-storage effect.

snap diode—This is a kind of diode which has an abnormally rapid discharge of stored charge. It can be used effectively as a frequency multiplier.

trigger diode—This is a diode which breaks down or conducts at a reduced voltage after the application of a somewhat higher voltage. It is typically used to trigger silicon-controlled rectifiers.

point-contact diode—A point-contact diode is the modern version of the *galena crystal detector*, which was one of the first radio-signal detectors. It consists of a very fine, sharp wire point placed in contact with a small piece of semiconductor (typically germanium, silicon, gallium arsenide, galena, or other semiconductor material). These devices typically are used for the detection of radio frequencies up to the microwave and millimeter-wave regions of the radio spectrum.

zener diode—This is a diode which conducts normally in one direction, and does not conduct below some design voltage in the reverse direction. At and above the design reverse voltage, however, this diode conducts well enough that it acts as a voltage stabilizer at that level. It is normally used with a series resistance to limit the current flow.

half-wave rectifier—A half-wave rectifier is used in some power supplies. It consists of a single rectifier diode, and it only passes current during one half cycle of the applied ac signal waveform (Fig. 2-1). The lowest frequency component of the current passed by it is equal to that of the applied power.

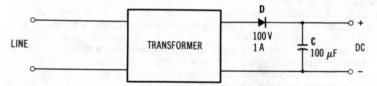

Fig. 2-1. A half-wave rectifier circuit.

full-wave rectifier—A full-wave rectifier is normally used with a center-tapped transformer, and consists of two half-wave rectifiers (Fig. 2-2). One of these rectifiers connects to one secondary output on the transformer, and the other rectifier connects to the other secondary. They are so polarized that conduction occurs first through one rectifier and, then, the other. At the common point between the rectifiers, the voltage always has the same polarity with respect to the center-tap. The lowest frequency component of the current out of it is equal to twice that of the applied power.

bridge rectifier—A bridge rectifier is a configuration of four rectifiers so con-

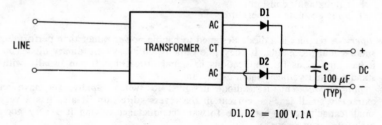

Fig. 2-2. A full-wave rectifier circuit.

nected that it can pass current from the full secondary winding on each half cycle. If the transformer has a center-tap, both positive and negative voltages may be obtained at the same time. If not, the output is taken across the remaining corners of the bridge. A connection diagram for a bridge rectifier is shown in Fig. 2-3.

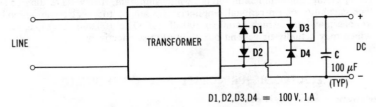

D1, D2, D3, D4 = 100 V, 1 A

Fig. 2-3. A bridge rectifier circuit.

positive rail—This is a term referring to the most positive voltage applied to an electronic computing system.

negative rail—This is a term referring to the most negative voltage applied to an electronic computing system. It may be the ground line.

inverse peak voltage, peak reverse voltage—Abbreviated PIV and PRV, respectively. This is the maximum voltage in the reverse direction that can safely be applied to a diode or a rectifier. Application of a higher voltage can be expected to cause a diode failure.

minority carrier storage—When a diode, rectifier, or transistor is conducting, there is a substantial amount of charge transported across the junction. When the voltage applied reverses, much of this charge is left on the wrong side of the junction. This is stored minority carrier charge. It must be drained off before the diode stops conducting.

temperature coefficient—This is a measure of the rate at which some parameter varies as the temperature of a device is varied. Most electrical and electronic devices, in addition to many mechanical devices, display nonzero temperature coefficients.

common-mode voltage—When it is necessary to measure the voltage between two points on a voltage-divider network, as between points A and B in Fig. 2-4, it is necessary to ignore the effect of the voltage from the ground or reference point to the points of measurement. This voltage that we have to ignore is the "common-mode voltage," and it is, in effect, measured to the

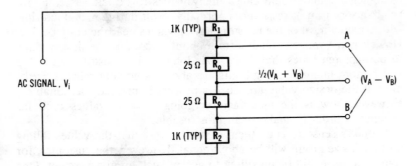

Fig. 2-4. Rejection of common-mode voltage.

midpoint between points A and B. What is required is the amplification of the voltage difference between points A and B without, at the same time, including the voltage from ground to the midpoint. The voltage to the midpoint is the common-mode voltage.

common-mode rejection ratio—The common-mode rejection ratio is the ability to reject the common-mode voltage while amplifying the voltage difference. The capability of operational amplifiers (op amps) to reject common-mode voltage may run as high as 60 to 100 db (1000 to 100,000 times), a value which may actually be beyond our ability to use effectively.

ELECTRONIC PRINCIPLES AND COMPONENTS

Resistance and Resistors

It is assumed that you, the reader, know enough about electricity and electronics to understand Ohm's law and its application to simple dc and ac circuits. It is also assumed that you know a good deal about what resistors, inductors, and capacitors are and how they function in general. It is also assumed that you have at least a superficial understanding of what a transformer is and what it does. (We are not particularly concerned about how one is made, however.) Therefore, the discussion on these topics will be limited to some special ideas which should prove useful.

The way values are selected for resistors and capacitors initially strikes one as quaint, to say the least. Fortunately, however, there is very sound logic behind the way these values are selected. It is regrettable that the basic philosophy behind this selection method is not more generally used in electronics. If one chooses resistance values based on numerical or integral steps, one quickly notes that the first step, from one to two ohms, is a much larger proportional step than that from nine to ten ohms. It is not economical, when resistors are manufactured, to make all the resistors in a single batch so carefully and so precisely that they will have the same value within 10% of a chosen value. Their selling prices would be exorbitant and customer demand would cause many values to be left as scrap. Instead, a spectrum of resistances normally result during manufacturing, with possibly most of the resistors having values near the design value. However, the range spread is from about a tenth of the design value to perhaps ten times that value. (As long as the resistors at the top and bottom manufacturing limits are about as equally reliable as the ones at the design value are, such a process is perfectly reasonable.) However, how is the manufacturer going to code values onto his product in order to make them most usable?

It makes sense to code them in such a way that the values falling into each size group will be about equal. However, the customer, for some reason, might be unwilling to accept such an arrangement. One effective way that coding can be achieved is to divide each design range into a number of equal-ratio steps. This will make it possible to take the lot of resistors and group the ones within ±10% of the design value together, with the remaining groups scaled proportionally. This, in fact, is just what is done. Based on a 100-ohm design center, the nominal values and their corresponding ranges are:

Value (ohms)	Range (ohms)	Value (ohms)	Range (ohms)
10	8.2-12	100	82-120
15	12-18	150	120-180
22	18-27	220	180-270
33	27-39	330	270-390
47	39-56	470	390-560
68	56-82	680	560-820

Such an arrangement serves as an excellent means of supplying the user with convenient resistance values and, at the same time, it enables the manufacturer to make the most of his production. If the user is satisfied with 20% resistors, he can choose values from 10, 22, 47, or multiples thereof. (That usually is all he needs.) With a 5% tolerance, there is a similar series of resistors, only with additional values, etc.

There are other features that are of importance to us with respect to resistors. The most common resistors on the market are sometimes called "mud" resistors. Others include carbon-film resistors, metal-film resistors, and wire-wound resistors. Needless to say, the price placed on these is dependent to a large extent on the precision with which they are made. The so-called "mud" resistors are likely to contain a pellet of a material that consists of carbon mixed with some kind of an insulating clay. Whereas the resistance of a piece of wire usually increases with increasing current and the concurrent heating, these typical resistors may very well have a decreasing resistance with an increasing current. As a user, you should find out about some of these properties, as they could be important in your circuits. Resistors also may have a voltage coefficient of resistance, in that the resistance may change as a function of the applied voltage independently of its temperature coefficient.

Inductances and Inductors

The principal way that inductors are used in digital electronics is as elements of filters or tuned circuits and as components of transformers. Each of these applications will be described briefly here, but the transformer application is the principal use that requires consideration in our experiments. One of the experiments will demonstrate the resonant effects in an inductor, and several will use tuned circuits, in addition, to several experiments involving transformers.

An inductor is formed from a group of several turns of wire placed close together in such a way that the magnetic field resulting from a current flow will thread through all of the turns in a more-or-less circular fashion. This threading action causes the field from each turn to affect that of the other turns. Any change of the field, that occurs due to a change in the current, attempts to prevent a further change from occurring through an induced voltage that is opposite and opposing to the change. When the current is increasing, the voltage opposes the applied voltage, and when the current is decreasing, it aids the applied voltage. Because of this property, an inductor is able to store energy in the magnetic field. The energy is again withdrawn as the current decreases.

Capacitances and Capacitors

A capacitor is a device in which it is possible to bind an electric charge through the force of attraction of unlike charges. A capacitor consists of plates or layers of a conductor material separated by very thin layers of an insulating material. Capacitors are vital elements in digital electronics because they can store the charge needed to supply very short pulses of current.

Many of the capacitors used with digital electronics are required for the control of these switching pulses. They keep the pulses from transferring from one circuit to another, and they keep them from getting onto power distribution lines. For example, the digital logic units called *transistor-transistor logic* (*TTL*) draw very short but heavy current pulses when they switch from one voltage state to another. It is important that a capacitor be able to store enough of a charge that the supply voltage will not decrease below a safe operating range when such a current pulse occurs. With TTL circuits, the voltage should not be allowed to dip more than 0.2 volt, in a typical situation. Since a current pulse lasting only 0.1 microsecond is entirely possible when switching states in a TTL unit, it becomes necessary to provide a capacitance in accord with the formula:

$$\Delta Q = C\Delta E = i\Delta t \qquad \text{(Eq. 2-2)}$$

where,
   ΔQ is the charge which must be stored,
   C is the capacitance,
   ΔE is the allowed voltage change,
   i is the peak current,
   Δt is its duration.

Typically, a capacitance of at least 0.1 μF is connected across each small group of TTL units for exactly this reason.

Tuned Circuits

A tuned circuit, typically consisting of an inductance and a capacitance, generates what is sometimes called a *coherent oscillation*. It is capable, when properly excited, of generating a sinusoidal signal following a well-defined pattern. The coherence of the signal is a consequence of the exchange of energy, potential to kinetic and back, twice each cycle. The quality of this coherence may be measured by the ratio of the energy stored and exchanged to the energy dissipated per cycle or period. The balance wheel of a watch or the pendulum of a clock is a good analogue of a tuned circuit, with the spring or the lifting of the pendulum mass corresponding to the capacitance, and the mass-momentum corresponding to the inductance. If all other properties are adequately controlled, then the timekeeping of the pendulum is dependent on these two properties alone. The coil in a tuned circuit stores kinetic energy in its magnetic field when a current is flowing through it; the associated capacitance stores charge as an electrical stress as the current stops flowing.

Transformers

The magnetic field associated with a coil or inductor is critical to the operation of a transformer. When the current through a coil changes, it generates a "back-pressure" called "back emf." It will also generate an emf in a coil that is so situated that the varying magnetic flux also penetrates it or links with it. The result is that a voltage may be measured across the terminals of the second coil. To be strictly precise, one should say that the "back emf" is a function of the rate of change of the flux linkages that are threading through the coil. (Flux linkages can be measured in terms of ampere-turns.) Anything which will change the effective total flux linkages will produce or induce a voltage.

A transformer must be used on a power source which changes polarity rapidly. The higher the rate at which this polarity changes, or the higher the frequency of the alternating current flowing through the coil, the higher is the rate of change of the flux and, as a result, the smaller and lighter that a transformer of a given power rating can be. This is one of the principal reasons that aircraft and similar vehicles, on which weight limitation is particularly critical, generate their electrical power at 400 Hz or higher frequencies. In applications where lightness and compactness are not so critical, it is more economical to generate power at 50 or 60 Hz. On some electric railroads, power is provided at frequencies as low as 25 Hz or even 16⅔ Hz because traction motors do not operate efficiently with higher-frequency power. In addition, the higher rotating inertia and the greater total weight can both be advantageous in these kinds of applications.

The transformers you will use will largely be components of power supplies, and will transform 120 volts or 240 volts to voltages as low as 5 volts. You will see later that in some situations, it is desirable to use as low a voltage as possible. Overall efficiency at low voltage can be limited by the characteristics of the available rectifiers, and may lead to a limited return to the use of germanium rectifiers.

Transformers can be built with either single or double primaries (for single- or double-voltage ranges), and they can have secondaries that are with or without taps. The tap most commonly found on transformer secondaries is a center tap. This tap, in effect, divides the available output voltage in half. Some of your power-supply circuits will make extensive use of these taps.

The reason for a double primary on a transformer is so that it can be used on equipment for either North America, where electrical distribution is predominantly at 120 volts with respect to a common ground, or elsewhere in the world, where, by and large, power is delivered at 240 volts. Each system has its advantages and disadvantages and it is unlikely that a common standard will become available.

Transformers are usually designed to be "constant-voltage" devices. The reason for this is that it has been found more economical to distribute power on a constant-voltage basis rather than on either a matched-impedance or a constant-current basis. The principal drawback to the constant-voltage system is that, in principle, it requires a large "floating reserve" of power. Transmission systems are so extensive these days, however, that statistically the magnitude of the floating power reserve required is much smaller than it used to be.

A constant-voltage system is almost a necessity for electronic equipment. You will find in later paragraphs that most of your solid-state electronics is operated from some form of voltage regulator. This will be discussed in more detail when power sources that are helpful for your laboratory experiments are discussed.

LINEAR AND NONLINEAR RESISTANCES

It is necessary to draw a sharp line between linear and nonlinear resistances, because each kind has properties which are distinctive and of major importance to users. The linear devices are not strictly ideal, as you have already noted, since the values of resistance are typically functions of temperature and, sometimes, voltage as well, and they have both inductance and capacitance associated with them as undesirable, or parasitic, parameters. The range of variation of the resistance is rather small unless the devices are badly overloaded, however. Typically, the change in value will not exceed 10% unless they are abused. The devices which are least likely to be damaged or to have significant changes in value (with either time or modest overloads) are the metal-film and wire-wound varieties. Their resistances are most likely to be stable, and the noise generated internally is also likely to be minimal.

You will be concerned primarily with nonlinear resistances which are nonsymmetric in their properties. That is, their behavior is dependent on the terminal that is selected as the positive terminal. One way, the device will conduct; the other, it will not. There are also some symmetric nonlinear resistance materials, like thermistors, varistors, thyrite, and related materials. They are mostly used for surge suppression or special sensors; however, they can be used either for measurement or protection. They are too specialized for further consideration in this book.

Solid-State Nonlinear Devices

The most important kind of nonlinear device we wish to consider here (chosen because of their importance in active or amplifying configurations) have essentially an exponential relation between voltage and current. As has been noted in the definition of a diode, they obey, to a greater or lesser extent, Equation 2-1:

$$I = I_s \exp (qV/kT)$$

where, the exp (qV/kT) means that the number 2.71828 has been raised to the (qV/kT) power. As it must be physically, (qV/kT) is dimensionless, with the result that (q/kT) of necessity must be an inverse voltage, or a transconductance per unit current. (All further discussions will assume that you are familiar with exponentiation.) This equation is the basic equation governing the behavior of semiconductor diodes and rectifiers, and it is also extremely important when dealing with bipolar and field-effect transistors as well as electron tubes. This chapter will be concerned with its application to solid-state diodes and rectifiers, while later chapters will be concerned with its application to bipolar and field-effect transistors. It is important to note that if the applied signal is small enough, *even the most highly nonlinear device will appear linear!* With bipolar transistors and solid-state diodes, signals less than 5 mV will appear as if they are linear. With field-effect transistors and with tubes, the behavior is usually linear if the applied signal is less than a tenth of a volt. *However, each of these devices does have regions of operation where they will behave nonlinearly unless the signal voltage is less than 5 or 10 millivolts.*

Linear Time-Variant Devices

All nonlinear devices can be made to behave as linear time-variant devices. This is accomplished by using two signals—one small enough to operate the device in question in a linear manner, and the other large enough to assure that it does not operate in a linear manner. As a result, on an instant-by-instant basis, the device will look linear to the smaller of the two signals, but its value will appear to be changing continuously in response to the action of the larger signal. This phenomenon is of extreme importance in modulation and mixing (two related phenomena). Both of these phenomena are important in communications. You will explore these properties of solid-state devices later by doing some experiments.

Historical Notes on Diodes

It has already been noted that the ideal diode is the basic nonlinear element with which you will be concerned and that many other devices are closely related to diodes in their operating properties. The next few paragraphs recount the development of these and later devices.

As most readers probably know, Thomas Alva Edison was the first person to make a successful evacuated incandescent lamp. His lamp used a carbonized strand of material which would glow when a voltage was applied to it. Edison experimented with a wide range of materials before he finally succeeded. He had to remove most of the air from his bulb in order to prevent oxidation of his filament. Nonetheless, as his lamp "burned," the inside of the bulb did darken and, eventually, reduced the light output of the bulb excessively. (This was not often noted by the users, as the operators of the electric generating plant would reverse the polarity of the line voltage periodically, frequently leading to lamp failure.

This phenomenon caused Fleming to wonder just what was happening, and he introduced a small piece of metal (which he called a "plate" because of its form) into the bulb. He found that when the filament had a negative potential with respect to the plate, current would flow, but not otherwise. Thus was born the vacuum diode. Later Lee deForest placed a wire structure between the filament and the plate, and found that he could control the flow of current to the plate. And so, radio and electronics were born. Later, Dr. Julius Aceves showed how to operate these "tubes" with alternating current, and the home radio became a practical reality. A few years later, Dr. Albert Hull introduced the "screen" into the bulb, and multigrid tubes were born.

At the time that Hertz made his initial experiments with radio waves, no sensitive detectors were available, so he used coupled spark gaps to demonstrate his "action at a distance." Marconi devised a detector composed of carbon granules like Bell used in his microphones, and found that if he tapped a small tube containing these granules between two electrodes, he could get detection of signals. (We today know how wise that selection was, as carbon belongs to the same group of atoms as silicon, tin, and germanium.) The device was called a *coherer*. The granules would pack when an electrical signal was applied to them, and would change the sound coming from an earphone that was connected to the granules. It is really remarkable that Marconi ever managed to get any transmissions across the Atlantic Ocean using these crude means. Then, the development of the Alexanderson alternator made limited radio communication possible during World War One. By the early 1920s, power transmitting tubes had been developed from the deForest Audion tube, and commercial radio broadcasting was born. People used crystal sets, at first. The crystal was the equivalent of our present-day switching diode except for one thing—it required only millivolts to activate. However, a person had to hunt out a sensitive spot on the crystal with a cat whisker (a fine-pointed wire). Undoubtedly, the spot that was found was the equivalent of a semiconductor grain boundary, which will be discussed briefly in the next chapter.

Modern Solid-State Diodes

It is important in developing an understanding of diodes as we know them today that Equation 1-6 be studied in considerable detail. You will be referring back to this basic equation in all of the later chapters of this book. In starting this discussion, you should note that the parameter (q/kT), (which is also identified with the symbol Λ, capital Greek letter Lambda) has the approximate value of 39 inverse volts, or 39 mhos per ampere, at room temperature. Its reciprocal is 0.026 volt, a number which some readers may recognize as the Fermi potential. For that reason, (q/kT) will be called the Fermi parameter. The Fermi potential is often divided by the diode or the emitter current to get the diode dynamic resistance at a specified current level, or for a transistor, the effective emitter input resistance. The product of the Fermi parameter (which is an inverse voltage) by a voltage produces the dimensionless product which is required for the exponent in Equation 1-6.

A very important question which we must now answer is a simple one. How much of a change of voltage is required to cause a two to one change in current through a device obeying this equation? It does not matter whether the ratio equals two or one half, the magnitude of (qV/kT) for this change is the natural logarithm of two. Solving for V, after substituting the value of q/kT, gives 0.018 volt required to double or halve the current. With the possible exception of high-vacuum diodes, this equation is nearly universal in application.

We still need to determine the I_s parameter in this equation, as it is as important as the Fermi parameter. It determines how much bias must be applied to a diode or a transistor to get it operating at a point that will have a usable amount of current available. With silicon and germanium diodes, this also determines the maximum safe operating temperature for the device. The current, with $V = 0$, is essentially the leakage current in the device (for these materials). It turns out that with germanium, between 0.1- and 0.2-volt forward bias is required to establish proper operating conditions; you will measure this voltage in your experiments. With silicon, it is between 0.5 and 0.6 volt. Clearly, it is easy to understand why it took a while for the semiconductor properties of these materials to be recognized.

The typical semiconductor diode, at first, was a point-contact diode, having a very sharp needle point pressing against a small piece of silicon. The silicon used for this purpose had to be pure silicon, but it did not need be as well refined as is common practice with integrated circuits. These diodes were initially used as radar detectors, and were very effective when properly biased. Their development led to the development of germanium diodes. Germanium diodes were very popular just after World War Two and are still used today for some applications. These were followed by the development of junction diodes and, then, a wide variety of specialty diodes. Eventually, people learned how to control a stored minority charge, either to immobilize it fast for use with switching diodes, or to cause it to dump in a controlled manner which led to the snap diode. The zener diode and the trigger diode both use a form of avalanching to lead to some kind of breakdown; in the one case, with a precisely controlled voltage level, and in the other case, with a sharp voltage drop. With the latter, the momentary reduction of the voltage causes a recovery of the nonconducting mode of operation until the threshold is again exceeded. The operating properties of diodes can also be varied by both the kinds of dopants used and their placement with respect to the junction, and the kind of variation of the doping level with respect to the junction. (A junction, as explained in Chapter 3, is the boundary between two layers of semiconductor which react with opposite electrical polarities. This boundary is obtained by introducing different impurities on the two sides of the boundary.)

The main reason that silicon is used in ordinary diodes is the significantly higher temperature at which they can be operated with a low failure probability. Carrier mobility (the charge carriers) in silicon is much lower than in germanium. Other materials are also used for diodes, such as gallium arsenide, gallium arsenide-phosphide, and other special alloys. It has also been found possible to make diodes of different materials if the crystal structures are compatible so that a single-crystal structure can be maintained across the boundary.

It is usually found that in a diode it is best to have only one of the layers, either the positive or the negative side but not both, doped rather heavily. The other side may be prepared in a way that will trap minority carriers when the polarity of the field is reversed, thereby, enhancing the available switching speed. This can also minimize the heating in the diode.

Optimum Diode Load

The forward and the back resistance in a signal diode, as well as the features previously mentioned, can be of considerable importance to the user. When the diode is conducting, you want its resistance to be small enough, compared to the impedance of the load, that negligible loss will occur due to its presence. At the same time, when the diode is in its nonconducting state, its resistance should be high enough that reverse current does not interfere with its normal operation. It can be shown that for a diode having a forward resistance (r_f) and a reverse resistance (r_r), the optimum load resistance can be expressed in terms of the equation:

$$R_{Lopt} = (r_f \times r_r)^{1/2} \qquad \text{(Eq. 2-3)}$$

(This equation also applies to rectifiers.) Needless to say, the optimum diode has as high a reverse resistance as possible and, at the same time, as small a forward resistance as possible.

The Diode Bessel Function Relation

The voltage applied to the diode generally takes the form of either a sine-wave signal or a complex of such signals. It is easily shown that, to get a rigorous solution of the current-voltage regulations with such a combination, it is necessary to expand in terms of modified Bessel functions. These Bessel "I" functions might be called the trigonometric Bessel functions, inasmuch as they are based on an equation of the form:

$$I = I_s \exp [U(t) + A \sin \omega t] \qquad \text{(Eq. 2-4)}$$

This equation can be expanded in terms of a Bessel function series, the resulting functions not having the characteristics encountered with the conventional Bessel $J_n(x)$ do. They are distantly related to the functions used in the solution of skin-effect problems. The Bessel expansion form of this function will not be used, but it will be "piecewise linearized" when necessary in the solution of the problems to be examined.

Piece-Wise Linearization

The process of piece-wise linearization of an input-output relation is extremely useful, particularly when the mathematical functions which represent rigorous solutions are complex, and difficult to manipulate. (The relation may not be expressible in closed form also, making the use of this technique mandatory.) The technique is rather commonly used, but when it is used, one finds that the nonlinear function is often represented by two or at most three segments. It has been found that it is best to use a series of segments, each one having a slope value differing from its neighbors by at least 10 to 20%. In this way, it is possible to get a rather clear picture of the behavior of the device, which will give results closely approaching the Bessel results (where they do apply), and give a good representation, even if the Bessel solution for some reason does not apply. This leads to an inverse kind of representation which has been reduced to computer programs that are ideal for use with highly nonlinear relations. Where the relations are more linear, either the usual Fourier or the Legendre/Chebyshev techniques can be counted on to provide the required information.

The fact that is of paramount importance with the diode equation is that a doubling or a halving of the current flow in any one of these solid-state junctions may be achieved by a junction voltage change which may be as small as 0.018 volt. It is this unique property which draws a clear line of demarcation between linear and nonlinear elements. There are other kinds of nonlinear devices, like magnetic cores, which can lead to distorted waveforms as you will see in the tests on transformers, but these devices are weakly nonlinear, not strongly, as is characteristic of solid-state diodes and active devices.

It is possible to further refine an understanding of these devices by noting that there are conditions under which a multiplicative factor must be included with the exponent (q/kT), changing it to the form (q/nkT), where the value of n is typically a ratio of small integers having a value between 0.5 and 2.0. Even in these instances, the high order of nonlinearity still exists. This matter will not be investigated further as an approach to the use of the devices will only require you to be aware of it.

DIODES AND RECTIFIERS

As you have seen, the varieties of diodes available at this time are so extensive that all that can be done is to touch on some of the more important kinds and to say a few words to at least put you in a position that you have an idea what they are and what they can do. You must remember that, in the final analysis, "getting your hands dirty" by testing devices is the most effective way to get a clear feel of how to use them.

Most of the diodes you will encounter are well behaved. That is, when you put a forward voltage on them, they obediently draw current, yet when you reverse the voltage, they do not. There are a number of different kinds of diodes which do not behave in this way, however, and some of them are rather important. Some of these diodes conduct and hold their voltage level when a reverse voltage in excess of some arbitrary value is applied to them. Some of them switch to a low-voltage, high-current condition. Some can be switched off, and some cannot. Some will oscillate under specified conditions. The list of special properties is quite extensive. Since the basic properties of ordinary diodes have already been reviewed, this section will discuss some of the more unusual properties of both ordinary devices and special devices.

Triggerable Rectifiers

The silicon-controlled rectifier (SCR) is the best example of this kind of device. They are strange beasts in that they will not start to conduct unless you tell them to and, then, you cannot turn them off without removing the voltage from them. Your circuit has to be designed so that it will turn the SCR off by the applied voltage either going to zero or reversing or by getting that same effect with the help of an inductive kick (a high-voltage reverse pulse that is generated by stopping the flow of current in a coil). Since this is an important but not-too-well understood phenomenon and since it is simple to demonstrate, you will find a circuit for demonstrating it in Appendix E. These SCRs are used extensively in lighting and for use with power tools. The light dimmer you may have on one of your lights is likely to include a form of one of these as its operating element. They are used in power supplies for electronic systems, as they offer a means of adjusting the voltage output from the power supply without wasting power in resistance. Their main problem, in that application, is the amount of radio-frequency noise they tend to generate.

The typical silicon-controlled rectifier has three terminals—a cathode, an anode, and a gate. The circuit used with them is so connected that the current to be controlled will pass through from cathode to anode (anode positive). A trigger voltage is applied to the gate. If more current is demanded of the circuit, the time at which the gate turns the SCR on is advanced so that the SCR conducts for a larger part of the half cycle, and vice versa. These devices are rather sensitive to the effects of lightning but, otherwise, they are very useful.

There is a variant of the SCR, called the *triac,* which can conduct in either direction. This device also has a gate, and its gate can be triggered on either half cycle. It is the preferred device if it is necessary to pass both halves of a sine wave. They can be placed before a transformer, whereas, a single SCR cannot, as transformers do not behave well with pulsating dc applied to their inputs.

A type of diode commonly used with SCRs when they are to be initiated from the power line is the trigger diode. This diode behaves more or less like a low-voltage SCR without a gate lead, in that it will turn itself on when a certain voltage level is reached, and the voltage drop across it decreases to a fixed value on initiation. These diodes are placed in series with the gate of the SCR and they return to the tie point between the capacitor and the resistor which provide the phase shift for the gate voltage (Fig. 2-5). A decrease in voltage across this diode sharply pulses the gate of the SCR, causing it to trigger into full conduction. It again turns off when the supply voltage approaches zero. In a sense, it can be said that this diode possesses a negative resistance characteristic at the transition point. The distinction between negative resistance and negative conductance will be considered shortly.

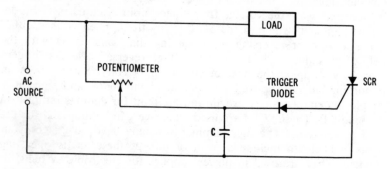

Fig. 2-5. A typical SCR circuit using a trigger diode.

There is a modification of the trigger diode that can be used with triacs. These devices differ from ordinary trigger diodes only in that they demonstrate the same trigger characteristics with either a positive or a negatively applied voltage. This is a necessary property for controlling triacs, as they must be able to be triggered with either polarity of signal. With minor changes, you can use the test setup that is described for use in demonstrating the characteristics of tunnel diodes, for examining the characteristics of either trigger diodes or diacs.

The Principles of Negative-Immittance Diodes

Probably one of the most interesting of all of the class of specialty diodes is the tunnel, or *esaki,* diode. This device shows a pronounced break, or reduction, of the current it passes as the voltage applied to it increases. It is a member of the negative-immittance diode class that is known as a negative conductance diode.

You must be very careful in the application of these specialty diodes as you can easily make serious errors in describing them. There are two basic kinds of negative immittance obtainable. Negative-conductance (or negative-admittance) diodes reach the decreasing-current-with-increasing-voltage region by going through a short region of zero change of current, and they leave in a similar way. On the other hand, negative-resistance (negative-impedance) devices reach a region of decreasing voltage with increasing current by passing through a region of zero change of voltage with increasing current and, then, leave in a similar manner. Negative resistance is commonly called "S-type negative resistance" in Russia and many other countries. Similarly, negative conductance is commonly called "N-type negative resistance." We will use negative resistance and negative conductance to refer to the two types of devices. The distinction, although largely ignored in this country, is an important one.

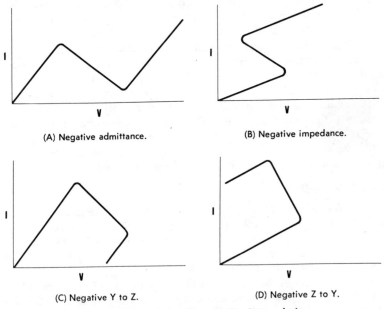

(A) Negative admittance.

(B) Negative impedance.

(C) Negative Y to Z.

(D) Negative Z to Y.

Fig. 2-6. Characteristics of negative-immittance devices.

As has been noted above, all of these devices exhibit positive-resistance characteristics over a major portion of their operating range. The important distinction is *how* they shift from positive to negative, a transition which must happen at each end of the negative-slope region. The characteristics of each are shown graphically in Fig. 2-6. Fig. 2-6A shows how the transition will take place if the transfer is through zero conductance (infinite resistance), and Fig. 2-6B shows how it will take place if the transfer is through zero resistance (infinite conductance).

One can ask, "What happens if the transition goes from positive to negative through zero conductance at one end and through zero resistance at the other?" Can you plot the kind of characteristic which might result? The first segment, as always, must have a positive slope, as in either Fig. 2-6A or 2-6B. But the transfer back to positive has to occur as shown in Fig. 2-6C or 2-6D if it is to be through the opposite kind of transition. The two possibilities are then as shown in Fig. 2-6C for a zero conductance to a zero resistance, and as in Fig. 2-6D for a zero resistance to a zero conductance. The author does not know of any devices having the properties shown in either Fig. 2-6C or 2-6D so, at least for now, Figs. 2-6A and 2-6B will have to serve as typical presentations.

Why is all of this so important? It is because negative resistance and negative conductance can be used to cancel out the corresponding positive values, and to reduce the effective resistance value for a circuit to zero. The result is a device that can be used with an appropriate kind of tuned circuit to make an oscillator. This oscillator will oscillate at almost any frequency you desire; just connect it to an appropriate kind of tuned circuit. In fact, unless care is taken, the oscillator will find its own circuit (made up of wiring inductance and stray capacitance) and it will oscillate on merrily in the ultrahigh-frequency or microwave part of the radio spectrum. If a circuit, which might have one of these beasts in it, behaves peculiarly, it is worth while to see if it might be oscillating. However, your scope may not show you the oscillations in this situation!

Tunnel Diodes

Tunnel diodes and many other solid-state devices behave as shown in Fig. 2-6A; they may be called negative-conductance (or admittance) devices, or N-type negative-resistance devices, as you choose. These devices always reach the negative segment by going through *zero* conductance, or infinite resistance.

A negative-conductance device like a tunnel diode only needs to have a parallel-tuned circuit coupled to it in the proper way for the combination to take off and oscillate strongly. This can be illustrated based on a curve like that shown in Fig. 2-6A. In this instance, it is assumed that a voltage is applied to the device which varies from zero to the value V_{cc}. See Fig. 2-7. The lines A, B, and C, which we will call load lines, show how much voltage will appear across the diode if it and the load resistance are connected as shown in the sketch of Fig. 2-8. As you can see, there are several different lines designated by A, several by B, and several by C. Each line carrying the same literal designation has the same slope and, therefore, represents a resistor of the same value, but in each case, the supply voltage, V_{cc}, is different. With line A, there is only one intersection with the

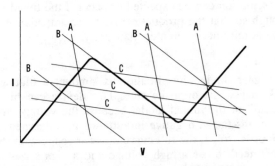

Fig. 2-7. Negative-admittance curve with load lines.

diode curve for each line. With line B, two lines cross as with A, but one is tangent to the curve at the midpoint of its negative-resistance region. With line C, each one of the load lines crosses the diode curve *three* times.

Load line A represents a load resistance of 150 ohms, B a resistance of 450 ohms, and C of 1933 ohms. If you set up your diode test circuit with resistor A, your scope will show you the full curve. This curve is illustrated in Fig. E-12. With resistor B, you will have roughly the full curve, but as you increase the resistance by just a little bit, the center section where it contacts the negative section of the curve will start to disappear. The device is now switching. You can repeat this experiment with varying values of resistance and a variable dc power supply and see all this going on if you wish. (Be sure the circuit is not oscillating, however!)

As long as the load line only crosses the negative conductance region and does not touch the diode curve more than once, oscillation probably can be avoided. But if a parallel tuned circuit is placed across the diode, using a decoupling capacitor to prevent shorting

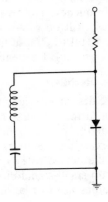

Fig. 2-8. A low-resistance, series
tuned circuit.

the diode, oscillation can be expected to occur if the tuned impedance is high enough so that the product of the tuned impedance (Z_t) with the negative admittance is greater than unity.

Trigger Diode

With a trigger diode, the situation is similar to that above, but things work in just the opposite way. This device can be called the network "dual" of the tunnel diode, because the voltage and current characteristics, as far as negative immittance is concerned, are interchanged. In order to get the full curve on your scope with this device, you will now need to use a high voltage and a high series resistance, precisely the converse of the requirement for a tunnel diode. As you reduce the resistance and readjust the supply voltage as required, the load line again will cross the curve more than once and, again, switching will occur at the appropriate voltage levels (really current levels here).

With a tunnel diode, therefore, you use a carefully selected voltage which will get you into the middle of the negative-conductance section of the operating contour (at about one volt), and make sure that your source resistance is small enough so that its load line will pass directly upward through about the middle of the active region, Fig. 2-7. Then, you use a parallel-tuned circuit to couple the source of voltage to the tunnel diode, as shown in Fig. 2-8. The circuit will oscillate at the frequency of the tuned circuit if the amount of negative conductance is sufficient to cancel out the positive resistance in the tuned circuit.

With a trigger diode, you select a supply voltage which will enable you, with an appropriate source resistance, to cross the characteristic curve in the negative-resistance region at one point, and use a low-resistance series tuned circuit to generate oscillations. A possible circuit is shown in Fig. 2-8. If you use the wrong kind of tuned circuit, you are likely to get a switching action which may not be at the right frequency. However, if it does turn out to be at the right frequency, it will be rather rich in harmonics. There may be times when it is useful to choose the wrong circuit, as long as it does control your operating frequency properly!

Diodes Vs. Rectifiers

We have described briefly the differences between diodes and rectifiers, but a further discussion may be helpful. Superficially, it would appear that the principal difference between the two might be the junction area. It certainly is one difference, but it is also necessary to design the device to have a higher peak reverse voltage (prv or piv) than do small-signal diodes. The junction area, and the contact and wire areas *need* to be larger also. All parts must be capable of carrying the peak current and standing off the peak voltage. Increasing the prv involves such things as reducing the conductivity of at least one region adjacent to the junction, and may also require increasing its thickness. This tends to increase the internal resistance of the diode, increasing heating, and possibly causing other problems, such as charge storage effects.

Inductive kick can be particularly brutal to these rectifiers. When the rectifier stops conducting, any inductance carrying that current will attempt to make the current continue to flow, and will create a powerful voltage spike which can easily destroy a rectifier. A capacitor properly placed to take up the charge will usually correct the problem. You will find the inductive kick experiment in Appendix E instructive in convincing you that you really can get 100 volts out of a 1.5-volt battery this way.

APPLICATIONS OF RECTIFIERS

The rectifier configurations you will want to be able to use in the definitions in this chapter have already been discussed. Now, it is important to learn more about their properties in terms of what you have just learned about the characteristics of the basic devices. The configurations which will prove to be of most use are the half-wave rectifier, the full-wave rectifier, the bridge rectifier, and the ladder rectifier.

You may wonder why silicon rectifiers are so important, particularly since they do not start conducting much until over one half a volt of forward bias has been applied. It takes less voltage to start a selenium, a copper-oxide, or a high-vacuum rectifier to conducting. However, at full current, the voltage across the silicon rectifier is much lower because of the exponential relation, and its piv is also much higher. (Only the mercury-vapor rectifier may be able to compete here.) Also, much larger amounts of current can be passed by a silicon or a germanium rectifier for a modest internal power dissipation.

Half-Wave Rectifier

Power supplies based on this arrangement, shown in Fig. 2-1, are only used when very modest amounts of current are required. With them, a pulse of current flows only once every cycle, meaning that on one half cycle, no current flows at all. The charge required to provide the necessary load current must be stored by that pulse, meaning that the stored charge must be adequate to provide load current for a full cycle, typically $\frac{1}{50}$ or $\frac{1}{60}$ second. Calculation of the size of capacitance required to store the necessary charge is explained in Chapter 7. The total charge stored must be sufficient to provide the required current with a limited change of voltage. This means that the total charge stored can be as much as five to twenty or more times that withdrawn each cycle. Equation 2-2 may be used to calculate the size of the capacitance. The equation must be used twice; first to determine the total charge being withdrawn each cycle and, the second time, to estimate the peak current which the rectifier must provide. If the charging current flows for 36 electrical degrees, and must provide the total charge withdrawn, the peak current through the rectifier must be at least ten times the average load current, a lot of current in some instances. A halving of the time between recharging periods by using a full-wave rectifier can be a big help. Typical waveforms are shown in Fig. 2-9.

Full-Wave Rectifier

Power supplies based on this rectifier use a center-tapped transformer and recharge the storage capacitor twice each cycle, first through one rectifier, then through the other (Fig. 2-2). Since the time between recharges is half that needed for the half-wave rectifier

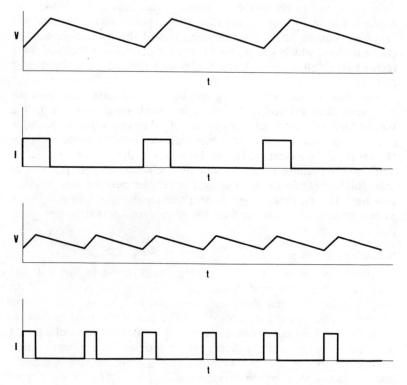

Fig. 2-9. Typical rectifier waveforms

for the same frequency and current level, the peak current in each device may be half as much as for the half-wave power supply. This configuration is better in another respect. The very fact that a current pulse flows every half cycle means that the transformer is better utilized, as transformers must be designed with small air gaps in their cores if they are to work effectively with a half-wave rectifier load. The shorter storage period means that a smaller capacitor can be used also. Although a series resistor may be desirable to limit peak current with all silicon rectifier supplies, it is not quite as important with full-wave or bridge rectifier configurations.

Bridge Rectifier

The power supply that is based on a bridge rectifier is in effect a pair of back-to-back full-wave rectifiers, one providing a positive voltage output, the other a negative output. The proper mode of connection is indicated in Fig. 2-3.

The bridge rectifier circuit can be used either to provide a single output voltage (in which case the transformer need not have a center-tap), or it can be used to provide both a positive and a negative voltage of about the same magnitude (in which case, the transformer center-tap is required). With the single-voltage arrangement, there are two rectifiers, in series, between the transformer and the dc output at all times, whereas, with a balanced output with respect to the center-tap, there is only one on each side. (There are still two rectifiers overall.) The bridge rectifier, like the full-wave rectifier, supplies the required charging current to the smoothing capacitor every half cycle, leading to smaller peak currents and a smaller capacitor for a given amount of smoothing compared to the half-wave rectifier.

Power Rectifier Configurations

A variety of operating configurations can be obtained from the use of a bridge rectifier and a center-tapped transformer. The four most useful circuits are given in Figs. 2-10 through 2-13. The first two circuits do not require the use of a center-tapped transformer winding; the first circuit provides a positive regulated voltage and, the

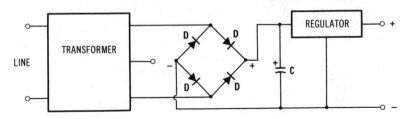

Fig. 2-10. A positive-regulated power-supply circuit.

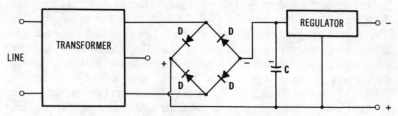

Fig. 2-11. A negative-regulated power-supply circuit.

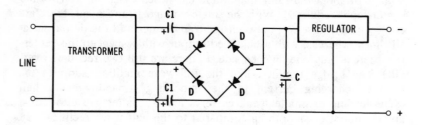

Fig. 2-12. A positive-regulated doubler circuit.

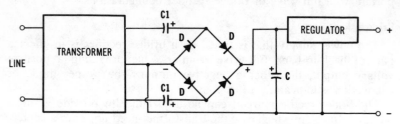

NOTE: CAPACITORS ARE AT LEAST 100 μF

Fig. 2-13. A negative-regulated doubler circuit.

second, a negative regulated voltage. Interesting enough, with minor changes, this circuit can also be used as a full-wave bridge voltage-doubler. It doubles the peak voltage that is available from the transformer center-tap to one side of the standard center-tapped winding (Figs. 2-12 and 2-13). It can also put out either a positive or a negative voltage, and it can use any appropriate solid-state regulator. (Note: All capacitors used in power supplies have a value of at least 100 μF. Their size can be computed.)

A single circuit board can be used to provide both a positive regulated voltage and a negative regulated voltage, as shown in Fig. 2-14. This board can be used with any two of the boards just described to provide combinations of voltages that are useful for transistor testing. These voltages are typically $+2V_m$, $+V_m$, and $-V_m$, and

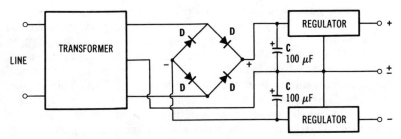

Fig. 2-14. A dual-voltage power-supply circuit.

$-2V_m$, where V_m is the voltage obtained from half of the winding using a half-wave rectifier. A variant of this board, to be used with it, has been designed to give the variety of voltages useful for testing transistors. This board can provide both positive and negative voltages, all regulated, in values of 5 volts, 12 to 15 volts, and a variable-voltage that is adjustable between 2 and 12 volts.

Voltage Doubler Rectifiers

It is important to know how a voltage doubler works. There are two forms of this configuration, the half-wave doubler and the bridge doubler. Circuits for these are shown in Fig. 2-15. The discussion here will deal only with the half-wave doubler since the bridge doubler is the full-wave equivalent, in all respects, except that both

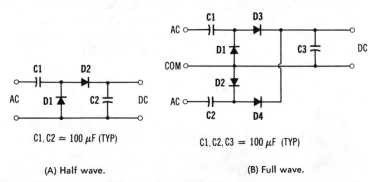

(A) Half wave.　　　　　　　　　　(B) Full wave.

Fig. 2-15. Voltage doubler circuits.

doublers draw charging current on both half cycles. The full-wave doubler charges the output capacitor on both half cycles, whereas, the half-wave doubler does it on one half cycle only. In each case, two rectifiers are needed in the doubler for each single rectifier in a conventional circuit. The convenient feature of the full-wave doubler is that the conventional bridge rectifier is ideally adaptable for use in it.

The half-wave doubler uses two rectifiers and two capacitors. The input rectifier causes a charge to be stored in a capacitor, connected in series with the transformer output, by closing a conducting path to ground (Fig. 2-13A). When the polarity of the transformer reverses, this rectifier turns off, and the sum of the transformer voltage and the capacitor voltage is applied to the second rectifier. This rectifier conducts, and permits part of the charge on the first capacitor, augmented by the transformer voltage, to be deposited in the second capacitor. This cycle repeats itself and, in the process, the voltage across the second capacitor approaches twice the peak value available from the transformer winding (Fig. 2-16). It takes a number of cycles for the voltage to build up to its operating value. One interesting feature of this supply is that the transformer is required to provide current pulses every half cycle, but the output is charged only once every cycle. With the full-wave bridge doubler configuration, however, both parts of the winding provide current every half cycle and, thus, the output capacitor also charges every half cycle. Can you explain why?

The reason is that while one input rectifier is charging its capacitor in series with the transformer, the output rectifier on the opposite side is forward biased and is charging the output capacitor. Since the output capacitor is common, it is first charged by one output rectifier and, then, by the other. You will find this voltage doubler circuit much more convenient than the half-wave doubler circuit.

The Ladder Rectifier Supply

The ladder power supply is an extension of the voltage doubler circuit and it uses the same principle two or more times to achieve high voltages with very small currents. It is probably the most convenient supply to use for testing rectifiers and high-voltage zener diodes. The ladder supply is based on the half-wave doubler, and uses both a ladder of capacitors and a series of cross-connected diodes. Care must be taken so that the diodes are installed with proper polarity! (See Fig. 2-17.) The respective capacitors charge on the appropriate half cycle, and the charge gets transferred up the ladder step-by-step until the peak voltage is approached at the output. The current capacity of this arrangement is rather small, but the *VOLT-AGE IS DANGEROUS!* Be very careful with it. As you can see, there are current-limiting resistors from the supply to the main capacitor, but it still can give you a nasty shock. When you observe the

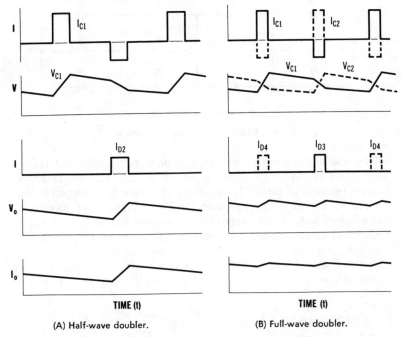

(A) Half-wave doubler.   (B) Full-wave doubler.

Fig. 2-16. Waveforms for voltage-doubler power supplies.

range of output voltage obtainable from the supply, you will under-stand why such a supply as this is convenient to have.

You probably already know that it is possible to encounter some extremely high-voltage rectifiers, particularly in oscilloscopes and in black and white and color television receivers. If it is necessary to make temporary repairs, this can be accomplished by the use of ordinary high-voltage silicon rectifiers. However, you must be cer-tain that you have a resistive and capacitive balance across all of the elements of the chain (Fig. 2-18). If the rectifiers have individ-ually different values of back resistance and different internal capaci-

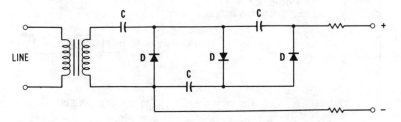

Fig. 2-17. The ladder multiplier power supply.

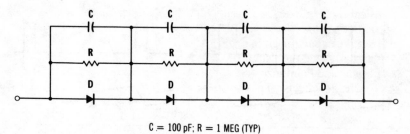

C = 100 pF; R = 1 MEG (TYP)

Fig. 2-18. Load balancing on rectifiers.

tances, the devices having the highest back resistances and smallest capacitances will prove to have the highest applied voltage. To prevent failures because of this, it is necessary that care be taken to balance the resistances and the capacitances. This makes them individually less efficient but, at the same time, it assures that they will not fail because of over-voltage. The internal elements in special rectifiers that you can buy, especially for this purpose, have been matched, with the result that they can take the full voltage without compensation. However, failures can and do result from an internal imbalance even then.

Rectifier Storage Effects

This brings up the question of just what happens when a rectifier is turned off very rapidly. You can get a hint of what happens by putting a small capacitor across a rectifier (under test) to make it "misbehave." Otherwise, the action is likely to be too fast to see on most scopes. For this test, you will want a small current-metering resistor in series with your rectifier (Fig. 2-19). When you start up the circuit, you will find that there is a short but significant reverse-current

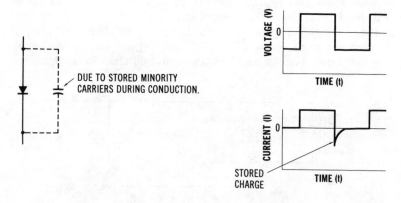

DUE TO STORED MINORITY CARRIERS DURING CONDUCTION.

Fig. 2-19. Equivalent charge-storage circuit.

50

transient as the capacitor charges while the rectifier ceases conduction. (Strictly, storage is a result of the charge that remains during turn-off, but the capacitor gives a similar effect.)

The design of the "snap diode" is such as to maximize the peak amplitude and minimize the duration of the current pulse resulting from storage effects. The resulting device can "dump" the stored charge into a tuned circuit, exciting it, and at the same time, generate a relatively stable output frequency which is a multiple of the source frequency. There can be a substantial phase instability in such a circuit, particularly when the multiplication ratio is high. These devices use storage as a harmonic generation aid.

This energy-discharge phenomenon is a serious problem with high-speed switching diodes, which must be designed to minimize both the total charge and the duration of the pulse. At the same time, it is desirable to have the forward conductivity of the diode as high as possible. Fortunately, both of these functions can be accomplished at the same time by introducing the right kind of dopants that are capable of trapping minority carriers on reversal, but at the same time are able to aid forward conduction. You should keep these points in mind when selecting diodes for switching uses.

COMMON-MODE PHENOMENA

Common-mode voltages can be a severe problem in some kinds of electronic circuits. When you need to measure a voltage difference between two points without disturbing them and without contaminating your reading as well, you have to reject an apparent voltage that exists with respect to ground. As long as you are using conventional electromechanical indicating instruments, you often do not have to use a ground reference. However, such is not necessarily the case with electronic instruments. The voltage to the midpoint of your measuring points is called a "common-mode voltage." Often, you will need to make a measurement of the voltage difference when the common-mode voltage may be ten or a hundred times the desired signal voltage. The ability to reject this common-mode voltage is called the "common-mode rejection ratio."

You will be confronted with common-mode voltages and the problems of common-mode signal rejection rather often while trying to get viable measurements on active devices. You will also find that it is a common problem. The difficulty is that it is impossible to measure a current without inserting a resistance, and it is impossible to measure a voltage without placing at least a small load across it. This is why you will be concerned with this problem and will need some special measuring techniques for measuring the direct voltage and the direct current used by your devices. You will find in the next

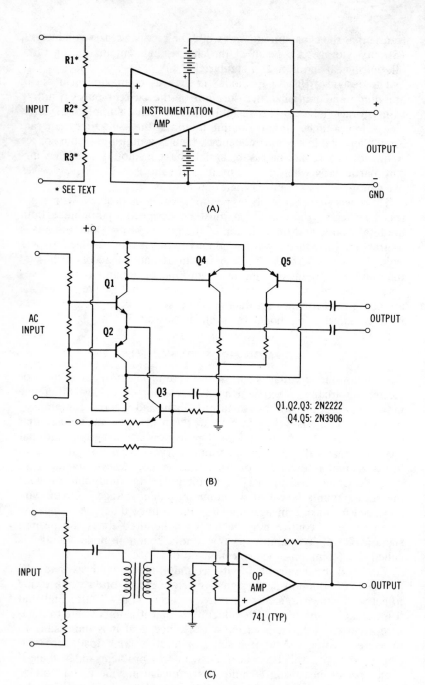

Fig. 2-20. Some techniques for isolating common-mode voltage.

chapters that the feedback which can develop across a metering resistance can be intolerable when using solid-state devices.

Ideally, you need to amplify the differential signal voltage sufficiently so that you can separate it from the common-mode voltage. Then, perhaps you should either establish an independent reference for a meter or use a transformer to complete the isolation. Some of the possible circuit arrangements are shown in Fig. 2-20. For calibration purposes, use 1000 ohms for resistors R1 and R3 and 25 ohms for R2. When calibrating, operate off the amplifier power supply. If you can sufficiently amplify the differential voltage (independently of the common-mode component), you can largely eliminate the effect of the common-mode component, but if you are required to accept a common ground for both circuits (such as a circuit across an emitter current-measuring resistance and a base voltage-metering circuit), then it may be that a transformer might be almost indispensable. If you are providing the signal to a meter (which by its nature may be independent of ground), this reference problem usually will not exist.

EXPERIMENT 1
Verification of the Relative Linearity of Resistors

You can conveniently use your variable-voltage power supply (see Chapter 7) for this experiment. The things you will observe include linearity, the effect of temperature, and some of the differences in the properties of various kinds of resistors (wire-wound, metal-film, carbon-film, carbon, etc.).

Step 1.

Connect your circuit as shown in Fig. 2-21. A conventional voltmeter and a conventional milliammeter will be adequate for this test. (A digital voltmeter would be equally as good.) Select a value of resistance which will cause approximately a full-scale current reading

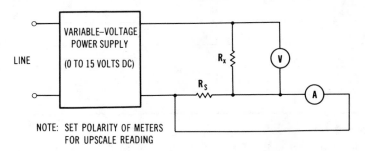

Fig. 2-21. Resistance-measuring circuit.

at the maximum voltage from your supply. (As an example, with a maximum of 12 volts and a current of 2 mA, you might choose a 5600-ohm resistance.) For your first series of tests, use a conventional carbon resistor. Make your calculation in the following blank spaces to determine the proper value of resistance:

$$R = V/I$$
$$= (\underline{\hspace{1cm}})/(\underline{\hspace{1cm}})$$
$$= \underline{\hspace{1cm}} \text{ ohms}$$

Step 2.

Set the supply voltage at its minimum value, and read the value of the current flow through the resistance. Then, read the voltage across it. Record the values in Table 2-1. Note that the table has space for measurements of more than one resistor. Now, take a hot high-wattage soldering iron and bring it close to the resistor. Note any change in the current and record the reading in the appropriate ΔI row of the table. Do the same with an ice cube or, if you know how to handle dry ice and have some, use it. You will fill in the last items, the temperature coefficients, in connection with Step 6.

Table 2-1. Resistance Data Table

Resistor type				
V				
I				
R_N				
ΔI_h				
ΔI_c				
ΔI_{vc}				
R_h				
R_c				
R_{vc}				
α_h				
α_c				
α_{vc}				

Step 3.

Repeat Step 2 at different voltages (up to the maximum you can get from the supply). Record your data in the remaining columns.

Step 4.

Calculate the nominal resistance values from the values of voltage and current that you read for normal temperature, heated, and cooled conditions, and record them in Table 2-1. Then, prepare another table for use with additional resistances. How much have the resistances changed? Record your data in Table 2-2.

Table 2-2. Test Data

R_N				
ΔR_h				
ΔR_c				
ΔR_{ve}				

Step 5.

If you have a can of component freezing spray, repeat the tests. This time, freeze the resistor with the spray and record your normal and chilled resistances, as calculated, in Table 2-3.

Table 2-3. Test Data

R_N				
R_{ch}				

Step 6.

Using the temperatures of both boiling water and ice, establish the resistance values. Calculate the temperature coefficient of resistance for your resistor using the equation:

$$R = R_o\ (1 + \alpha \Delta T) \qquad \text{(Eq. 2-5)}$$

where,

α is the temperature coefficient of resistance,
ΔT is the temperature change.

How much difference did you observe between the temperature coefficients when you heated the resistance and when you cooled the resistance? Explain what you think causes this.

The temperature coefficient is what we call an empirical coefficient, in that normally there is no analytic basis for its use. With pure metals, its value is roughly the reciprocal of the absolute temperature at which the measurement is made, e.g., at the melting point of ice, it is approximately 1/273, or 0.00366. However, it differs widely from this value with alloys and semiconductor-type materials, of which carbon is one. With the latter, alpha may be either positive or negative, and it may be quite large or small when compared to 1/T.

Step 7.

You need to test some other kinds of resistors, such as wire-wound, metal-film, carbon-film, boron-carbon, or whatever you can get, in the same way that you tested the resistor chosen in Step 1. Include them in the table that you prepared for Step 4. Also, use a coil of insulated wire and repeat the tests. Do you think you could make a thermometer from any of these? Explain why you think so.

The properties required for making a reliable resistance thermometer are: first, a stable resistance, second, a uniform value of temperature coefficient (over the range of resistance used), and third, an adequately large value of coefficient to make the resistance change measurement useful. You should make your decision based on how satisfactory you find the properties of the resistances that you tested.

EXPERIMENT 2
The Relation of Current to Voltage in a Diode

In this experiment, you will compare the behavior that you observed in diodes with the behavior you observed in resistors. It is suggested that you use one or more 1N914 diodes, as these are quite excellent and readily available.

Step 1.

You will need either a sensitive digital voltmeter or a high-sensitivity milliammeter and voltmeter for this experiment. (See Chapter 7 for the description of a "home-brew" unit that will do the job.) If you are using a digital voltmeter, either as a voltmeter or as a milliammeter, it is important that you find the voltage sensitivity of the unit when it is used as a milliammeter. Since it is also desirable to test any analog milliammeters you may plan to use, the first step

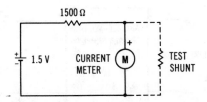

Fig. 2-22. Circuit for calibrating meter sensitivity.

is making this test. Refer to the circuit given in Fig. 2-22. Use a 1.5-volt dry cell (either a C cell or a D cell) and wire the circuit as shown. Assuming you are using a 1-mA full-scale meter and have a good battery, your meter should read full-scale with this setup. (If your meter has a different sensitivity, scale your resistance accordingly.) Then, connect the parallel resistance as indicated in Fig. 2-22, and adjust its value until your meter reads half scale. The parallel resistance then equals the internal resistance of your meter. For a meter suitable for the work you are undertaking, a 1-mA meter should have an internal resistance of at most 5 ohms. (No commercial analog meters are this sensitive.) When using meters having different sensitivities, the product of the full-scale current reading divided by the measured resistance should not exceed 5 mV.

When using a digital voltmeter, particularly a 3½-digit unit, it may be possible to choose "full scale" as being the 5 or 10 mV reading. Your precision may not be better than 20% in that case, but the loss of precision is less severe than is the loss from the effects of using a higher value of full-scale voltage. (If you are fortunate enough to have a digital voltmeter that reads 3½ digits with a full-scale sensitivity of 200 mV, millivolts, you are fortunate as your reading can be within a few percent.) A simple way of boosting the sensitivity of any meter is by the use of an LM4250 operational amplifier. It is connected as a "times-ten" booster either with a conventional vom or with a dvm that has an inadequate sensitivity or an inadequate number of digits. A suggested circuit for this purpose is included in Appendix B. You will want to be sure that you have adjusted your *null balance* as well as checking the sensitivity on your meters before you proceed.

Step 2.

You can connect your meter(s) in the appropriate configuration of the circuit shown in Fig. 2-23, depending on what you have. Your voltmeter must be usable as a differential voltmeter, so that you can get as much precision as possible, because you are going to be measuring values under 50 mV in most instances. You need the very-high-sensitivity milliammeter so that the voltage changes across the current-metering resistance cannot introduce voltage and amplification errors.

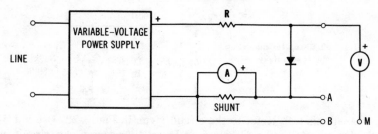

NOTE: DVM RETURN MAY BE CONNECTED FROM M TO A.
VOM RETURN IS CONNECTED FROM M TO B.

Fig. 2-23. Diode static characteristic measurements.

To clarify this in your mind, assume that your initial measurements caused 5, 10, 20, 40, and 80 millivolts across your milliammeter. Computation of the error you would encounter in measuring the change in voltage across the diode (in series with this meter) is enlightening. If you assume that the diode current changes by 2 to 1 for every 18 mV change in voltage (from anode to cathode), and if you use the above mentioned milliammeter, your 18 mV will look like a larger voltage if the current meter is also included in the voltage measurement. What will you find?

The true voltage change across the diode will be 18 mV, as will be shown when the meter is directly connected. When the meter is connected across the diode and the milliammeter to keep the milliammeter from reading the voltmeter current, however, the apparent voltage will have values of 23, 28, 38, 58, and 98 millivolts of change. As a result, the apparent diode resistance values will be 27, 56, 111, 222, and 444% high. With anything over about 20 mV connected across the meter, you will almost be measuring the meter alone, not the diode.

What all this means is that you must measure the voltage directly across the diode with a very-high-input-impedance meter, and you must be able to measure *small changes* with good precision. Fortunately, the static value of the voltage (which *is* temperature sensitive and may be as much as half a volt) is much less important. A convenient way of handling the problem is the introduction of a "zero reference" which is adjustable. Then, you can measure changes with respect to it. That is the technique used in the high-sensitivity meter that is described in the discussions on special instrumentation.

Step 3.

Now, you apply voltage to your diode and measure the diode current in the anode lead with an analog meter. Measure the voltage changes from anode to cathode on the diode, adjusting your zero offset to zero just prior to each change in operating conditions. Adjust your series resistance to double the current through the diode, and read the change in the voltage across the diode. The change should be about 18 mV if the current is metered external to the voltage measurement but, otherwise, it will be the 18 mV plus the milliammeter voltage. Your current meter needs to be very sensitive so you can use it in the loop if necessary.

You will want to perform the experiment with your meter in both positions, and when using both normal-sensitivity and high-sensitivity meters. Thus, you can verify that the meter-voltage-loss problem is real. It will be convenient to estimate the effective diode resistance at 1 mA and at 10 mA diode current. Based on these data, you can calculate the theoretical value of the effective resistance, and can take data to enable you to estimate the effective value, using both the normal and the high-sensitivity meter. (Hint: use values of current that is 20% more than the test value and that is 20% less, and determine the ratios of ΔI to ΔV at each point.) Enter your results in Table 2-4.

Table 2-4. Test Data.

I_{nom}				
ΔI				
ΔV				
Y_t				
Y_c				

Data for both the test conditions, with voltage measured across the diode alone and, then, across the diode and meter, should be recorded and calculated separately. The Y_c data are the values obtained by taking the ratio of the change of I to the change of V, and the Y_t data are obtained from the use of equation:

$$Y_t = (q/kT)\ I \qquad \text{(Eq. 2-6)}$$

Step 4.

Repeat the same experiment with other silicon diodes and, also, with germanium small-signal diodes. Do they behave in essentially the same way? Record your data in Table 2-5.

Table 2-5. Test Data

I_{nom}				
ΔI				
ΔV				
Y_t				
Y_c				

The germanium diode requires much less voltage to initiate conduction but, otherwise, it behaves in exactly the same way as the typical silicon diode. Various diodes may require different voltages to establish an initial conduction level, but the voltage change required for the 2-to-1 current change (either an increase or a decrease) is approximately 18 mV as noted previously.

Step 5.

Using your original diode, double the current through it in several successive steps. What can you change besides the source voltage to do this? Tabulate the voltage change each time and plot the data on a graph. (It is usually convenient to rezero the differential voltmeter at the start of each step by readjusting the reference voltage.) Tabulate and plot your data on the graph given in Fig. 2-24, and explain your results.

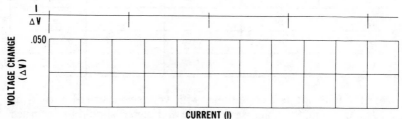

Fig. 2-24. Graph of I versus ΔV.

The author found that the voltage change stayed rather close to the 18 mV value until the current got rather high. It is assumed that it was a consequence of the internal resistance of the diode becoming important. As long as the voltage change is less than about 25 mV for the 2-to-1 change, your diode should prove satisfactory.

Step 6.

Repeat the above steps for both increases and decreases in current for your test diodes, and determine whether you find any irregularities. If so, try to explain them. Add your data to the above tables and graph.

Behavior of diodes at very low currents can be rather unpredictable. At high currents, at previously noted, the internal resistance can become important and can limit efficiency. In addition, you may find a diode which will actually oscillate. Those diodes can be quite a bit of a surprise. It means that they have a negative-immittance region of some kind.

EXPERIMENT 3
Further Diode Tests

So far, you have tested principally one kind of diode. The next question to be answered is plain—do all diodes behave this way? Also, do rectifiers behave this way?

Step 1.

You can identify the material from which the diode that you are testing is made by noting the voltage that appears across it when 1 mA of current flows through it. Typically, a silicon diode will show a voltage drop of about half a volt, whereas, a germanium diode will show a voltage drop that is nearer a fifth of a volt. Repeat your tests of Step 6 in Experiment 2, using several other diodes of the 1N914 type. Then, test some diodes made with alternative materials such as germanium, gallium arsenide (LEDs will do), and any other materials you might have. Mark these diodes so that you can later identify them for further testing. Tabulate your data in Table 2-6.

Step 2.

As you test the above mentioned diodes and record your data, you will want to prepare the data for plotting. In this step, you should plot these data, but instead of using the 2:1 scale for current as above, you should use a logarithmic scale as indicated on the graph in Fig. 2-25. This will cause an exponential relation like that in Equation 2-1 to give a horizontal line for the relation of ΔV in terms of the

Table 2-6. Diode Test Data

Diode				
I				
ΔV				
Y				
Diode				
I				
ΔV				
Y				
Diode				
I				
ΔV				
Y				

logarithm of I. Plot your data for the above diodes, and note particularly where any devices may start showing larger values of voltage change. These data will give you an indication of the range over which the devices will obey Equation 2-1.

Step 3.

Select a diode which when plotted in Fig. 2-25, has a curved characteristic in some region. Then, replot the data for that region of the curve on conventional graph paper. Does the new graph appear to approach a straight line when plotted as voltage (V) vs. current

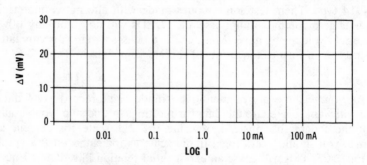

Fig. 2-25. Plot of ΔV vs. log I for typical diodes.

(I)? Can you explain the meaning of this? (The curve can be plotted as ΔV vs. ΔI if you prefer.)

It was found that where semilogarithmic plotting seemed to yield a curved contour, the use of linear scales tended to approach a straight line. This indicates that as long as the small-signal impedance of the diode is large compared to its series resistance, you will get an exponential characteristic, whereas, when the internal impedance of the diode is sufficiently small, the series resistance controls device behavior.

EXPERIMENT 4
Diode Small-Signal Characteristics

In this experiment, you will repeat some of what you did in Experiment 3, but with one significant addition. You will introduce a small ac signal on your dc bias and will actually estimate the effective small-signal conductance and resistance of the diode at each of a series of operating points. The points correspond to points at which you have already tested a particular diode. The ac signal that you superimpose on the dc operating conditions should not be more than 10 mV. It may be either a 60-Hz signal from a transformer and voltage divider as shown in Fig. 2-26, or it may be a signal obtained

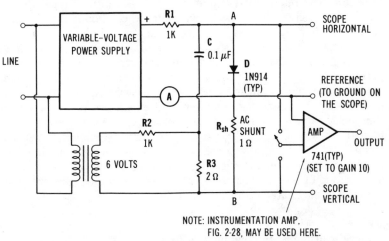

NOTE: INSTRUMENTATION AMP,
FIG. 2-28, MAY BE USED HERE.

Fig. 2-26. Circuit for measuring diode impedance using a 60-Hz test signal.

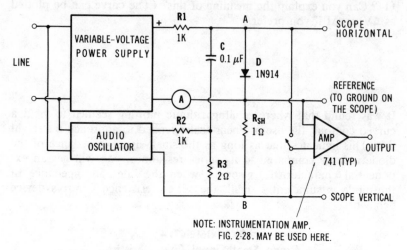

Fig. 2-27. Circuit for measuring diode impedance using an audio signal.

from an audio oscillator or an audio signal generator (Fig. 2-27). (The latter is a more precisely controlled and more precisely calibrated example of the former.)

Step 1.

Wire up your circuit in accordance with either Fig. 2-26 or 2-27, depending on which type of signal source you are using. The test signal is introduced to the diode through a voltage divider and a capacitor. The operational amplifier, which is mounted on a solderless breadboard, may be used to monitor the amount of ac signal across the diode or across the current-monitoring shunt. (The fact that you must measure your signal current in the cathode return shows why the previous discussion on the effect of series resistance was important.) A single-pole, double-throw switch is convenient for switching from one location to the other. The measurement has to be done in this way in order to minimize measurement errors. In some instances, a capacitance-isolated transformer may be required to extract the signal voltage.) This is only the first example you will encounter in which a common-mode rejection problem exists. The difficulties in making such measurements, in the face of this problem, may be one of the reasons we do not have really clear presentations of many of the characteristics of these solid-state devices.

Step 2.

After you have your circuit working, replace your diode and its metering resistor temporarily with the instrumentation amplifier and

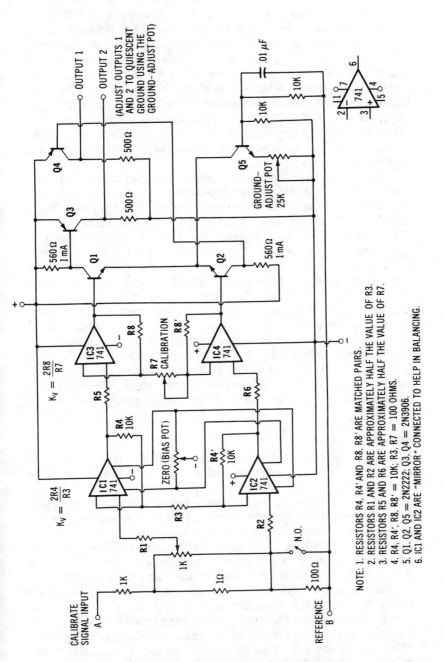

Fig. 2-28. A dc-ac amplifier for common-mode rejection.

NOTE: 1. RESISTORS R4, R4' AND R8, R8' ARE MATCHED PAIRS.
2. RESISTORS R1 AND R2 ARE APPROXIMATELY HALF THE VALUE OF R3.
3. RESISTORS R5 AND R6 ARE APPROXIMATELY HALF THE VALUE OF R7.
4. R4, R4', R8, R8' = 10K: R3, R7 = 100 OHMS.
5. Q1, Q2, Q5 = 2N2222; Q3, Q4 = 2N3906.
6. IC1 AND IC2 ARE "MIRROR" CONNECTED TO HELP IN BALANCING.

65

its calibration circuit shown in Fig. 2-28. (Points A and B in the schematic shown in Fig. 2-28 are to be connected to Points A and B, respectively, in the schematic of Fig. 2-26 or Fig. 2-27, depending on which circuit you have chosen to use.) This arrangement will enable you to balance your op amp configuration, which really is an instrumentation amplifier, to be sure it is operating properly. The combination of the 1-ohm resistor and the potentiometer will enable you to tap off a small fraction of the signal for use in the adjustment of the balance of your amplifier. (After calibration, resistors R1 and R2 shown in Fig. 2-28 are disconnected from the calibration circuit and used as the inputs.)

The first step in balancing is to adjust the input trimmer or "bias pot" so that the direct voltage is balanced, as indicated by a zero output by a voltmeter connected from Output 1 to Output 2. Then, the zero or "bias pot" can be adjusted to give a minimum reading. The "bias pot" is connected across pins 1 and 5 of the two input op amps and can be used for this adjustment. Then a common-mode signal is introduced from the potentiometer across the 1-ohm resistor, and the signal balance pot is again adjusted to minimize the ac output from the overall amplifier. (If it does not trim to balance, it means that the feedback network on the input op amps is not properly balanced.) If you cannot zero the output, connect the voltmeter across pins 6 of IC 1 and IC 2 and adjust for zero output. If you cannot get a zero output, replace the 741 op amps and try again. Once there is a balance across IC 1 and IC 2, repeat the process across IC 3 and IC 4, and then across each pair of transistors in the circuit. Once you have all good devices, you can trim the balance at the output with the bias pot.

Step 3.

Once the preceding adjustment is accomplished, it is necessary to calibrate the amplifier array so that you know its sensitivity to the applied ac signal. This may be accomplished with the ac calibration resistor R7 shown in Fig. 2-28. Then, a known source voltage may be applied to the input and the output calibration established on the scope. You will need the calibration arrangement shown on the left side of Fig. 2-28 to get this circuit set up and working. Once the circuit is working, you can measure the ac signal across the shunt that is in series with the diode and be reasonably confident that your measurements are valid. Once your amplifier is calibrated, disconnect the calibration circuit from resistors R1 and R2. (The calibration circuit consists of the 1K resistor, the 1-ohm resistor in parallel with the 1K potentiometer, the 100-ohm resistor, and the adjacent switch.) Then, use resistors R1 and R2 as your inputs. The Reference terminal (point B) should be connected to the Reference connection in the circuit that you use (Fig. 2-26 or Fig. 2-27). The vertical connection on the scope can be connected to either Output 1 or Output 2.

Step 4.

When both of the preceding calibrations have been completed, you can start varying the current through the diode. You will also be observing the signal voltage across it, and the signal current across the metering resistor. You should tabulate both the dc and the ac values of voltage and current, setting the dc values to those points at which you already have data. Then, you can determine the small-signal admittance or impedance (Y or Z) of the diode by using the equation:

$$Z = \frac{1}{Y} = \frac{R_s \Delta V}{V_s} \qquad \text{(Eq. 2-7)}$$

where,

R_s is the metering resistance,
ΔV is the diode signal voltage,
V_s is the shunt signal voltage.

Enter your data in Table 2-7. (I is the diode current.)

Table 2-7. Data Table for Step 4

R_s				
V_s				
ΔV				
Z				
I				
Y				

Step 5.

As you tabulate all of these data, you will wish to examine how the value of Y varies with the current through your diode. Try this on a number of the diodes that you have tested. Can you find a region in which it is given by Equation 2-8?

$$Y = (q/kT) \ I \qquad \text{(Eq. 2-8)}$$

Our results showed a rapidly increasing value of Y in the exponential operating region, but a much less rapid variation as the linear region caused by internal resistance was entered.

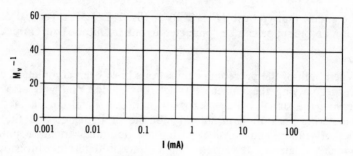

Fig. 2-29. Graph of Y/I as a function of log I.

Step 6.

Repeat the previous steps with some typical rectifiers. Tabulate and plot your results in Fig. 2-29 showing the relation of I to diode dc bias, and the relation of Y/I to I. What do these results tell you? Enter your data in Table 2-8.

Table 2-8. Data Table for Step 6

R_s			
V_s			
ΔV			
Z			
I			
Y			
R_s			
V_s			
ΔV			
Z			
I			
Y			

We found that rectifiers behave very much as do the regular small-signal diodes. They have exponential regions of operation and, when the current through them is large enough, they do show some evidence of an internal series resistance.

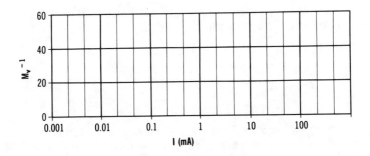

Fig. 2-30. Graph of Y/I vs. log I for Step 7.

Step 7.

Repeat Step 5, but using germanium diodes. Tabulate your results in Fig. 2-30 and plot the curve of Y/I against log I. Are there any significantly different results? Compare these results with those for Steps 5 and 6.

Table 2-9. Data Table for Step 7

R_s				
V_s				
ΔV				
Z				
I				
Y				

The principal differences with germanium diodes are the relatively poor (low) back resistance, and the relatively low breakdown voltage. The forward voltage required to start them to conducting is also lower than for silicon diodes and silicon rectifiers.

Step 8.

Compare the results of all of the above tests, and note what features proved to be common to all of these devices. What does this tell you about the properties of solid-state junction devices? Explain why you believe these properties are significant.

The very close parallel between the measured and the calculated values of the conductances for all of these diodes indicates that this property is a prime factor in controlling the current flow in all of these devices. You will have found this in silicon diodes, in germanium diodes, and in rectifiers, as well, and you can possibly find it in crystals of gallium arsenide, galena, and a variety of other materials that have proven effective as diodes.

EXPERIMENT 5
Charge Storage Effects

One of the serious problems in the use of solid-state diodes is the discharge of the minority charge that still lingers after the reversal of the applied potential on a diode. This charge causes a momentary continuation of conduction *after* the diode is reverse biased and, in some circuitry, this conduction can cause severe problems. In this experiment, we shall attempt to observe this phenomenon in some conventional rectifiers, and shall attempt to simulate it with switching diodes by the use of a shunt capacitor. (The excess minority charge acts as if it were stored in a capacitor at the instant of switching.)

Step 1.

In order to observe this phenomenon, it is necessary to switch voltage polarity very rapidly as, otherwise, the charge will be swept out as the voltage decreases. You, therefore, need a square-wave generator as your signal source, and it should be able to generate square waves up to the limit of TTL (transistor-transistor logic) switching speeds, which are the equivalent of over 25 MHz. Since TTL devices switch from about 0.4 volt to about 2.5 volts, it is necessary to establish a reference at about 1.5 volts and filter it well. Use a 10-μF electrolytic capacitor that is bypassed by a 0.1-μF ceramic capacitor and a 100-pF mica capacitor, all with the shortest leads possible (the smaller the capacitance, the more critical the lead length). Be certain that the leads on the mica capacitor are just as short as you can get them. As is evident in Fig. 2-31, the diode has a 10-ohm current-metering resistor in series with it, with this re-

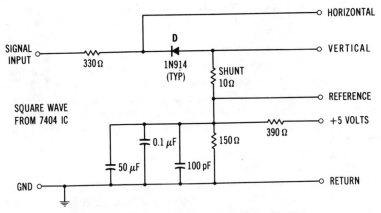

Fig. 2-31. Circuit for testing charge-storage effects.

sistor returning to the 1.5-volt supply. *The diode is connected to conduct when the voltage is low, because of the higher current capacity of the 7404 IC chip when the voltage is low.* The resistor, in series with the diode, should limit the peak current to about 10 mA.

Step 2.

After your circuit is wired, you can insert a conventional silicon rectifier as the test diode. Then, apply a square wave at a frequency of about 1 MHz to the input of the 7404 inverter being used and observe your current waveform. Sketch it on the graph given in Fig. 2-32.

Step 3.

Estimate the time duration of the spike and also the peak current. Then, estimate the total amount of charge by multiplying the peak amplitude of the current pulse by ½ and, then, multiplying this by the time duration of the pulse. Remember that your current probably will be in milliamperes or even microamperes, and your time may be

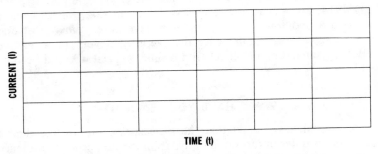

Fig. 2-32. Rectifier current waveform graph for Experiment 5.

in fractions of microseconds. How many picocoulombs of charge do you estimate were stored?

If your peak current was about 0.5 mA and the approximate pulse duration was 0.5 microsecond, the total charge was about 125 picocoulombs. This is the right order of magnitude, as the junction capacitance might be about 5 pF and the voltage change about 2 volts, indicating a probable charge magnitude in the low hundreds of picocoulombs.

Step 4.

Now, substitute a 1N914 diode for the rectifier. This diode by design should have very little stored-charge effect, as it is designed to "lose" its minority carriers very rapidly on reversal. Can you detect the charge storage effect?

If the writing speed of your oscilloscope is sufficiently high, you may see a small trace of a charge storage in your diode. If you do not, don't be alarmed, because it can be extremely small and very fast; in that case, simulate some charge storage by placing a 100-pF mica capacitor in parallel with the diode. Then, repeat the test. What have you learned from this experiment?

You should have observed that ordinary rectifiers, which make very poor switching diodes although they are excellent for their purposes, do have a charge storage which must be dissipated on the reversal of voltage. You probably have found that silicon switching diodes, like the 1N914, are much superior for switching in this respect (unless you tried a poor one) and, unless you have an extremely fast and sensitive scope, you should have discovered that you could not see it. But, by simulating the effect with a shunt capacitor, it did become visible.

WHAT HAVE YOU LEARNED?

You should have found that the properties of solid-state diodes and rectifiers differ in substantial ways from the properties of ordinary resistors. You should have found that, by nature, the current-voltage relation for these devices is exponential in form rather than linear and, in addition, you should have found that the devices normally conduct significantly in only one direction. You should have found that rectifiers behave much like diodes, but that they can block a much higher voltage than can diodes, and the rectifiers are much more "sluggish" (they switch much slower) than are signal diodes.

This detailed evaluation of the properties of solid-state junctions is of an overriding importance to you, as these junctions are the basic elements which contribute to the operation of bipolar transistors. They, also, have great importance in various kinds of field-effect transistors. In fact, one can also say that the same phenomena are of more than a casual importance even with electron tubes.

You are now ready to see how the placement of two diodes in a back-to-back configuration can lead to the bipolar transistor and, also, how the operation of these devices can be developed from what you have learned about the properties of diodes.

CHAPTER 3

The Bipolar Transistor

In this chapter, we will study and measure the characteristics of one specific type of bipolar transistor, in order to gain some understanding of how bipolar transistors function, what is important in their operation, and what must be known to use them most effectively and economically. You will find that some parameters, normally understood to be of prime importance, are important in more different ways than is often thought, and you will find that there are other parameters that are of overriding importance. You will also find that these devices are extremely useful if they are applied correctly. By the end of this chapter and the next, you will understand what these parameters are, how they are important, why they are important, and how you can use them most effectively.

OBJECTIVES

After studying this chapter and performing the experiments, you will understand the highly nonlinear characteristics of the npn silicon bipolar transistor. You will also be able to verify the following concepts and properties through your studies and your measurements.

1. Definition of a transistor.
2. Relation of the transistor to earlier active devices.
3. Current-voltage relations in bipolar transistors.
4. Significance of transistor current gain.
5. Undependability of current gain.
6. Dc characteristics of good and bad transistors.

7. Small-signal characteristics of transistors.
8. Effects of spreading resistances.
9. Controlling the characteristics of transistor amplifiers.
10. Comparison of voltage-gain and current-gain operation.
11. Effects of emitter degeneration on operation.
12. Statistical variations among transistors.

DEFINITIONS

The next chapter will explore in greater depth the characteristics of bipolar transistors and their use in amplifiers. The discussion will include pnp silicon transistors as well as germanium transistors. But, first, a list of definitions, some of which will be used significantly later but most of which must be understood now, will be reviewed. In the following discussion on bipolar transistors, these definitions will be useful to you:

semiconductor material—This is a material which normally conducts electricity very poorly. However, its ability to pass current increases as its temperature is increased. It also can be made to conduct more freely by putting certain kinds of contaminants into it. These contaminants are called doping materials or dopants.

intrinsic semiconductor material—This is a semiconductor material in its purest and most perfect state. Because of its purity and the care with which it is prepared, its crystallization is as perfect as can be achieved.

single-crystal material—A semiconductor which has been prepared in a form in which there are almost no significant crystallization defects. These crystals can be either intrinsic or doped with a dopant if the process has been sufficiently carefully controlled.

p-type semiconductor material—A semiconductor containing a contaminant or dopant which causes it to have a deficiency of electrons and, as a result, shows an abnormal affinity for them.

n-type semiconductor material—A semiconductor material containing a dopant which causes it to have more than the normal number of electrons. These electrons are extremely mobile.

bipolar transistor—A device made up of at least three layers of semiconductor material, with a layer of one type situated between two layers of the opposite type. The outside layers are called the emitter and the collector, and the inner layer, which must be extremely thin, is called the base.

emitter—The emitter of a transistor is a region that is able to create a flow of charge, either positive or negative, when properly polarized by an electric field. It is usually doped rather heavily with an appropriate kind of impurity to make additional carriers available. (The equivalent region in a field-effect transistor is called the source.)

base—The base of a transistor is the layer which can draw the charge out of the emitter by means of a forward bias (which can be called an accelerating potential). It is the control electrode in the bipolar transistor. All presently known kinds of active devices have a region which serves the function of the base. (In a field-effect transistor, this area is called the gate.)

collector—The collector is the region which finally collects (or drains) the charge that is being released by the emitter. It is biased with such a polarity of voltage that it can draw the charge through the base and collect this charge. (The corresponding region in a field-effect transistor is called the drain.)

beta—The beta of a bipolar transistor is the current gain. It is defined as the ratio of a small change in collector current to a corresponding small change in base current. It is most easily measured in terms of the ratio of two sinusoidal currents. (Normally, the currents are measured in terms of the voltage generated by passage of the currents through resistances of known value.)

dc beta—The dc beta of a transistor is the ratio of its collector current to its base current.

diffusion—Diffusion is the process by which the charge leaves the emitter and becomes available for attraction to the collector of a bipolar transistor. A variation in the concentration of carriers through a region is necessary for diffusion to take place.

drift—Drift is the motion of charges as a result of an accelerating field. It is quite distinct from diffusion, which results from a variation of carrier density, and can occur under an essentially negligible field. Drift is more predominant in devices having a decreasing density of dopant in the base in the direction of carrier movement.

recombination—Recombination is the neutralization of a carrier by its being trapped by a charge of opposite polarity. It is one of the causes of current being required in the base lead of a transistor.

majority carriers—The kinds of carriers that are in the greatest concentration in a semiconductor region are called the majority carriers. They are the ones that are the most free to move, diffuse, or drift.

minority carriers—When a material is doped with an excessive concentration of one kind of carrier, it tends to suppress the number of the opposite kind of carrier available. However, there are always some free negative and positive charges in any material capable of conducting electricity. The charge carriers having the lower concentration are the minority carriers.

base-spreading resistance—The base-spreading resistance is the resistance present between the external base terminal of a transistor and the active boundary between its base layer and its emitter layer (its junction). The existing electric fields present in an operating transistor can cause some variation in this parameter.

emitter-spreading resistance—The emitter-spreading resistance is the resistance between the external emitter lead and the active part of the emitter-base junction. It may be orders of magnitude less than the base-spreading resistance.

collector-spreading resistance—The collector-spreading resistance is the resistance between the collector junction at points where current is flowing and the external collector terminal. It is very similar to emitter-spreading resistance and, with some types of transistors, may have roughly an equal value. (In fact, some transistors can be used equally well with their collectors serving as emitters.) Almost all bipolar transistors exhibit some inverse beta, or inverse current gain.

input capacitance—The input capacitance for a bipolar transistor is the total capacitance as seen at the base of the transistor. Typically, it comprises the capacitance from base to emitter, and the capacitance from base to collector. Its main component is the diffusion capacitance, although there is also a transition capacitance.

diffusion capacitance—The diffusion capacitance is the capacitance across the input of a transistor that results from the interaction of charge carriers crossing the base with the charge on the emitter boundary. This capacitance is much larger than the static capacitance that might be measured across the junction (without an activating bias applied).

transition capacitance—This is a component of capacitance at the input that is voltage dependent. It is a function of the voltage from base to collector.

output capacitance—The output capacitance is the combination of the base-to-collector capacitance, the collector-to-emitter capacitance, and the collector-to-ground capacitance. It is quite small compared to the diffusion capacitance, but can be magnified in apparent magnitude by the Miller effect.

Miller effect or *Miller capacitance*—This is the apparent multiplication of the value of the base-to-collector capacitance as a result of the voltage amplification in the transistor. The value of the capacitance as seen at the base will appear to be:

$$C_{cb}(1 + |K|)$$

where,

C_{cb} is the collector-to-base capacitance,

$|K|$ is the magnitude of the voltage amplification measured at the edge of the junction.

(When a significant amount of collector-spreading resistance is present, the value of K at the junction may be significantly larger than the value measured at the device terminal.)

header capacitance—The various capacitances not resulting from the semiconductor chip itself, but which are present from wire-to-wire, wire-to-case, etc., can all be lumped together as header capacitances. They can be particularly important at high frequencies.

low-injection operation—Under ordinary conditions, relatively small numbers of carriers are drawn from the emitter into the base of a transistor. These carriers are of such a sufficiently small number that the balance of charges in the base is not seriously upset. This condition is known as a condition of low-injection operation.

high-injection operation—Under high-injection operation, the number of carriers injected into the base region may approach (or exceed) the number of majority carriers normally present in the base region. Under these conditions, additional majority carriers must be drawn into the base through the base lead, and a diffusion-type of variation of charge density will exist in the base between the active region and the base lead. This sets up the equivalent of a thermocouple voltage across this region, and causes a change in the behavior of the device. It is of extreme importance at high currents, and it can cause significant changes in the behavior of the transistor. In fact, it changes the transconductance-per-unit-current value which will be measured in the device.

graded base or *diffused base*—A graded base is one in which the concentration of the doping material varies across the base region. Normally, this grading is in such a direction as to reduce the doping level as the collector is approached.

transconductance—The transconductance of any active device, governed by the solid-state diode equation, is the ratio of the change in value of output current to the change in the input voltage causing it. If a small value of sine-wave voltage is applied at the input and the sine-wave current measured (in terms of the voltage) across an appropriate value of output resistance, the value of the transconductance g_m is given by the equation:

$$g_m = \frac{v_o}{v_i R_L} \qquad \text{(Eq. 3-1)}$$

where,

 g_m is the transconductance,
 v_o is the output-signal voltage,
 v_i is the input-signal voltage,
 R_L is the output-current metering resistance.

transadmittance—When a transistor is operated at a sufficiently high frequency, the inductive, capacitive, and time-delay effects can become important. Then, the transfer function for the device is no longer resistive but may include capacitive and other components as well. Under these conditions, it is necessary to talk about transadmittance rather than transconductance.

alpha—The common-base current gain for a bipolar transistor, or alpha, is the current gain from emitter to collector. It is closely related to transistor beta, which is defined as $\alpha/(1-\alpha)$. Since with best-quality transistors, the alpha differs only slightly from unity, it is clear that the denominator is a small difference of two numbers.

point-contact transistor—The first solid-state device made that proved capable of amplifying was the point-contact transistor. This device was made by placing a pair of very sharp needles extremely close together and in contact with a piece of semiconductor material. Then, when a current flowed from one needle into the semiconductor, it was often possible to get a somewhat larger current to flow from the second needle into the semiconductor. These devices were not very stable, and it was impossible to predict whether one would work or not, so they remained laboratory curiosities.

small-signal transistors—These transistors were designed to handle small amounts of current and voltage and to provide voltage or current amplification.

power transistors—These transistors behave like large numbers of small-signal transistors connected in parallel. They are designed to handle significant amounts of power.

germanium transistors—These transistors are made from the semiconductor element *germanium,* and were the first devices successfully marketed. They have been largely displaced because they fail at temperatures that are near the maximum which is normally encountered in more stringent applications, and because the leakage current encountered with them is large enough to create application problems.

silicon transistors—These devices are based on the semiconductor element *silicon.* Silicon has a higher maximum operating temperature; one that is more compatible with the normal operating conditions of most electrical and electronic equipment. It also has a much lower leakage current and, consequently, has proven more satisfactory for industrial and commercial application. Its high-frequency characteristics are less favorable than those of germanium, however.

gallium-arsenide transistors—These transistors and a variety of other element combinations are only now becoming important. They are devices that you will want to be alert for.

npn transistors—These transistors consist of three layers of semiconductor material, with the outside (or emitter and collector) layers being doped with an "N"-type material which creates an excess of electrons, and the inside (or base) layer being doped lightly with a "P"-type doping material which gives it an affinity for electrons. Because of this affinity, the base layer must be extremely thin to assure that the electrons will largely penetrate through it into the collector region.

pnp transistors—These transistors consist of three layers of semiconductor material, with the outside (or emitter and collector) layers both being doped with a "P"-type material which creates layers that are deficient in electrons. These layers can be considered as having an excess of "holes" or "vacancies." The inside layer (or base) is then doped with an "N"-type dopant. Conduction results from the motion or movement of the holes. These devices are the inverse of npn transistors.

electrons—Electrons are negatively charged carriers which are responsible for most of the current flow in electrical and electronic circuits. They are the lightest-mass charged particles that are normally encountered. Many of them are bound to atoms. Only a small percentage are free to move about.

holes—Holes are the positively charged equivalents of electrons. In most semiconductor materials, they are generated by what might be called "electron-catching sites," that is, places which attract and catch electrons. The movement of electrons from site to site creates the effect of motion of a positive charge, or the effect of a positive current flow.

vacancies—Another name for holes.

diffused base—Another name for a graded base. It applies an accelerating field, in addition to the diffusion fiield, to carriers in the base region.

carrier injection—Carrier injection is the process of moving free electrical charges from one region into another region of different electrical polarity. This injection is accomplished by applying a small forward bias across the boundary or junction.

junction—The region, including the boundary between two layers of semiconductor, which exhibits significantly different electrical properties.

beta cutoff frequency—The beta cutoff frequency is that frequency at which the common-emitter current gain for a bipolar transistor begins to decrease to half its previous value for each doubling of frequency. It is usually defined as the frequency at which the current gain has decreased to 70% of its value at low frequency.

alpha cutoff frequency—The alpha cutoff frequency is that frequency at which the common-base current gain has decreased to 70% of its low-frequency value. It is near the upper operating frequency for the transistor.

maximum oscillation frequency—The maximum oscillation frequency is that frequency above which insufficient energy can be generated by an amplifier to sustain oscillation. The output energy is less than that needed at the input to sustain oscillation. The power gain at this frequency is unity; above, it is less than unity.

unit under test (uut)—When the characteristics of some device or component are being measured, the device is often referred to as the *unit under test* or *UUT*.

device under test (dut)—Another identification for the unit under test.

integrated circuit (IC)—An elaborate configuration of transistors, diodes, resistors and, possibly, capacitors created on a single piece of semiconductor substrate such as silicon.

input admittance (y_i)—The ratio of the change of input current to the corresponding change in input voltage, for a device having separate input and output terminals or ports.

EIA Code—The alphanumeric code assigned to an active device by the Electronics Industries Association to identify it specifically. This code also can be used for referring to a set of specifications that present the characteristics for a device that has been registered with EIA by a manufacturer. Diodes are typically identified by codes like 1N914 and 1N4007, while transistors use codes like 2N34, 2N2222, 2N3055, 2N3906, etc.

THE NATURE OF ACTIVE DEVICES

An active device is a device that can take energy in one form and, under the influence of a control of some kind, convert the applied energy into a duplicate of the controlling source, but at an increased power level. An active device can take on any of a number of forms.

The first active devices built were triode electron tubes. The first tube of this kind was made by Lee deForest in the early part of the century. Then, Dr. Albert Hull in the early 1920s introduced the screen grid into the deForest "Audion," and the electron-tube industry was on its way. At about the same time, Dr. Julian Aceves showed how tubes could be operated successfully on alternating current. This was followed by the cathode-type structure, which used a filament (inside a sleeve) that was coated with a material capable of emitting electrons. This permitted the operation of radio receivers and transmitters on alternating current. Finally, just after the end of World War II, Dr. Shockley and his co-workers invented the point-contact transistor. This was followed by the invention of the bipolar transistor and the field-effect transistor. These inventions eventually made it possible to build a pocket calculator which would do the work of ENIAC. (ENIAC, a room-size calculating monster, was the first high-speed electronic digital computer to see service.)

The only kind of "one-port" devices presently known which will amplify are devices like the *tunnel diode*. A tunnel diode, over a modest range of either the voltage or the current, will exhibit a decreasing value for one variable as the value of the other variable increases. These devices are known as *negative immittance* devices. The word immittance indicates it may be either an impedance or an admittance value. The tunnel diode and certain types of gas-discharge tubes are perhaps the best-known examples of these devices. For technical reasons, they are relatively unimportant in most of the branches of electronics that one will encounter. All of the active devices you are likely to use extensively are two-port devices; these are devices with a pair of terminals for introducing the control signal and a different pair for removing the processed signal. The first of these terminals is known as the input, or control, port and the second terminal is known as the output port.

Theoretically, there are at least four kinds of two-port active devices. They are based on the kind of variable that provides the control action on the output, and on the kind of the variable that is controlled. This condition is a result of the existence of two basic kinds of variables, namely, *across variables* (like voltages), and *through, or flow, variables* (like currents). Either kind of variable may be the input or control variable, and the input variable may theoretically control either kind of variable. However, it can be shown that all presently known two-port active devices that are used in electronics consist of devices having the control function performed by an across, or voltage, variable. This control action affects a through, or current, variable. This is readily recognized from an examination of the basic Ebers-Moll equations, which control transistor operation and, also, from the diode equation, which controls the basic action of semiconductor junctions. In both of these sets of equations, the independent variables are junction voltages while the dependent variables are current. You can examine the basic equations and their significance in more detail in Appendix A.

One of the important advantages of the approach (to circuits using active devices) that you are studying is that it applies to *all* devices that meet relatively loose specifications. As long as a bipolar device has a beta value that exceeds 10 to 20, the voltage and power ratings are adequate, and the maximum operating frequency is high enough, *any* bipolar device should prove satisfactory. You simply set the output current level at the value required to assure that the transconductance requirement of the device is met, and your desired operating conditions will almost always be available. A similar rule applies to field-effect transistors and electron tubes.

One happy consequence of this is that you have a wide range of options in device selection with both bipolar transistors and field-effect transistors. Your only problem is making certain that the selections are all in a common class; for example, npn-silicon or p-channel JFET devices. You can even cross those boundaries if you are careful to take the basic differences into account. (One of the reasons this book is organized as it is is to help you to recognize how to cross those boundaries successfully.) If the device you are considering tests good in either a TeeDeeTester or a FeTester, and it is properly matched to its environment, it will usually function as planned. The result is that while we have suggested the 2N2222 transistor for much of your npn-transistor work (they are readily available), almost any npn silicon or germanium transistor will prove satisfactory in most of the circuits. A similar situation is true for pnp transistors, and any other devices other than a 2N3906 transistor, of your choosing, may be substituted. Feel free to substitute any similar devices that you may have or may find easier to get. Simply adjust the input bias circuit to provide your selected output current level and you will be "in business."

Similarly, where an operational amplifier (op amp) is required, most applications will function without problems if a 741 or a 4250 is used. If you find the input loads your circuit excessively, you can use an FET-input op amp or insert a FET or IGFET source-follower in the circuit as an isolator. A suitable FET-input op amp is the NE536; however, there are many others that will prove satisfactory.

The result is that you *do* have ample alternatives, and so the "suggested" types are just that. Further, advances in technology are such that you will find the devices are improving to the extent where it is almost useless to recommend specific devices. In short, don't be afraid to try substitutions!

SOME INTRODUCTORY NOTES

There is a wide variety of devices which have been given the name "transistor," and they all have their own properties and peculiarities. Some obey equations that are related to the Ebers-Moll equations given in Appendix A, and some do not. There is an even larger group of devices which carry the "2N" designations that are used to identify transistors. The result is that you cannot be sure without looking up the specifications if you have a transistor or something else.

In this chapter we will discuss the npn silicon bipolar transistor. As in previous chapters, the discussion is directed toward providing the knowledge that is necessary to understand what you are doing as you perform the experiments—so that you will understand the results that you will observe. A very brief review of some of the more mathematical considerations is included in Appendix A, but it need not be studied unless you wish a better mathematical understanding of the behavior of the devices. These mathematical considerations support the results you will observe during your experimentations.

Transistors are closely related to solid-state diodes in that they are made of the same type of materials and they use junctions and regions of differing electrical properties just as diodes do. In fact, they obey the same kind of exponential current-voltage relation that you have observed so consistently with diodes. Needless to say, that is one of the reasons we were so careful to document the existence of this current-voltage relationship in Chapter 2.

The physicists working on the diode projects (the phrase "solid-state physicist" had not yet become an important phrase in our language) concluded that they should be able to develop the theory of an amplifier based on back-to-back diodes, and thus was born the Ebers-Moll model of the junction transistor. But the device did not exist. Why not?

Careful examination of the equations showed that in order for an effective action to result, the current deflected into the base terminal had to be kept to a minimum. But how did one do that? Further consideration showed that there were several possibilities. One way was to keep the conductivity of the base layer as low as possible so as to minimize the recombination of the charges entering the base from the emitter. It was necessary, on the other hand, to have the emitter conductivity high so that carriers would be available to be injected into the base. A further thing which could be done was to make the base as thin as possible so that the carriers could escape to the collector before they had time to recombine. Also, it was necessary to make the semiconductor material into as perfectly crystallized a crystal as possible, because it was found rather quickly that grain boundaries are excellent carrier generators and were able to initiate recombination action. This emphasized even further the importance of growing single crystals as perfect as possible and, then, doping them as uniformly as possible, based on the requirements at hand.

In the final analysis, all of these things had to be accomplished before the first transistors were successfully made. These devices were known by various names, such as alloy and grown-junction and surface-barrier transistors as well as others, and they were made from both germanium and silicon materials. Because of the imperfections still existing in the methods of preparing the basic crystalline material and because of the many problems in achieving consistent and reliable results, most of the early transistors were rather unreliable devices

The growth of perfect crystals (called single crystals), with an absolute minimum of flaws and imperfections, has become of increasing importance as more and more elaborate device configurations have been developed. The production of tens to thousands or more perfect chips is essential to keeping manufacturing costs low, while a flaw under the site of just one device on a chip is enough to make the chip useless. (A chip is a piece of semiconductor material on which a transistor or array of devices is developed.) Consequently, the development of optimum methods of growth, including growing crystals in outer space, has become of increasing importance. The development of planar and ion-implantation techniques is equally important.

THE TRANSISTOR (DIODE) EQUATION

The diffusion of carriers in any transistor is dependent on the applied voltage across the base-to-emitter junction, with the number of carriers that are introduced being controlled by the diode equation:

$$I = I_s \exp (qV/kT) \qquad \text{(Eq. 3-2)}$$

where,
   I is the current through the junction at the operating voltage and temperature,
   q is the electron charge,
   V is the voltage across the junction,
   k is the Boltzmann constant,
   T is the absolute temperature,
   I_s is the junction saturation current.

The ideal objective with a transistor is to have all of the current "swept up" by the collector, leaving only the input-voltage, output-current shown in the equations of Appendix A. The goal with the bipolar transistor, then, is to make sure that as many of these carriers as possible *do* reach the collector, with a minimum lost through either recombination or diversion in the base region. The more carriers that get through, the higher will be the current gain (or beta) of the transistor. The number of carriers available are controlled strictly by the voltage "V" given in Equation 3-2.

Some of the carriers, which in the base region are minority carriers, will recombine with the majority carriers in the base. Some, also, will find their way out of the base lead to the base terminal. In addition, some carriers will enter the base from the collector and will recombine. Each of these factors lowers the overall transport efficiency for the emitter charges entering the base and, thereby, reduces the overall available current gain. This loss of carriers can be important to the device user. However, the prime importance is to keep the loss from exceeding a certain maximum value. Lost carriers must be supplied by the input signal voltage and this will cause an increased flow of signal current in the input, with the resulting loss of input impedance for the device. As long as the number of carriers combining prematurely is kept less than a specified design number, however, it is possible to minimize the consequences of this leakage. This will be demonstrated later in the experiments.

Device Characterization

The characterization of the properties of any active device, in a way that is most satisfactory for the user, is a problem of more than trivial importance. In some respects, it is well done by device makers, but in others, it is very poorly done. We will now examine this problem briefly, as it is of overriding importance in the use of active devices, and it is of nearly equal importance in both circuit design and circuit troubleshooting. It is, therefore, of significance to the worker who has to make any kind of changes to computing hardware. The following discussion will be divided into two principal aspects, namely, the static configuration and the small-signal configuration. Of particular importance are their interrelations, as there is no way of using any active device which does not involve a properly coordinated combination of both kinds of operating conditions.

The original way of presenting the static data for bipolar transistors was to plot contours of constant value of emitter current as a function of collector voltage and collector current. This arrangement was selected as a consequence of the apparent similarity to the point-contact transistor and to the form taken by the basic back-to-back diode equations (not the modified form used in Appendix A) for the bipolar transistor, and what was then believed to be the overriding importance of current gain. Since the collector current is very nearly equal to the emitter current with these devices (when operating in the active region), these curves appear pretty much as shown in Fig. 3-1.

Emitter current was selected rather than base-to-emitter voltage principally because the voltage is very sensitive to temperature. This is particularly true with germanium transistors because of the rather small voltage required to place them in their active operating region. (Typically, between 100 and 200 millivolts is required for this; in addition, a current change in the collector of as much as thirty times could be caused by a further change of 100 millivolts.) Actually, the junction voltage change per degree Celsius change in temperature is approximately 1.5 to 2 millivolts with both germanium and silicon materials, easily sufficient to justify the choice of silicon as the basic material.

Device users quickly found that the plot of contours of the constant base current as a function of collector-to-emitter voltage and collector current was significantly more useful for routine circuit work. That is the form used for the equations given in Appendix A and for the curves shown in Fig. 3-2.

More information is needed, however, for a person to either design a circuit or to easily troubleshoot one. The basic diode equation (Equation 3-1) clearly shows that the collector current is likely to be primarily a function of base-to-emitter voltage, and the experiments you will make will show that such is indeed the case. As a consequence, the input-voltage characteristics, which are typically

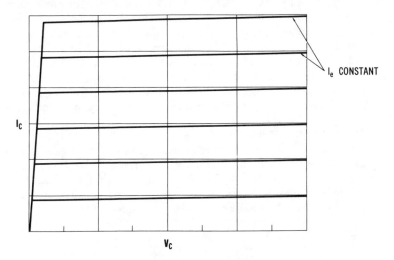

Fig. 3-1. Typical common-base curves for a transistor.

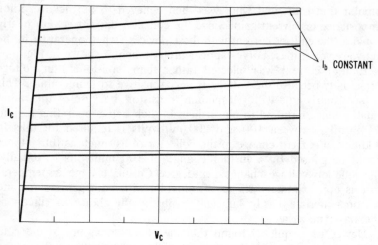

Fig. 3-2. Typical common-emitter curves for a transistor.

contours of the constant base current as a function of the base-to-emitter voltage and the collector-to-emitter voltage, can be very helpful in many difficult applications. Unfortunately, coordinated sets of curves for input and output, with a common collector-to-emitter voltage axis, are only provided by a few foreign device manufacturers.

An optimum set of static contours is only part of the story, however. Any change in operating conditions for a device means that there has been either a small or a large change in the voltage and the current at the device input. As a result, it is important to be able to at least make an estimate of how fast things will change in a circuit when such an input change takes place. This, incidentally, is equally true with switching circuits, since in some way voltage changes must be developed to carry the circuit from state A to state B. The critical condition required to assure that the switching *will* take place is that, in the critical circuit, the voltage amplification around the feedback loop *must exceed unity*. Current gain alone just will not do it.

Transistor Voltage Gain

The data provided for calculating voltage gain, particularly with bipolar transistors, are unfortunately very meager. It is shown by the equation in Appendix A that the voltage gain for a typical transistor amplifier may be simplified to the following approximate form (neglecting such things as the spreading resistances for now):

$$K_v = - g_m R_L \qquad \text{(Eq. 3-3)}$$
$$= - (q/kT) \ I_c R_L$$
$$= - y_f R_L$$

where,

g_m (or y_f) is the transconductance,
I_c is the collector current,
R_L is the load resistance,
K_v is the voltage gain,
q is the electron charge,
k is the Boltzmann constant,
T is the absolute temperature.

The transconductance, which is nominally $(q/kT) \times I_c$, is shown as g_m or y_f. The circuit configuration is as shown in Fig. 3-3. Later, we will see that it can be necessary to include a transconductance-per-unit-current efficiency symbol, κ, in addition to the (q/kT) term. We will use the Greek letter kappa as the symbol for this efficiency. As long as the beta for the transistor being used exceeds a minimum value, this equation is an extremely effective one for calculating the voltage gain. For some special circuit configurations, modifications of the equation, along with diagrams are included in Appendix A. The subject is also discussed extensively in the author's book, *Handbook of Transistor Circuit Design*.[1]

Transistor Beta

The measurement of the beta of a transistor requires the measurement of small-signal currents, both in the base and the collector circuits. The measurement of the signal current in the collector is relatively straightforward, but the measurement for the base is somewhat more difficult because of the fact that the test-signal voltage must be applied to this terminal through a current-limiting resistance. As long as this resistance has a large value when compared to the input resistance of the transistor, it is possible to determine the base current with the aid of this voltage. (You can determine this by finding how much of the applied voltage is available at the base of the device and correcting for it, if you wish.) The equation you can use for this, based on Fig. 3-3, is:

$$\beta = \frac{\left(\dfrac{V_o}{R_L}\right)}{\left(\dfrac{v_s}{R_s + (y_i)^{-1}}\right)} \qquad \text{(Eq. 3-4)}$$

which can be simplified, when $y_i R_s \ll 1$, to:

$$\beta = \frac{R_s v_o}{R_L v_s} \qquad \text{(Eq. 3-5)}$$

where the voltages and resistances are as required in Fig. 3-3.

---

1. Pullen, K. A., *Handbook of Transistor Circuit Design*, (Englewood Cliffs: Prentice-Hall, Inc., 1962).

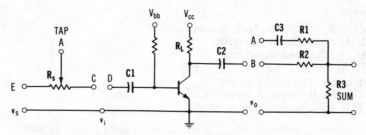

NOTE: FOR BETA MEASUREMENT, JUMP C TO D AND SUM A AND B.
FOR K_v MEASUREMENT, JUMP E TO D AND C TO COMMON. SUM A AND B.

Fig. 3-3. Circuit for a simple transistor amplifier.

Establishment of voltage and current levels, as well as resistances, is not critical. However, they must be established in the experiment. If you wish to include some suggested values, however, make R1 = R2, with values between 10 kilohms and 100 kilohms. Choose resistance R3 so that it has a tenth of the resistance of R1. A typical value of v_s, for current gain, would be about 100 millivolts and, for voltage gain, about 5 millivolts. Also, the collector current may be set to 1 milliampere by changing the value of resistor R_b. A value of 1000 ohms should be a good trial value for R_L. The value of R_b would be between 50 kilohms and 200 kilohms for most transistors. The value of potentiometer R_s might be about 25 kilohms.

Clearly, one of the easiest ways to measure the beta is the adjustment of either R_s or R_L, so that the two voltages are equal. We suggest changing R_L in large steps, and varying R_s for balance. The matter of phase error will be discussed in a later section.

Referring to Fig. 3-3, the two points marked "A" are connected together when making small-signal current- or voltage-gain measurements. When this is done, the segment of resistance R_s between points A and C is used in Equation 3-5 as the value of R_s. When current gain is being measured, point C is connected to point D, the two points marked "B" are connected together, the two points marked "A" are connected together, and the two points labeled "v_o" are connected together. When this is done, resistors R1, R2, and R3 form a summing circuit, and the sum of the signals that are applied to points A and B will appear across R3. It may be examined using an oscilloscope.

A point that needs to be made is that in Equation 3-5, the value of R_s is that part of R_s that lies between the wiper arm of the potentiometer (point A) and point C. With voltage gain, point E is also connected to point D. When voltage gain is being measured, point C is connected to ground, or common.

The actual measurement may be made by measuring the values of

v_o and v_s or, as noted above, by varying the tap point on R_s for balance. (The sum of the two voltages will then be zero.) It is important to hold the actual signal and the direct current into the base constant while doing this. The adder, or summing, circuit off the output and the tap provide the signal to the op amp array, and potentiometer R_s is adjusted for minimum signal. (An op amp may be used to amplify the voltage out of R3, which then goes to a scope.) The beta then is a function of the resistances R_L and that segment of potentiometer R_s that is from point "A" to point "C," as noted for Equation 3-5 and Fig. 3-3.

If the frequency of the test signal is high compared to the beta cutoff frequency, a true null cannot be obtained without using the Lissajous technique, which is fully described later. For that reason, the test signal frequency should be no more than a tenth of the beta cutoff frequency, f_β.

The Lissajous technique is explained in Experiment 3, Chapter 3. With it, you place the input signal on the horizontal plates of the scope, and the differences of the voltages, as obtained from R3, on the vertical plates. Then, adjust for minimum vertical amplitude of the elliptical part of the signal for balance. The presence of a quarter-moon appearance, either up or down, indicates the presence of distortion.

The best scale calibration probably can be obtained if R_s is a calibrated variable resistance and if at least three different values of R_L are available. This configuration will then prove rather effective as a means of determining the values of beta as a function of both the collector voltage and the collector current.

Transistor DC Beta

At the same time that the small-signal beta of a transistor is being measured, it is convenient to measure the dc beta. This is easily done by measuring the currents in the base and the collector leads, and taking the ratio.

$$\beta_{dc} = \frac{I_c}{I_b} \qquad \text{(Eq. 3-6)}$$

The value of this parameter will vary rather widely from that of the small-signal beta, although both will be approximately equal at very small currents. The small-signal beta increases much more rapidly than the dc beta, and then decreases much more rapidly. The reason is that the dc beta is a running average of the value of the small-signal beta. Because of this, the range of variation of the dc beta is significantly less than that for the small-signal beta. It is comparatively easy to estimate the dc beta from the small-signal values, but rather difficult to make the reverse conversion.

The curve for the dc beta as a function of either base current or collector current is of some use to us, however, as there appears to be some evidence that the point for maximum dc beta is at least near the operating point at which high-injection conditions become significant. For device currents which place the dc beta below this point, low-injection conditions apparently apply; above, high-injection conditions apply.

High-Injection Conditions

The existence of the high-injection condition has a significant effect on the transconductance efficiency of a bipolar transistor. With npn transistors, high injection causes an increase of the effective value of kappa, to values which may be as large as 1.6 times the value obtained with low injection. With pnp transistors, the effect is to reduce the effective value, which may then approach 0.6 times the nominal value. The cause of this phenomenon is the large number of majority carriers which must be drawn into the base region to maintain electrical neutrality. This is discussed in more detail in Chapter 4.

The beta value of a transistor is important to the device user, but in a somewhat different way than most people realize. If a circuit is designed so that, for the range of beta values encountered with a given type of transistor, the load placed on the previous amplifier does not seriously affect the voltage gain of that stage, any transistor having a beta greater than that minimum value in the subsequent stage is likely to lead to satisfactory operation. Installation of a transistor with an abnormally high value of beta will show little if any effect. The result of the use of design procedures when complying with this condition will be a substantial improvement of circuit reliability and stability, and a substantial reduction in repair problems.

Transistor Modes of Operation

Since the diode equation is really the one controlling the flow of current in the output circuit of a bipolar transistor, it would appear that making extremely small changes in the base-to-emitter voltage should cause surprisingly large changes in device output current. Your experiments will show that this is in fact the case. The change in base-to-emitter voltage required to double or halve the output current is once again in the neighborhood of 20 mV. One important consequence of this observation is that there really are two modes of operation of bipolar transistors as amplifiers. The first mode is essentially a voltage-amplification mode, in which a signal voltage as small as a microvolt is built up to a value approaching 100 mV, more or less. The second mode is one in which the device is operated in a way to increase the signal current level, with little change in the base-to-emitter signal voltage level. This mode of operation is one method of using your transistor as a true current amplifier.

One cannot say that there is a firm and fixed line of demarcation between these two modes of operation, as internal feedback can make significant changes in the exact transitional voltage level. For example, if an emitter return resistance is used which will develop a voltage gain to the emitter of 0.9, then, the stage can be used as a voltage amplifier with an output level of about ten times the nominal voltage level indicated above. Since the voltage gain to the emitter of an amplifier, using emitter feedback, is less than unity, the safe output level as a voltage amplifier is increased by approximately $(1 - K_{ve})^{-1}$ times the value without the emitter feedback resistance in use.

The behavior of a transistor amplifier having large peak-to-peak input signals compared to 5 millivolts is of more than passing importance to you. For this reason, we have designed an experiment that you can perform, in which you will examine the variation of small-signal amplification of a transistor amplifier as a function of bias changes ranging to 100 or more millivolts. You will quickly observe that with a typical transistor, the voltage amplitude of the output signal *does,* in fact, change very substantially with just this small change of bias, and you can determine the significance of this change.

Your test will show you that when you reach signal levels in the neighborhood of 100 millivolts, you must expect to have to make some changes in the way you handle your circuit arrangements for devices. You will find it convenient to select some peak-to-peak signal voltage level, and then devise your circuits in a way to keep the voltage level from exceeding the chosen level. You will be building amplifiers having voltage gains in the neighborhood of unity, input to input, increasing the current level substantially in each successive stage. As noted above, these are true current amplifiers which have been designed to have minimal voltage gain. They even differ from the amplifiers using transistors which are usually considered to be current amplifiers.

This does not always mean that the gain from input to output on an individual stage is limited to unity. Where the device is operating in the common-emitter mode, the load impedance required will normally be much less than the input impedance of the following amplifier, and you will, in fact, have a stage gain of roughly unity. In the common-base mode, however, it is necessary that the signal be introduced into the following amplifier at an extremely low-impedance level, and some kind of impedance transformer is commonly required as an interstage. In this case, emitter-to-collector voltage amplification must be sufficient to make up for any voltage loss due to the impedance matching network. Voltage gains between ten and one hundred, emitter to collector, are sometimes required. Because of the extremely low input impedance of the amplifier, phase instability and oscillation usually can be avoided in spite of the increased voltage gain. Typically, at least part of the power introduced into the emitter is available as output from the amplifier as well.

Parasitic Elements of the Bipolar Transistor

With any device, passive or active, which functions at a junction of currents, any shunt impedances or admittances may be combined into the device as part of its interelement admittances. As a consequence, the parasitic elements which can degrade such a device's behavior most severely will be primarily series elements, typically resistances and inductances. It is helpful for this reason to introduce simulated series resistances and inductances into the transistor model. Then, you can observe what the effects are that you must anticipate from these elements.

With bipolar transistors, the parasitic elements are principally the spreading resistances (Fig. 3-4), the base-spreading resistance, the emitter-spreading resistance, and the collector-spreading resistance. The effect of each of these is sufficiently different that a set of experiments is well worth while. Further, these experiments will shed light on how the transistor can best be operated under some rather diverse conditions.

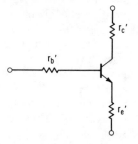

Fig. 3-4. The spreading resistances of a transistor.

The base-spreading resistance acts as an additional impedance in series with the base when the transistor is used as a current-input device. As your experiments will show, when the frequency is much lower than the beta cutoff frequency, you can completely ignore this resistance. Unfortunately, the beta cutoff frequency for many transistors is so sufficiently low that it is essential to operate them from a voltage source instead and, then, you can no longer ignore the base-spreading resistance.

Device manufacturers frequently give an indication of the value of the base-spreading resistance for those of their devices which are intended to operate either as switches or as high-frequency (rf or if) amplifiers. On most other devices, however, the user is left to find out on his own anything at all about the value of the base-spreading resistance. It is for this reason that device users must have some idea of the nature of this resistance and how one might estimate its value. We will discuss this parameter in more detail in the next chapter.

Probably the most troublesome of these resistances is the emitter spreading resistance. It functions to limit the overall transconductance of a typical device through the introduction of a term of the approximate form of Equation 3-7 in the denominator of the voltage-gain equation:

$$[1 + y_i r_b + (y_i + y_f) r_e] \qquad \text{(Eq. 3-7)}$$

In this equation, the value y_f is a phasor admittance which is measured inside of the base-spreading resistance, and the value y_i is also a phasor admittance measured inside, the former being a transadmittance, and the latter an input admittance. In this expression, the $y_i r_b$ term corrects for the signal voltage lost across the base-spreading resistance, and the remaining term corrects for the effect of any resistance in the emitter lead, either external or internal to the transistor. This third term might not appear to be of major importance at first glance, but with an emitter current of 5 mA and an r_e value of 5 ohms, the additive term will have a value of approximately unity, reducing the voltage gain to half the expected value. You will observe this in your experiments.

One very important effect of the collector-spreading resistance is that it increases substantially the minimum dissipation for a device at saturation. The same resistance which increases the minimum voltage at saturation additionally increases the overall amplification generated at the transistor junction for any given set of operating conditions. Under adverse conditions, it might cause an amplifier that should be stable to become marginally stable, possibly even to the point of oscillation.

In a strict sense, the diffusion capacitance is not a parasitic parameter, but it is extremely important in the way it can affect device operation. When you operate a device as a current amplifier, the effect of this capacitance is in parallel with the input, and it is responsible for the loading which helps produce the low beta-cutoff frequency. When you operate a device such as a voltage amplifier from a low-impedance source, the frequency at which loss of response sets in is much higher. The prime cause of loss of gain with frequency, then, is a combination of the base-spreading resistance and the effective value of the input capacitance instead of the total effective input resistance and the input capacitance.

EXPERIMENT 1
Plotting the Typical Operating Characteristics of an Npn Silicon Transistor in the Common-Emitter Mode of Operation

The purpose of this experiment is to plot the typical characteristic curves of several npn transistors of the same general type and designation. We suggest use of the 2N2222 silicon device. From your measurements and the curves which result, we hope to learn much about the properties of these devices, and to be able to compare them with the theory that is described briefly in this chapter.

Special Equipment Required

The following list of equipment is needed for this experiment.

1. A power supply having both positive and negative outputs. Outputs are to be 5 volts, 12 to 15 volts, and variable from zero to about 12 volts.
2. A high-sensitivity multirange milliammeter, with ranges from 100 microamperes full-scale to 100 milliamperes full-scale, and with an internal voltage drop at full-scale of less than 5 millivolts.
3. A differential voltmeter, with a 50 millivolts full-scale range.
4. A digital voltmeter that is adequately sensitive may be substituted for one or the other of the above meters.
5. A solderless breadboard with related hardware.

Procedure

Take your power supply, and test each section of it with your volt-ohm-milliammeter (vom) to be sure that it is fully operational. Then, take your solderless breadboard and mount it on a chassis or a box lid. Connect each of the outer rows of the breadboard, on either side, to its own individual binding post. Connect at least one of the component cross-connectors on each side to a ground terminal. You will then have a flexible arrangement for powering your board. This arrangement is sketched in Fig. 3-5. You will find that this configuration is extremely convenient for much of the testing you will be doing.

The actual wiring diagram you will use is shown in Fig. 3-6. You will notice meter shunts indicated, which may be installed by placing an appropriate resistance from one component strip to another, and connecting the meter across the strips.

If you install your meter shunt on the board, you can change ranges by use of several shunts, and by moving only one meter wire and one circuit wire each time you change ranges. Remember to keep the total shunt resistance in the circuit small enough that it does not affect the circuit operation. This is discussed in more detail in

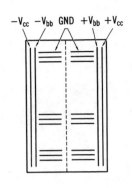

$-V_{cc}$ $-V_{bb}$ GND $+V_{bb}$ $+V_{cc}$

(A) Solderless breadboard configuration.

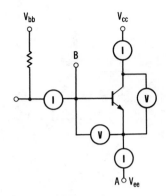

(B) Circuit diagram.

(C) Meter equivalent circuit.

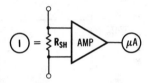

Fig. 3-5. Powering layout for use with breadboard.

Chapter 7. Some of your metering options are shown in Fig. 3-7. The switching arrangements show how you can minimize the effect of contact resistance.

Step 1.

The magnitude of the base current is adjusted with a potentiometer so that a fixed value of voltage can be used for this source. The variable supply can then be used for the collector-voltage source so that a variety of voltage points may be taken at each level of base current. First, make a curve with the base lead open (base current zero). Then, starting with a base current of about one microampere, read the collector current for a 2N2222 transistor (or other npn silicon device) at collector voltages such as 0.5, 1.0, 2, 4, 6, 8, 10, and 12 volts. It is also desirable to get readings between 0.05 volt and 0.5 volt, if you can. Record your data in Table 3-1, starting with the open-base contour. Plot your data on the graph in Fig. 3-8. First, plot your curve for $I_b = 0$ (base lead open). This curve is very important as it will tell you a great deal about the quality of the device you are testing.

Figs. 3-9 through 3-11 show typical curves of the sort that you are likely to observe from time to time during your testing. You should

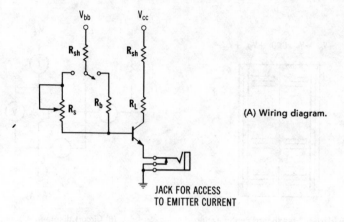

(A) Wiring diagram.

JACK FOR ACCESS
TO EMITTER CURRENT

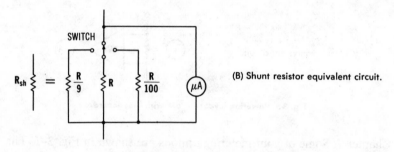

(B) Shunt resistor equivalent circuit.

Fig. 3-6. Transistor tester circuit diagram.

attempt to describe the conditions under which you might expect to get each type of curve that is shown in these figures.

The ideal transistor would have a zero base-current line corresponding to the zero-current axis, as indicated in Fig. 3-9. This device has essentially zero leakage current, and would not begin conducting over the normal operating range of collector voltage. When it is drawing current, at any given base-current level, the contour above

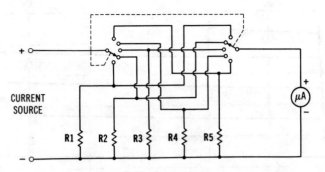

Fig. 3-7. Suggested meter shunt switching arrangements.

Table 3-1. Data Table for Experiment 1, Step 1

I_b				
V_c				
I_c				
I_b				
V_c				
I_c				

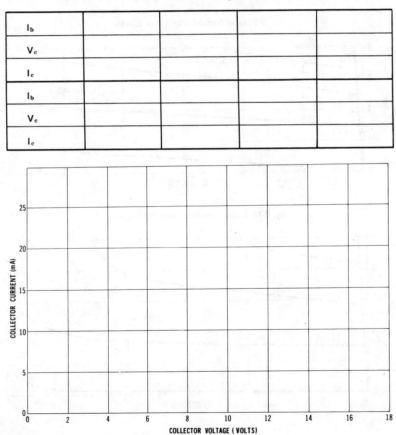

Fig. 3-8. Graph for Experiment 1.

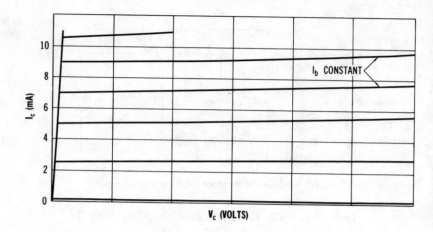

Fig. 3-9. Ideal transistor collector curves.

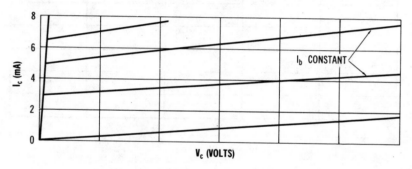

Fig. 3-10. Lossy transistor collector curves.

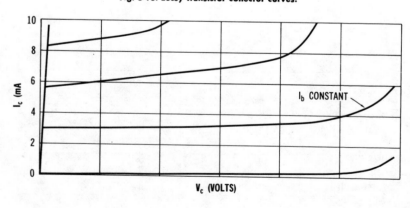

Fig. 3-11. Low-breakdown transistor collector curves.

saturation voltage slopes only a little as the collector voltage is increased. The contour is nearly a straight line. The only factor which can cause a positive curvature is heating. Under high dissipation conditions, the contour will "loop" toward high current as the supply voltage is increased, and will return toward zero collector voltage along a higher current contour than the initial one.

High values of capacitance in the transistor can cause some opening of the contour at all levels of current but, in this instance, as the amount of opening tends to become larger, the larger the overall current. Also, the contour appears to be relatively constant in slope in this instance. (This characteristic is more likely to be noted on a curve tracer like the Tektronic 575 series.)

If the transistor has a high value of resistive loss, then the zero-current contour and all other contours will slope upward, with the slope at low values of current being particularly severe. Transistors displaying this characteristic should normally be discarded as they are very unlikely to give satisfactory results during operation.

When a transistor has an abnormally low breakdown voltage, the contours bend at abnormally low voltages as shown in Fig. 3-11. If these devices are to be used at values substantially below the level of collector voltage, where the curvature is serious, and if there is no inductance in the circuit, the transistor may prove to be acceptable. Nonetheless, they normally should be avoided.

The characteristic curves in the saturation region should all be nearly vertical, and they should remain so as the scope sensitivity is increased and the applied voltage reduced. If there is a substantial break in the slope at increased currents, one can suspect the possible presence of either emitter- or collector-spreading resistance, or both. The problems resulting from this are explored in later experiments.

Naturally, you won't find many typical examples of any of these kinds of devices. As a result, you have to decide whether a given amount of ohmic resistance will or will not give you a problem, and whether the form of the curve as it leaves the region of zero current is acceptable. (The subject of second breakdown is touched on in the next chapter.) The presence of an inductance in the collector load circuit of the transistor can introduce voltage spikes which can have a severe effect on the choice, particularly if you plan to cut off the current in the device.

Transistors which behave similarly to those whose curves are shown in Figs. 3-10 and 3-11 are often called leakers when the supply voltage and current are sufficient to cause problems. Ideally, the search for leakers should be made at the most extreme conditions your transistor is likely to encounter, as the operation of the circuit under adverse conditions can be particularly informative. Since temperature is one of the particularly severe parameters, the use of both heat and cold can prove useful. But, be careful that you do not over-heat your device with your soldering iron.

Step 2.

Now, you can compile a full set of data on your device. Adjust your base series resistance to give you a series of base current values having a ratio of either 1.4 to 1 or else 2 to 1, and record the data in Table 3-1. It is probably convenient to make your base-current adjustments at the same value of collector voltage each time; a convenient value is 2 volts. Note the values of base current, collector voltage, and collector current. Then, change the collector voltage to the next value, leaving the base current fixed. Read and enter your data in Table 3-1. Use the values of collector voltage suggested earlier for your measurements. (For example: 2, 4, 6, 8, 10, and 15 volts, etc.)

Step 3.

After you have taken your set of data, plot the data on the graph in Fig. 3-8. Hopefully, you will obtain a set of curves like those shown in Fig. 3-9. Compare the curves carefully, noting any differences that you find.

The factors that you will want to look for are the same as those discussed in connection with Figs. 3-9 through 3-11. You can detect the presence of either emitter- or collector-spreading resistance by the appearance of the saturation region of the curve, but you cannot tell yet just where it is. Also, you can detect the presence of ohmic leakage by the slope of the zero-current contour and the other contours. You can get an idea about the value of beta by noting the current difference between two contours and dividing that difference by the amount of base-current change which generated it. In short, you can tell a lot about your transistor if you are observant.

Step 4.

There is one additional important piece of information which you need to collect on your device, and that is the typical values of base voltage along the various contours that you just plotted. You were not asked to do that in Step 1 because these curves are seldom presented, and the measurement involves voltage changes. Fortunately, however, the exact reference point is not important in this measurement as long as it is stable.

You will want to tabulate the voltage change, with respect to your selected reference voltage, at the same points that you measured the data in Step 1 so that you can plot your data in a set of curves like those developed in Fig. 3-8. With this new plot, however, you will have both the curves shown in Fig. 3-8 and a new set, based on your new data. You can, in fact, get a very complete set of curves from these data and, as a result, will have an excellent basis for deciding for yourself what configuration can be the most useful to you. The example graphs shown in Fig. 3-12 are especially set up in the way that they are shown. This is for your convenience. In the upper right-hand corner, you can insert standard collector curves like those plotted in Fig. 3-8. Directly below the collector curves you can plot constant-base current contours on a graph of collector voltage vs. base voltage. Then to the left, a set of contours arranged as a function of base voltage and collector current can be plotted. Record your data

Table 3-2. Data Table for Experiment 1, Step 4

I_b				
ΔV_b				
I_c				
V_c				
I_b				
ΔV_b				
I_c				
V_c				

in Table 3-2. How much does the base voltage change from base-current contour to base-current contour? What can you conclude about the relationship of this to what you learned about diodes?

You should have found that between successive contours of constant-base current, the collector current will nearly double. (This is not exact, as the beta does vary.) The change in base voltage from contour to contour will be around 20 millivolts. The change *along* any given contour will usually be only a few millivolts.

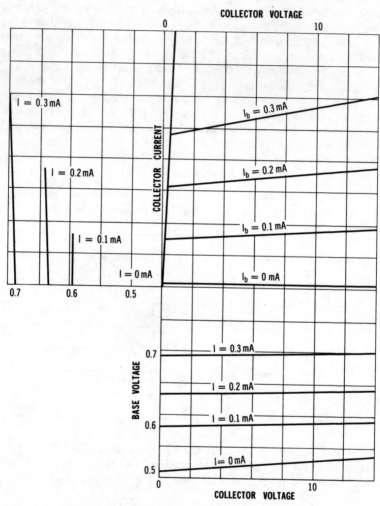

Fig. 3-12. Set of characteristic curves for a typical transistor.

Step 5.

You can now plot your data on graph paper in the format described in Step 4, so that you have both the base-voltage, collector-voltage field and the base-voltage, collector-current field plotted. Examine the graphs after you have them drawn. Where on these curves do you normally operate your transistor? Which pair of curves present data in an open way in the region of operation?

It is clearly evident from these curves that the normal operating area for a transistor involves the curves shown in the right-hand portion of the curve set. Even though some manufacturers provide the graphs shown in the upper half of Fig. 3-12. the author has not been able to find any very significant use for the set involving base voltage and collector current.

Based on these measurements and the resulting curves, you can see why there is a problem of sorts in keeping track of the base-to-emitter voltage, and why a voltage offset can be helpful in understanding what is happening. You want to remember that the magnitude of the offset voltage is a function of temperature, so that precision in its measurement is much less important than is the measurement of the changes in the base voltage. That is why a way has been set up to avoid having to take the difference of two large numbers when measuring the offset voltage. This gives you almost all of the information you need. Since you can easily calculate the dc beta of your device with your data, and you do not now require any great precision in your answer, you have all but the direct measurements of transconductance directly available to you.

Step 6.

Now you want to try to push your device somewhere near its dissipation limit. Look up the dissipation value for your transistor. (The dissipation rating of the 2N2222, at room temperature, is 500 milliwatts.) For this test, you can apply up to at least 200 milliamperes as the peak rating for the 2N2222 is 800 milliamperes. You want to examine the operation of your device in the low-voltage area to find out how it behaves with large currents. You need to find both the location of the saturation line at this value of current, and how high you can run your current when staying approximately one half a volt above the saturation line, and still staying below one half of the rated dissipation level.

It is best to limit the range of the collector voltage rather than the collector current, since we will find that the transconductance is controlled by this current even in practical devices. It is also helpful to keep a current-limiting resistance in series with the collector, since there are sound reasons why the voltage from the collector to the emitter should not exceed one half of your supply voltage. (There are a few exceptions, but usually this is a wise rule.) What do you observe?

When a transistor heats, it is normal for its collector current to increase. If half of the voltage is consumed across a stable resistor, the increase in current will cause a larger decrease in voltage across the transistor than the increase in current, with the result that the power input will decrease. If less than half of the voltage appears across the series resistor, the power input to the transistor will first rise, then level off, and, finally, it will decrease, but only after the voltage across the transistor is less than one half the applied voltage. The curves you have plotted show that for all practical purposes the behavior of the device is independent of collector voltage as long as you have at least one half a volt more supply across the transistor than the saturation voltage is at maximum current. The result is that if you want to minimize heating and maximize reliability, you select the lowest possible collector voltage.

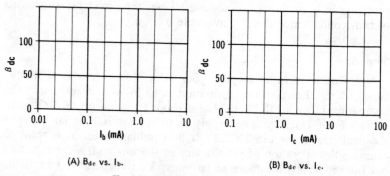

(A) B_{dc} vs. I_b.   (B) B_{dc} vs. I_c.

Fig. 3-13. Graphs of dc beta vs. current.

Step 7.

Now, you can plot the curves for dc beta as a function of the base current. You can, also, plot the curve of dc beta as a function of collector current. Simply take the ratio of the currents at some selected collector voltage (like 2 volts). Plot the curves on the graphs shown in Fig. 3-13. (Your curves should look something like those shown in Fig. 3-14). Later, you will sketch a plot of the small-signal beta

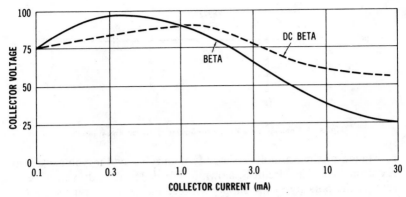

Fig. 3-14. Dc and small-signal beta curves.

on this same graph to see how they differ. (Let's hope the device survives its ordeal!) What can you conclude from these curves?

You are asked to plot these curves two ways since, while device manufacturers usually plot them one way, when you try to use them, you need them the other way. You will usually find the curves plotted as a function of collector current on the data sheets. Have you ever tried to use them in that form? If you want to make a Fourier analysis of the output waveform when a sine wave is applied to the base, how do you do it? You have to find out how the current output varies in terms of a sinusoidal input, and that means you have to know what is happening when going from a sine-wave input (as a function of time) to the output. If you plot it as a function of base current, you have what you need. The author has been looking for a method of plotting it in terms of the output for years, and still has not found one. The conversion back to equal-spaced inputs and then, replotting, has to be followed by another conversion to a sinusoidal plot in terms of time. To put it mildly, it's a real mess.

EXPERIMENT 2
Common-Base Characteristic Curves for an Npn Transistor

The purpose of this experiment is to test the transistor in the common-base configuration and to repeat Experiment 1. You will need to make some minor modifications to the circuit layout used in the previous experiment. The base of the transistor in this instance goes to ground, and the emitter is returned through a variable resistance to the negative supply instead of to the positive supply. The

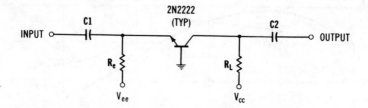

Fig. 3-15. Common-base circuit configuration for an npn transistor.

resistance of the variable resistor is about $\frac{1}{100}$ of the value used in Experiment 1. The circuit is shown in Fig. 3-15. Why do you have to reverse the polarity of the supply source, and why do you reduce the value of the variable resistance that is in series with the emitter?

With a bipolar transistor in the conducting mode, the base potential is always between that for the emitter and that for the collector. If the base is used as a reference point then, it is necessary that the emitter potential be the reverse of that of the collector. When the emitter is used as reference, the base-potential polarity is the same as that of the collector. Since the emitter current may be as much as 100 or more times the base current, the resistance in the emitter circuit cannot be over $\frac{1}{100}$ of that used with the base to absorb the same overall voltage. Actually, the emitter current is slightly larger than the collector current. Since the current equation for your device is

$$I_b + I_c + I_e \equiv 0, \qquad (Eq. \ 3\text{-}8)$$

you can see that the base current must be much smaller than either of the other currents when using a good transistor.

One conclusion you can draw about transistor curves that are plotted as contours of the constant emitter current as a function of the collector voltage and the collector current is that the curves really do not tell you very much about the characteristics of your device. The emitter-current contour is too close to coinciding with the collector current it defines.

Step 1.

Modify your wiring configuration to conform with Fig. 3-15. Then you can start collecting data for plotting a set of curves like those shown in Fig. 3-8. Measure the emitter current, the emitter-to-base voltage, the base-to-collector voltage, and the collector current, and record them in Table 3-3. Start with an open circuit on the input (or emitter) lead and plot the zero bias contour. Then, starting at about

100 microamperes, introduce an emitter current and measure the collector current at several values of collector voltage for each current. Then, double the emitter current and repeat the tests.

Table 3-3. Data Table for Experiment 2, Step 1

I_e				
ΔV_e				
I_c				
V_c				
I_e				
ΔV_e				
I_c				
V_c				

Step 2.

These data can be plotted either as collector sets or as complete sets involving the three fields as was done in Experiment 1. It is important to introduce a stable reference voltage so that the change in the emitter-to-base voltage can be tabulated. With silicon transistors, the offset voltage should be about 0.4 to 0.5 volt, and with germanium transistors, between 0.1 and 0.2 volt. As long as the differential voltmeter is calibrated so that it is reasonably precise, the reference zero can be reset after each measurement. Unless the calibration is good, however, a cumulative error can result. The use of a digital voltmeter of an adequate sensitivity is particularly good for this, it is normally calibrated very precisely.

When you are plotting an input family of curves, as you were in Experiment 1 and as you will in this experiment, it is best to leave your reference bias set at a known value, like 0.4 or 0.5 volt (for a silicon device), and measure all voltages with reference to that value (Fig. 3-16). When you are confirming that the voltage change re-

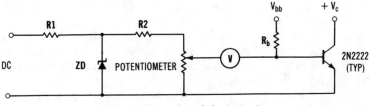

Fig. 3-16. Offset-voltage balance circuit.

quired for a two-to-one change in collector current is, in fact, near the 18-millivolt value, it is more convenient to rezero the bias each time so that you can observe the changes as they occur.

Step 3.

You will want to measure the "alpha," or current gain from the emitter to the collector, when you are measuring the beta directly. You, therefore, should determine the ratio of the collector current to the emitter current for a series of emitter current values (Table 3-4) at a typical value of collector voltage; for example, two volts. Later, you will add data of the small-signal alpha to these data, to see what differences, if any, you may find.

Table 3-4. Data Table for Experiment 2, Step 3

I_e				
I_c				
I_e				
I_c				

EXPERIMENT 3
Measurement of Small-Signal Alpha and Small-Signal Beta

The purpose of this experiment is to learn something about the most commonly used small-signal parameters for a bipolar transistor, namely, *beta* and *alpha*. As you know, these are the current gains from base to collector and from emitter to collector, respectively. Your initial efforts will be to measure the beta at a variety of points over the field of curves that you plotted for your transistor. Then you can do the same with alpha. However, plot its contour values on the common-base family of curves in this case.

There are two ways in which these curves can be presented. They can be presented as a function of either the input or the output current as you have already done with the dc beta, or they can be plotted as contours of constant value on a set of collector characteristic curves. You should do it both ways. You will find that you will obtain more useful information when the constant-value curves on the collector family are used. It is worth noting that constant-value contours of any small-signal parameter may be plotted on static characteristics in this way. The important thing is to determine the most useful and meaningful arrangement.

For this experiment, you need a low-frequency audio oscillator, whose frequency should be about 500 to 1000 cycles. This source must have a very low-impedance output, and it must have a good sinusoidal waveform. Why is it necessary that your source have a low-impedance output?

The reason is that transistors require some power at their inputs and you need to have the power available. The power demand is greatest in the common-base mode when you are putting your signal into the emitter, but it is significant in both configurations. You can use an EXAR 2206 signal source, following it with a low-impedance repeater such as an emitter follower if you wish. (The design of emitter followers is explained in the next chapter.) A possible repeater circuit is diagrammed in Fig. 3-17. It should be operated from +5 volts to −5 volts, with the base returned to dc ground.

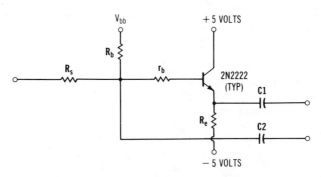

Fig. 3-17. An emitter-follower repeater circuit.

Step 1.

A circuit for measuring small-signal beta (or small-signal alpha with the base and emitter interchanged) is shown in Fig. 3-3. You will notice that the output of your signal source is connected to a series resistor with a potentiometer in parallel with it (but returned to ground rather than to the base). You will have to choose the size of this resistor based on the transistor and the current level you are using in your measurements. A good trial starting point is $(10kT/qI_b)$. (The value of (kT/q) is 26 ohm-milliamperes.) You will find your oscilloscope very valuable in these tests, so I hope you have it ready.

At this point, it may be well to digress briefly to the methods you can use in applying the scope most effectively for these tests. Basically, you have two ways you can proceed. Each will give you some useful information that is somewhat harder to obtain when using the alternative method. These are, first, the use of the scope as an amplitude-time device and, second, the use of the scope to plot an "X vs. Y" input-output analysis of the response. This latter method is sometimes called a *Lissajous figure* method.

When you use the scope as an amplitude-time device, you place the signal you wish to study on the vertical deflection amplifier, (the Y-input) and set the horizontal sweep rate to give you one or two full cycles of the signal across the face of the tube. The first step is to examine the waveform for application to your device under test (DUT) that is coming from your oscillator. This usually should be a good sine wave. You will next want to examine the corresponding voltage at the transistor input. Leaving the base current fixed, you can vary the value of the series resistance in the signal line. As you decrease the value of this resistance, you will discover that your waveform is becoming distorted compared to the sine wave that you had. Can you explain why?

The reason for this is that the input to the transistor does *not* present a constant resistance to a large signal, but varies exponentially (with some modification from the base-spreading resistance). When the signal amplitude is small enough, this variation does not matter. The result is that the waveform at the base will be distorted if the signal amplitude there is over ten millivolts.

Step 2.

Now, you should examine the waveform at the collector output. Even with a distorted waveform at the base of the transistor, the collector waveform may be good. Why is this?

The answer is that when the input admittance of the transistor increases, the forward transfer admittance also increases, with the result that the output-signal amplitude and waveform are essentially dependent on the source waveform, not on the waveform at the base. This will be true as long as the beta of the transistor is relatively constant over the waveform. It is also why the beta is often thought to be the primary basic parameter for the bipolar transistor. The actual value of the voltage gain from source to output is very sensitive to the value of the beta, however, and the value of the beta depends on the minority carrier lifetime, a value that even now is relatively very poorly controlled.

Step 3.

You can eliminate this nonlinear distortion and show better what you want to know by going to small-signal operation for your measurements. When you use the scope as a signal observation device, it indicates the signal voltage present at the test point. You find that when you reduce your signal amplitude enough, the transistor appears to be linear, and you will find no noticeable distortion at any location in the circuit.

To use your scope as an X–Y indicating device, you connect the input voltage from the source to the horizontal axis, and the amplifier output to the vertical axis. Under these conditions, you will find that the output signal will generate a sloping, almost perfectly straight, line across the face of the scope screen (in quadrants two and four). The trace may have some separation of the two directions of the traverse, but the figure should look like an ellipse. As a matter of fact, the higher the frequency, the wider the ellipse will appear to be. This is a *Lissajous figure*. Why does it slope this way? (See Fig. 3-18B.)

The backward slope is a result of the inversion of the signal polarity which takes place in a common-emitter transistor amplifier. This effect is very important in digital electronics.

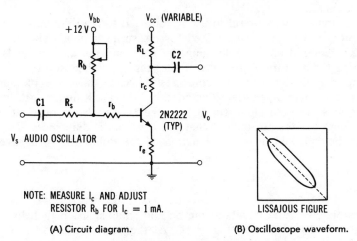

NOTE: MEASURE I_c AND ADJUST
RESISTOR R_b FOR $I_c = 1$ mA.

(A) Circuit diagram.

(B) Oscilloscope waveform.

Fig. 3-18. A basic transistor test circuit.

You may notice some imperfections in the ellipse at higher signal amplitudes which disappear as you reduce your signal amplitude and increase the sensitivity of your amplifiers in your scope. What causes these?

These irregularities are caused by variations in the value of beta over the operating path that is followed by the signal. The load contour may get too close to either the saturation line or the cutoff line for the device. They both can give lots of trouble!

Again, if you take the horizontal input directly from the base of the DUT, and your voltage amplitude there exceeds 10 to 20 millivolts, you will find that the plot of the output against the input is very distorted. Again, this is due to the nonlinear input resistance of the transistor. It largely disappears at small values of input voltage. You will find it useful to plot some of the curves that you observe.

These tests will tell you several things. First, you should operate your devices at small enough signal voltages so that you can largely get a linear operation, or else use compensating circuits if you wish to minimize distortion. The second thing is that you probably can get a better measurement by using a balancing technique in conjunction with the Lissajous figures. Your circuit diagram shows just such a summing-repeater amplifier. One input is connected to the arm of the potentiometer and the other to the amplifier output. You can adjust the beta potentiometer until you get approximately zero output and, then, read the beta directly off the potentiometer. With the Lissajous figure, the average slope of the combined signal will appear to be horizontal.

This circuit is very useful to use if you need to read the beta of a transistor, and it can also be used for measuring alpha. (The only changes required are interchanging the base and the emitter leads, reversing the polarity of the base/emitter supply, and substituting the appropriate smaller value of source resistance, at most 5% of that used in the common-emitter configuration.) You will find it instructive to measure both alpha and beta at the points where you plotted the dc beta and the dc alpha so that you can see the differences. This should be followed by an explanation of the differences you have found.

You probably found that both the dc and small-signal beta behave much as was indicated in Fig. 3-14. At the same time, you probably found you could hardly separate the two alpha curves. The small-signal beta curve crosses the dc beta curve when the latter is horizontal, and they both start out from the same point at zero current. The variation of the small-signal beta is much more rapid than that of the dc beta, since the latter is, in effect, the average of the former from zero current to the base current in question.

EXPERIMENT 4
The Measurement of Transistor Transconductance

In this experiment, you will use the configuration used in Experiment 3 to measure the transconductance of your transistor so that you can observe why this parameter is more important than normally recognized. You can use your small-signal test configuration in the common-emitter arrangement with only small changes.

Step 1.

You will apply signal voltages of about 5-millivolts amplitude to the base of your transistor. You will need to measure the voltage at that point. As long as the operating frequency is below the beta cutoff frequency, you can leave the signal source resistor in place as a means of reducing the available signal. The potentiometer may now be connected from base to ground. Also, you will want to be able to reduce your collector load roughly in half every time you double your collector current. Why do you want to do this?

It is helpful to do this since the nominal transconductance of your device is ideally proportional to the device current. To get a constant overall output to compare against the input, it is helpful to halve the load resistance every time the output current is doubled. That way, the value of the product $(q/kT)\, I_c\, R_L$ is kept constant. Since you are neglecting the effects of high-injection and the spreading resistances for now, this will not be exactly constant but it will be surprisingly stable, and the variations can easily be taken care of with the potentiometer.

Step 2.

Any variations caused by any of the parameters that tend to degrade the constancy will show in your measured data. You should plot the ratio of the output voltage (for a constant value of the product $(I_c\, R_L)$ to the input voltage (at the base of the transistor) on your graph for small-signal beta. (With an ideal transistor having zero base-spreading resistance and no other defects, this will be a horizontal line.)

The equation for the transadmittance y_f or g_m, whichever you prefer to call it, is given by the expression:

$$y_f = v_o / (v_i R_L) = \kappa (q/kT) I_c \qquad \text{(Eq. 3-9)}$$

Since this ratio is actually the product of kappa and (q/kT), and you know that (q/kT) has a value of approximately 39 mhos per ampere (siemans per ampere), you can evaluate the variation of kappa directly. (This neglects r_b and the other spreading resistances.) We have chosen the npn transistor for this series of tests because the value of kappa breaks upward to a larger value in the high-injection condition, and this break should be detectable even in the presence of moderate values of r_b and r_e.

The equation for amplification of a transistor including the kappa and the base-spreading resistance, takes the form:

$$K_v = - y_f R_L / [1 + y_i r_b] \qquad \text{(Eq. 3-10)}$$
$$= - \kappa (q/kT) I_c R_L / [1 + (q/kT) I_b r_b]$$

where emitter- and collector-spreading resistances have been neglected. Interesting enough, the input admittance for the device may be stated in a similar form:

$$y_i = (q/kT) I_b / [1 + (q/kT) I_b r_b] \qquad \text{(Eq. 3-11)}$$

The kappa factor, which technically should be considered in Equation 3-11, normally can be neglected there. These equations can give you a rather good low-frequency representation for your transistor as a component of a simple amplifier.

What can you deduce about the values of these parameters, based on the data you will enter in Table 3-5?

Our experience indicates that you need to know something about r_b, and whether there is a significant magnitude of r_e in your device. Its value should be less that $0.1 [(q/kT) I_e]^{-1}$ at the maximum value of I_e that you expect to use to be sure that it has a negligible effect on the operation of the device. You also need to know where high injection begins so that you can correct for it. In addition, you will need to know what the minimum value of beta that you are likely to encounter is, so that you can design your source circuit to provide the required base current.

Table 3-5. Data Table for Experiment 4

v_o				
v_1				
I_b				
I_c				
R_L				
y_f/I_c				

Step 3.

It is interesting to attempt to estimate some of these data from the things you can measure on your transistor. The techniques described are approximate, as they assume things which really are not quite true. However, if they are used carefully, it can lead to some kind of an idea of what the correct answer is.

The value of r_b can be estimated at relatively small currents by the use of the equation:

$$r_b = \frac{(\Lambda I_c - y_f)}{\Lambda I_b y_f} \qquad \text{(Eq. 3-12)}$$

where,

y_f is the measured transconductance at a small current value (like 1% of the rated peak value),
Λ is always (q/kT),
I_b is the base current,
I_c is the collector current.

The value of Λ is approximately 39 mhos per ampere.

The value of r_e can be estimated at higher current values by determining the degeneration effects that it introduces. It has to be separated from the r_b component. The following equation can prove useful in its calculation. You should remember that it will be significant only if you can clearly see that the transconductance for your device has become roughly constant as you increase the collector current. The approximate equation for r_e is:

$$r_e = (\kappa \Lambda I_c - y_f\{1 + \Lambda I r_b\})/y_f \kappa \Lambda I_e \qquad \text{(Eq. 3-13)}$$

where the definitions are the same as for Equation 3-12.

You should make some trial calculations for r_e at several points of operation for your transistor. Measure all three currents with your high-sensitivity meter and your transconductance with the bridge configuration used earlier in this experiment. Does your resistance value given by the equation sound reasonable? Select collector currents in the region where the dc beta is decreasing with an increasing collector current, and the collector voltage at least twice the saturation value. Explain:

In the region where the transconductance contours seem to get far apart, and the value is relatively constant as the collector current is increased, you can expect that the emitter-spreading resistance is the limiting factor. Now, look at some typical transistor data sheets and see how much of the information, that you have just measured, could have been found on sample data sheets. What comments do you have on that?

Equation 3-13 shows why the transconductance curves spread due to r_e. You probably have found that much of what you now know about transistors is not included on device data sheets and, in fact, probably most of it is not even mentioned in textbooks. This discussion possibly can give you some guidance on how to proceed in finding out what you may need to know.

EXPERIMENT 5
The Transistor as an Amplifier

You want to learn as much as you can about the more general behavior of your transistor when used as an amplifying device. You have noted that under specialized conditions, it appears to have a current gain and, under other conditions, it appears to be a transconductance device. It is particularly important that you decide which is the more meaningful representation for your device, and in what way each representation is important in your use of the devices. This set of tests is best made with a simple amplifier, one in which you have complete control of the configuration. Then, you can make changes and observe the results, comparing them with what you have already learned, to see how they apply.

Step 1.

You will need your low-impedance source of an audio sine wave for this test. Also, you need a means of introducing varying, but known, amounts of resistance in series with each of the three leads on your transistor. You also need to be able to monitor the current in each lead without disturbing the operation of your device. In the output, you will actually want two resistances in series, one of them representing the load circuit, and the other simulating the parasitic internal impedance (called the collector-spreading resistance), within your device. You will also want to be able to substitute several different transistors of the same EIA code, and be able to adjust the collector current of each to the same value. A circuit that can be used for this purpose is shown in Fig. 3-18.

Your audio-signal source will need the special emitter-follower output discussed in the introduction to Experiment 3 and in later paragraphs. You, then, can use a series resistance to simulate whatever effective source impedance you may desire or require in either the base or emitter lead, as needed. If you find that the amplifier output-signal magnitude is almost constant for moderate values of this series resistance, you can conclude that your device input is probably voltage controlled; on the other hand, if you find it is not constant, you may be able to conclude that it is current controlled. What do you find?

You will find that as long as the dc base current is held constant, and as long as the product $(R_s + r_b)y_i$ is less than unity, the output will prove to be almost independent of the value of R_s. As the value of R_s does get larger than is permitted by the previous expression, however, the gain will be reduced. The input-signal current, then, is decreasing substantially. So far, all you can say is that the input appears to give a constant output as long as the signal voltage at the input is constant. Your test setup will show that if you do keep the input-signal voltage constant, the output stays constant. But, so does the input-signal current! All you have *proved* is that the input is lossy.

Step 2.

Examine the output waveform from your amplifier with your oscilloscope. Also, examine it both at the output of the oscillator and at the input to the transistor. Do you have as good a sine-wave signal at both the transistor input and output as you have at your oscillator output? What happens as you vary the amplitude of this sine-wave input from the oscillator? Describe what happens with your transistor. (See Fig. 3-19 for a good generalized test circuit.)

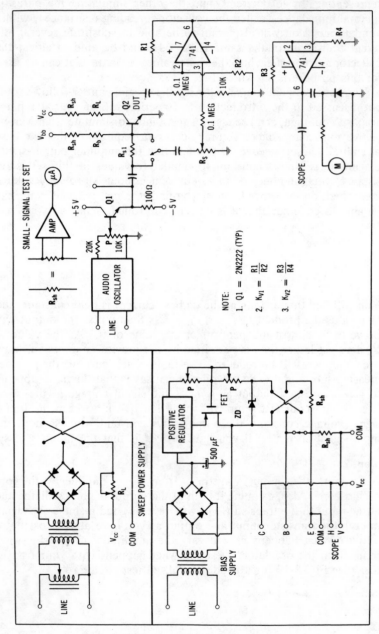

Fig. 3-19. Transistor generalized test network.

118

As we varied the signal voltage, it was found that the sine wave was quite good as long as there was less than 10 to 15 millivolts of signal (peak) applied to the transistor base. As the signal level was increased, the waveform at the base degenerated quite rapidly if we had a large input resistor, and quite badly at the output, if we did not. When the signal voltage at the base was set to about 10 millivolts, however, the waveform remained nicely sinusoidal at both locations. This can be explained by the use of Equations 10 and 11, both of which show that these important admittances change rapidly with base and collector current. Apparently, the relative constancy of beta is a consequence of the fact that these two parameters vary "in step" reasonably well. The input conductance and the transconductance both approximately double with a doubling of the collector current. But the relation is only marginally good enough for our purposes. With a voltage signal source, the output waveform flattens for reduced values of collector current; for a current source, the base-signal voltage flattens for increases of base current. The typically low value of the beta cutoff frequency limits our use of the current-mode linearization to very low frequencies.

In short, there are conditions under which you can cause your device to behave like a current amplifier, and other conditions under which it behaves like a voltage amplifier (strictly a transconductance amplifier). You definitely can get a better frequency response operating in the transconductance mode (as you will see directly), but you have not yet answered conclusively the question as to which kind of a device you really have. To settle this question, it is necessary to test several other devices having the same EIA code designation to see which parameters are more consistent from device to device.

Step 3.

To check which parameters are more consistent, replace your test transistor with another one having the same EIA code, and adjust your base resistance controlling the base current so that your new device is drawing the same value of collector current that the first device did. Is the base current the same? Is the series resistance for control of base current the same?

Your tests will show that the base current usually may be significantly different, and that the bias-control resistance will also usually be different. You will find, though, that with equal collector currents and minimum resistance in the signal input line, the output voltage for the new device will be almost identical with that for the first device. In fact, as long as you are operating under comparative voltage-input conditions, you can put any one of a number of transistors into the circuit and will get the same result. You only have to be sure that the collector currents are the same. They do not even have to have the same EIA code.

You will find it both interesting and instructive to repeat this experiment with a variety of transistors to verify for yourself that this is indeed true. But it definitely is *not* true if you operate your device in a current-input mode.

What all of this means is that if you want to get repeatable results with a series of devices with the same EIA code designation, you must use the transconductance mode of operation and adjust the circuits so that, in each case, the device will have the level of collector current you have designed it to have. This should be interpreted as indicating that the transconductance mode of operation for a bipolar transistor is more circuit independent or a more reliable mode of operation. It is interesting to note that you can predict this from the basic physics of the Ebers-Moll model for a bipolar transistor. (See Appendix A.)

Step 4.

Now, you can introduce a resistance in series with the emitter for the transistor and examine the consequences. Mathematically, one consequence of this is to modify the denominator term of Equation 3-10 (and, also, Equation 3-11) to the form:

$$y_f = \frac{\kappa \Lambda I_c}{1 + \Lambda(I_b + I_c)(r_e + R_e) + \Lambda I_b r_b} \qquad \text{(Eq. 3-14)}$$

where the standard definitions apply. You notice that the term in the first parentheses in the denominator is the emitter current, and the second term includes all of the resistance between the emitter side of the input junction and ground reference (or the source of emitter current at the point of bypass if part of the return resistance is bypassed). The final additional term in the denominator corrects for loss from the base-spreading resistance. (In this equation, R_s has been assumed to be zero; if it is not, its value should be added to r_b.) Set up your circuit so that you have $R_e = 10$ ohms and $I_c = 10$ mA. Now, plot a curve of the effective transconductance of the circuit for collector currents from 100 microamperes to about 50 milliamperes. (Use two-to-one ratios for your current.) What do you observe? Explain. Plot your curve in Fig. 3-20.

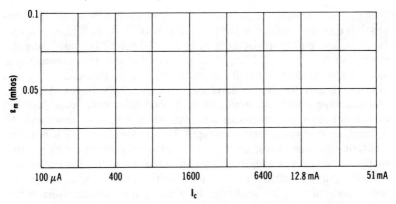

Fig. 3-20. Variation of effective transconductance with collector current.

The overall transconductance will appear to be approximately 0.1 mho until the collector current of the device is less than 2.5 milliamperes. Then, it will decrease roughly linearly with the current. The equation for determining the "turn-over" point is:

$$39 I_e R_e = 1 \qquad \text{(Eq. 3-15)}$$

or,

$$I_e = (39 R_e)^{-1}$$

Step 5.

At what value of R_e would you expect this new term to introduce a significant effect on device transconductance? To find this out, solve Equation 3-15 for R_e. For resistance values greater than those yielded by this equation for the emitter current in question, the transconductance for the device at that level of current will be largely controlled by the emitter resistance. For the equality condition, the transconductance should be half the nominal value. Test it and see. Is it?

Unless you have a transistor with large values for base- and emitter-spreading resistance, your results should check.

Step 6.

You can now make more extensive measurements of the transconductance of your transistor as a function of both collector current and collector voltage. It will be convenient to arrange your circuit so that you can introduce emitter resistance if you wish to do so.

One or more resistors with switches across them will serve that purpose. We suggest collector supply voltages of 2, 4, 6, 8, and possibly 10 volts, and emitter resistances of 1, 2, 5, 10, and 20 ohms. Using the graphs shown in Fig. 3-21, plot a curve for the transconductance against the collector current for each resistance value chosen. Use a different graph for the various collector supply voltages. What do you discover about the variation of transconductance as a function of the collector current, with and without the emitter resistors? As a function of the collector voltage? If you find a transistor whose transconductance characteristics are significantly affected by the value of the collector voltage, mark it and set it aside for further testing with the curve-generation tester used in Chapter 4. Also, when you are using emitter resistance, see how much input-signal voltage you can apply before distortion occurs at the output, and compare with

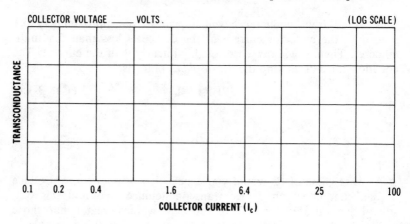

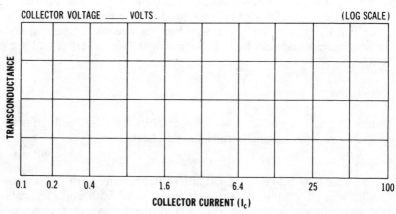

Fig. 3-21. Plots of transconductance

the results when the emitter resistance is effectively zero. Describe your results.

You will find that the value of transconductance begins to be limited at the value of current indicated by Equation 3-15 as the resistance value in the emitter return is changed. You will further find that when you are operating under conditions where the emitter resistance has a negligible influence, the distortion will set in at a rather constant milli-volt signal level. You will further find that if you multiply the signal-input level, with no emitter resistance, by the value of the denominator

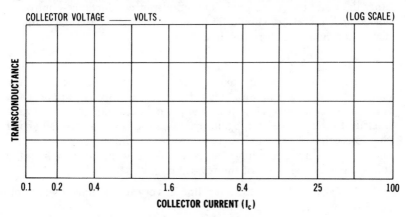

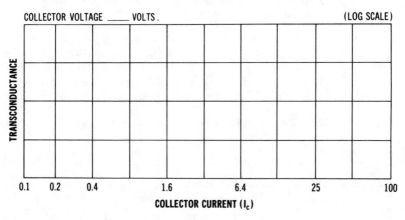

against collector current.

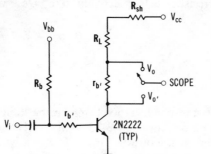

Fig. 3-22. Circuit for testing collector-spreading resistance.

of Equation 3-14, the result will be the approximate signal level for distortion with the corresponding value of emitter resistance. You will find it hard to come to any conclusion based on this evidence other than that the transistor is, in fact, a transconductance-controlled device and, therefore, should be basically treated as a voltage amplifier rather than as a current-gain device.

Step 7.

You can find out what the effect of an excessive resistance in the collector lead of a transistor might be. For this purpose, connect the small series resistance (Fig. 3-22) in the collector circuit as if it were internal to the device and, again, examine the static and small-signal characteristics of the device. This resistance will simulate the presence of collector-spreading resistance. You want to examine the following characteristics of the revised circuit:

1. The shape of the saturation line at very low voltage as the collector current is increased.
2. The comparative voltage gain to the collector on both sides of the simulated spreading resistance.

For this test, you can eliminate the emitter degeneration. The saturation data can be tabulated by passing a base current that is roughly equal to the collector current into the transistor and varying the collector supply voltage. The voltage and current values can be tabulated in the first section of Table 3-6, where the "primed" value is

Table 3-6. Test Data for Experiment 5, Step 7

I_b				
I_c				
V_c				
V_c'				

at the collector side of the "spreading" resistance, and the "un-primed" value is on the load side. The data for the nominal voltage gains in the active region (with the collector voltage greater than saturation, and the base current at its normal value) are tabulated in Table 3-7 for a collector supply voltage of about 3 volts where, again, the "primed" values are at the collector side of the simulated

Table 3-7. More Test Data for Step 7

I_c				
v_o				
v_o'				
v_i				
K_v				
K_v'				

spreading resistance, and the "unprimed" values are on the load side. In the circuit diagram (Fig. 3-22), a 100-ohm load resistance and a 50-ohm simulated spreading resistance are suggested. Describe and explain what you have found.

Regardless of the base or the emitter configuration, the voltage gain is linear with the load resistance. In fact, any "parasitic" series re-sistance, including a collector-spreading resistance, will increase the total amplification, even though the useful amplification at the output terminals is unchanged. This internal amplification will be the crit-ical factor for determining the effect of the Miller capacitance, and can lower the maximum operating frequency for the circuit.

Step 8.

Once again, apply your sinusoidal signal to the input of the tran-sistor, keeping the amplitude small enough so that the output is sinu-soidal. Then, vary the collector current by varying the value of re-sistance in the base lead, being careful to keep your resistance in the collector return small enough that the transistor does not go into saturation. (A good starting value is 100 ohms.) Keep your collector supply between 2 and 4 volts. As you vary the base bias, you are tracing out the operating load contour of your transistor as an amplifier. This procedure probably is the ideal way to test out a trial amplifier design. You can get data on small-signal amplification on a point-by-point basis, plot them, and determine any changes you may need to make before you select your operating point. The important thing to remember is that *all* of the parameters you are using are stable ones, parameters that do not change much from device to device, or within a given device.

In making this test, you can start with a very small load resistance, counting on linearity for scaling purposes, but you will then wish to substitute the selected value and rerun the test. You want to select the maximum value of collector current that you will wish to draw, and adjust your collector supply voltage so that your device will draw that amount of current just before it goes into saturation. Otherwise, you are wasting power and burdening your transistor unnecessarily. To verify this, vary the collector supply voltage and record data at typical points along your respective load lines, all for the same chosen value of load resistance. Before you take your data, however, remember one thing. The transistor that may be getting its signal from the amplifier you are designing will have to draw power from your amplifier. If you want to minimize this effect, you will want the loading *not* to load this amplifier. What do you do now?

Supposing your next amplifier is to draw 5 milliamperes of collector current, and you want it to be able to operate with any transistor having a beta greater than 20. The required input or base current for this transistor, then, is 250 microamperes, and that corresponds, by Equation 3-11, to an input admittance of about 0.01 mhos, or an impedance of 100 ohms. Your load impedance for the amplifier in question, therefore, should not exceed about 50 ohms. Surprised? Most people would be. If you have to have a higher input impedance and/or a better linearity, you will have to introduce emitter resistance to increase the input impedance of the following amplifier and improve its linearity. What did you find with your circuit?

When saturation occurs, the output signal will be severely distorted on the high-current part of the waveform. If the device is operated near cutoff, there will be flattening on the other end. If there is severe loading, the waveform will again be degraded, and the amplification will also be reduced.

To verify all of this, further measurements of the stage gain can be made with varying values of load resistance, and the stage can be coupled into a following amplifier. The easiest way to verify the loading effect is to vary the current in the output stage and to watch for variations of the amplification resulting from loading. You will want to use two amplifiers coupled together for this, and vary the base current of the second one. Use a 100-ohm load resistance for the first stage, and vary the collector current for the second stage from 100 microamperes to 10 milliamperes. In Table 3-8, record the voltage amplification of the first stage as a function of the collector current in the second stage. Set the collector current in the first stage at 1 milliampere. You will have found that up to a certain value of I_{c2}, the value of the first-stage voltage gain is unchanged. Above that

Table 3-8. Test Data for Experiment 5, Step 8

I_{c2}				
K_{v1}				
I_{c2}				
K_{v1}				

value, it degraded as the current increased. This shows the stage loading effect.

Step 9.

You have now observed the effects caused by the presence of spreading resistances within your transistor and know how to detect their presence. The base-spreading resistance affects the frequency response of your amplifier, and it also reduces the effective transconductance in an identifiable manner. The emitter-spreading resistance increases the amount of input voltage which can be applied without encountering distortion, and also has a clearly recognizable effect on transconductance. The collector-spreading resistance shows up primarily through its effect on the saturation voltage for the device. It does *not* affect the value of device transconductance as a function of collector current. You will find it interesting to test some transistors just to look for variations which may suggest the presence of these phenomena.

Step 10.

It is important to determine if there is any reason, ordinarily, to apply collector supply voltages above 2 or 3 volts when a device is operating in the common-emitter mode. (The presence of resistance in the emitter return can influence this decision.) You have already seen that you will get significant distortion if you apply more than

10 millivolts input signal. This normally indicates that there is less than a volt total output unless you are using emitter degeneration. Repeat your transconductance measurements at several of the collector voltages recommended in Step 6, and at several of the collector currents also recommended there. Record your data in Table 3-9. Do you have difficulties with waveforms if you limit your voltage gain to 10? To 20? To 50? What are your observations and interpretations?

Table 3-9. Test Data for Experiment 5, Step 10

V_c				
I_c				
K_v				
OK?				
V_c				
I_c				
K_v				
OK?				

Now, insert a 10-ohm resistance in the emitter lead and repeat the above experiment. Insert your data in Table 3-10.

Table 3-10. More Data for Step 10

V_c				
I_c				
K_v				
V_c				
I_c				
K_v				

You will discover that the operating point for the transistor will vary from near cutoff to near saturation in this test, so you will want to indicate what the basic collector supply voltage was for each series of data. It will be approximately the same as the voltage of the device when the collector current is only 100 microamperes, so you can use this as your guidepoint. What has happened to the amplifier characteristics with this arrangement? Explain.

At collector currents over 1 milliampere, you will have noticed that the expected linearization of the amplifier is beginning to set in. It is highly doubtful if you saw anything other than, possibly, saturation which might indicate that an increase in supply voltage is desirable. But if you estimate the total voltage change you can get out of the stage, you will find it is much more than you are likely to need for the following amplifier anyway. In other words, it would appear that only in unusual situations would you need more than 2 or 3 volts for your collector supply.

Step 11.

To document this point further, assume that you decided to use a 10-volt supply. What choices do you have? (You will want to set up the circuits and test them!)

One commonly used approach is to design for maximum output. If you select 10 volts, and decide you need 10 milliamperes of current with 8 volts across the load impedance, this would call for 8/0.01 or a load impedance of 800 ohms. Now, the base input resistance for the next transistor is likely to be about 500 ohms, coupled with a capacitor. The resulting effective load resistance is about 300 ohms, or less than a third of our chosen value.

Typically, you probably would choose your static operating point at about 5 milliamperes, giving an effective transconductance of about 0.2 mhos and a nominal voltage gain of about 60. Without the loading, it would have been 160. As a result, you decide to reduce the voltage to 5 volts and reduce the load resistance to 400 ohms. The combination now is 220 ohms, and at 5 milliamperes, the voltage gain will be about 44. Further decrease of the supply voltage to 2 volts leads to a load resistance of about 200 ohms, and a combination resistance of 140 ohms. Now, the voltage gain under similar conditions is 28, and the loading of the following input really has not done very much to the potential gain available. In short, the following amplifier stage has little influence on the stage you are working on.

You can, of course, still use the 200-ohm resistance in the output circuit for your transistor but, then, if you are using the full 10-volt supply, you have at least five times the power dissipation, and really don't have much to show for it. You will be dissipating about 90 milliwatts in your transistor when you need to dissipate only about 10. What have you really gained?

Keeping the voltage gain down is the only way to minimize the likelihood of oscillation or phase instability in a circuit. The voltage gain of 10 per stage is about optimum for that. So why the higher voltages unless there is a special circuit problem compelling its use? Record your experiences here, as you are actually running a simulated test, capacitively coupling a 500-ohm resistance to your amplifier stage.

Step 12.

There is one more interesting and important consequence to all this. We often need a voltage source which can give us a rather wide range of currents with extremely small changes of the voltage available as the current changes. Yet, we may not wish to set up a special regulator. We can use our transistor as an emitter-follower using a zener diode to set the base voltage, and we will need only a volt or so over the zener output to make the thing work. We can get up to a 2000-times current change through a transistor with less than a 200-millivolt change in the output voltage. Yet, at the same time, the total voltage across the transistor is less than 2 volts. The dissipation is much less than in a regulator.

Now, set up your transistor in the circuit shown in Fig. 3-23. You will notice that you have a variable resistance in series with the emitter-to-ground or reference point (the negative rail). You want to observe how the voltage from the base to the emitter on this transistor changes as you change the resistance in the emitter return. A 5- or 6-volt zener diode will be fine for stabilizing the base as indicated in the circuit. You can use an ordinary vom to measure the voltage across the load, and a dvm (or a high-sensitivity differential voltmeter) from the base to the emitter. Record your data in Table 3-11, and complete the graph in Fig. 3-24. We found that the voltage

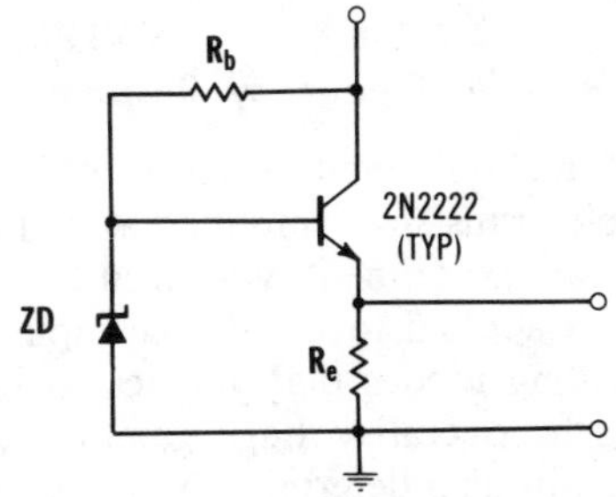

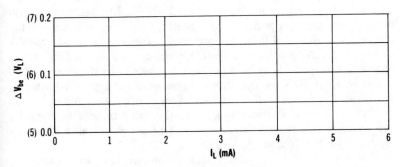

Fig. 3-23. A zener diode-transistor voltage regulator.

Table 3-11. Data Table for Experiment 5, Step 12

I_L				
ΔV_{be}				
V_L				
I_L				
ΔV_{be}				
V_L				

from emitter to return stayed very nearly constant (much less than a 100-millivolts change), and that most of that change appeared across the transistor from the base to the emitter. The base voltage itself was almost unchanged, with the small change that was there resulting from the very small change in current in the zener diode itself. Depending on the current level in the base circuit, there could be some loss due to base-spreading resistance.

Fig. 3-24. Graph of ΔV_{be} and V_L as a function of I_L.

EXPERIMENT 6
The Frequency Response of a Transistor Amplifier

You should want to learn more about the frequency response of a simple transistor amplifier as a function of its mode of operation. For this experiment, you need to vary the frequency of the test signal that is applied to the device input while using several of the typical operating modes that you tested in Experiment 5. You will observe how the overall voltage gain varies as a function of the frequency. You will also determine how the current gain varies as a function of the frequency so that you can use these data to help decide how to use your transistor most advantageously. During this experiment, you can ignore the spreading resistances except as they are built into the test devices.

Step 1.

As noted above, current gain is measured by applying a signal voltage to the transistor base through a known value of series resistance. From that, the input-signal current is determined by taking the ratio of the source voltage and the series resistance. You can assume that you can neglect any series inductance which may be in the resistor, and its shunt capacitance as well. You will want to keep track of your scope input capacitance, however. For practical purposes, these parameters (other than for the scope capacitance) are relatively unimportant except at very high frequencies. Once again, neglecting capacitances, the transistor beta equation is Equation 3-5, or

$$\beta = \frac{R_s v_o}{R_L v_s}$$

To measure the frequency response of your amplifier, you vary the frequency of the source sine wave, keeping its amplitude constant, and measure the input-signal voltage and the output-signal voltage. The test frequencies should be respectively doubled each time you make a new measurement. (If you prefer, they may be closer—each frequency is, then, 1.4 times the previous frequency, or approximately $(2)^{\frac{1}{2}}$ times.) As you increase your frequency, you will reach a point at which the ratio given by Equation 3-5 begins to decrease. When it has reached 70% of the steady value, you have reached what is known as the *beta cutoff frequency*. This is often designated by the symbol f_β. Somewhat above this point, the value of beta as given by the equation will be reduced to one half its previous value for every doubling of the frequency.

You will find this frequency disappointingly low. For audio-frequency transistors, it may be as low as a few thousand hertz, and it is usually less than a megahertz even with switching or very high-frequency devices. It is, in fact, roughly equal to the value:

$$f_\beta = \frac{f_{max}}{\beta} \qquad \text{(Eq. 3-16)}$$

where f_{max} is the frequency at which the maximum power gain available from your device has decreased to unity. This frequency is usually within a factor of two or three of the alpha cutoff frequency. The alpha cutoff frequency is the frequency at which the alpha for the transistor decreases to approximately 70% of its low-frequency value. It is immediately evident from this that the maximum operating frequency for a transistor must exceed several hundred megahertz if its beta cutoff frequency is to exceed 1 MHz by any significant amount.

Now, measure the voltages of your 2N2222 transistor as a function of frequency in the current-gain mode, and record your data in Table 3-12. Then, plot a curve of beta vs. frequency. (Use Fig. 3-25 for this purpose.) Give some comments on what you have learned.

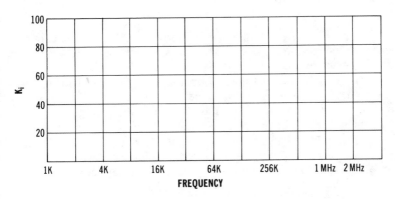

Fig. 3-25. Graph of beta vs. frequency.

Table 3-12. Data Table for Experiment 6, Step 1

f				
v_o				
v_s				
k_i				

$R_s =$ _____, $R_L =$ _____.

The approximate beta cutoff frequency of my transistor is _____ MHz. My comments on this experiment are:

You may realize that the 2N2222 is more than an audio transistor, since its minimum f_{max} is over 200 MHz. Unless your scope is a very good one, and your oscillator is useful to around 10 MHz, you may not get too much of a "rolloff" using it. If that does prove to be the case, repeat the experiment with an audio transistor and you will find that the slope is plainly evident with it. Also, the additional practice will be both interesting and fun.

Step 2.

Clearly, the current-gain mode of operation, at least in the common-emitter configuration, is not very satisfactory as a high-frequency mode of operation. This leaves you with two important choices, first going over to the voltage-mode of operation or going over to common-base operation, and using the stage as a current-mode device. It is worth your while to test both modes of operation. In this step, you can make the test in the voltage-gain mode of operation.

In the voltage-gain mode of operation, you can provide some charging current for the input capacitance for the device under test. It is true that the charging current from the signal source must pass through the base-spreading resistance for your device, but the hindrance to the flow of the charging current through it is very much less than with the current-mode of operation. An analysis of the circuit will show you that the source should be inductive for the best operation. In this case, the voltage-gain operation is limited only by the effect of the base-spreading resistance, which will lower the effective quality factor, or Q, of the tuned circuit. Under conditions where the frequency is high enough that the base-spreading resistance can affect the circuit Q, the input conductance of the transistor that is in parallel with its capacitance can usually be neglected.

Fortunately, you will find that with the better high-frequency transistors, and often with switching transistors, the value of $r_{b'}$ is small enough that the circuit will operate to a much higher frequency in this mode of operation than in the current-gain mode. This is because the product of $r_{b'}$ and g_i is normally much less than unity, and may be less than 0.1. It is vital in any of these circuits that you minimize the amount of limitation you place in the way of the flow of charging current, as a limiting resistance forces the circuit to become a current amplifier.

You can assure that your circuit is voltage controlled by providing your input signal from a source having a low enough impedance that it can provide the required charging current. In addition to a low impedance, however, energy storage in the source circuit can provide the circulating energy that, typically, is required. In short, a low-impedance tuned circuit can be optimum. Typically, this tuned circuit is located in the collector circuit of the previous amplifier stage. Once again, you will find yourself directed toward the use of low-voltage-gain, low-impedance, output tuned circuits for transistor amplifiers.

As you found in an earlier experiment, you can make your voltage-gain measurements with the same circuit that you used for current-gain measurements. In this instance, however, it is necessary to measure the input voltage at the base of the transistor. The series resistance in the source lead can be eliminated. The voltage gain equation now takes the form:

$$K_v = -\left(\frac{v_o}{v_i}\right) \qquad \text{(Eq. 3-17)}$$

$$= -\frac{i_c R_L}{v_i}$$

where,

i_c, v_o, and v_i are measures of signal current and voltages,
R_L is the output load impedance.

The minus sign, as usual, is introduced by the inversion property you observe in this device, a property you will use extensively in inverter and other IC circuits. This inversion property is an absolute essential in the development of digital inverters, and the NAND, NOR, Exclusive-OR and Exclusive-NOR circuits and devices used in digital electronics, whether they are RTL, DTL, I²L, TTL logic circuits or P-MOS, N-MOS, or C-MOS integrated circuits. It is in fact the most essential element one can find in all digital electronic circuits.

Step 3.

Now, you will measure the voltage gain for your transistor as a function of frequency, observing the input voltage at the base and the output voltage at the collector as you change frequency. You are now measuring the voltage gain, so the ratio of output voltage to base-signal voltage will remain relatively constant to a frequency significantly above the beta cutoff frequency. If you have significant resistance in the base lead, however, *both* the base and the collector voltages will begin to decrease at approximately the beta cutoff frequency, with the ratio of the two remaining relatively constant. You can increase the source voltage when this begins if you wish. But do not increase it enough to cause your device to distort. Remember also that the presence of the capacitance of the scope on the base can affect the frequency response, so use a probe to keep it to a minimum. If you do not have a ×10 probe, a resistive divider can be used instead. Connect two 10,000-ohm resistors in series and connect them directly across the signal input ahead of the coupling capacitor. (That way they will not affect the dc balance.) Then, tap the scope on the midpoint.

If you are using a high-frequency transistor like the 2N2222, you may find that gain degradation does not start until your scope itself has reached its operating limit. If that is the case, again substitute an npn silicon audio transistor.

You want to be careful about one other thing in making this test. Do not use an output load resistance large enough to give you a stage gain of more than about 2, or you will have to worry about the Miller capacitance. This will be discussed in more detail in the next chapter. Record your data in Table 3-13. Based on these data, my 3 dB frequency is approximately _____ MHz. My further comments on this experiment are:

EXPERIMENT 7
Tests for Statistical Differences

Silicon npn transistors of the same EIA code, and if you are so inclined, ones of different EIA codes, should be put through a series of tests to find out what parameters will be the same for the various devices, and where there are significant differences from device to

Table 3-13. Data for Experiment 6, Step 3

f				
v_i				
v_o				
K_v				
f				
v_i				
v_o				
K_v				

device. Only in this way can you learn what you can trust and what you cannot trust in applying these devices. It is particularly interesting to vary the collector supply voltage and to observe whether the amplification properties in particular are changed. You will, of course, note that the power dissipation will vary with supply-voltage changes. And it is also important to observe the magnitude of the input-signal voltage that you can apply to the base of the transistor before you notice a measureable degradation of the waveform. You will also find it useful to vary the frequency up to the point that marks the onset of rolloff at the various test points.

It is convenient for you to include both dc and small-signal beta, small-signal transconductance, and the base and collector voltages, as well as the approximate rolloff frequency so that you can tell how things are responding. Record your data in Table 3-14. The beta cutoff frequency is _____ MHz, and the voltage-gain rolloff frequency is _____ MHz.

You know that the beta cutoff frequency is $(1/\beta)$ times the maximum operating frequency, approximately. If the maximum frequency were 100 KHz, would you select a current-gain or a voltage-gain circuit for use with the transistor? Why?

Our tests showed that the most stable parameter for all of these devices has been the transconductance as a function of collector current. We have further found that the voltage-gain rolloff frequency was much more consistent than current gain.

Table 3-14. Data for Experiment 7

V_c				
ΔV_b				
I_c				
I_b				
v_o				
v_s				
v_1				
R_s				
β				
g_m				
$f\beta$				
f_v				
V_c				
ΔV_b				
I_c				
I_b				
v_o				
v_s				
v_1				
R_s				
β				
f_v				

EXPERIMENT 8
The Use of Emitter Degeneration

You have already observed the effect of the emitter-spreading resistance in limiting the value of the transconductance that can be obtained with a transistor. This naturally leads one to consider what can be done to improve circuit operation by introducing resistance deliberately into the emitter return lead. In this experiment, you will learn more about the effect of using such a resistance in the operation of an amplifier. Perhaps you can also learn from this something about how one goes about teaming up pairs of transistors to obtain power with a minimum of distortion. (A push-pull combination, for example, uses two devices driven with signals of opposite polarity. When one conducts, the other may be nonconducting. If the devices are properly matched, the required input-supply power can be reduced, and a substantially more linear and efficient amplifier results.)

It is first desirable to examine how a single transistor functions, both with and without this resistance (Fig. 3-26). Since a plot of small-signal voltage amplification as a function of static bias is a very sensitive way of indicating how such an amplifier is functioning, you can plot the small-signal amplification of your amplifier as a function of bias voltage or current with both modes of operation.

The voltage gain of the simple transistor amplifier is given approximately by the equation:

$$K_v = - y_f R_L \qquad \text{(Eq. 3-18)}$$
$$= - \Lambda I_c R_L$$
$$= - (q/kT)\, I_c R_L$$

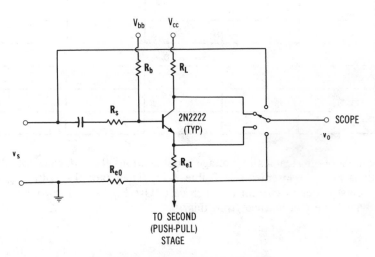

Fig. 3-26. An emitter-degenerative amplifier.

The equation, including emitter degeneration, for low-injection conditions, is:

$$K = \frac{-\Lambda I_c R_L}{1 + \Lambda I_b r_b + \Lambda (I_b + I_c) R_e} \qquad \text{(Eq. 3-19)}$$

where, ΛI_c, as before, is the effective value of the transconductance, and the last term in the denominator expresses the effect of the "emitter degeneration."

Step 1.

Assuming that R_e has a resistance of 5 ohms, plot the curves for both conditions of operation (both with and without the 5 ohms). For this test, use 200 ohms as your load resistance, and use $I_c = 10$ mA as your starting point. Calculate the voltage gain in Equation 3-19 with $R_e = 0$, and again with R_e equaling the specified value. Repeatedly double your current and make the same calculation. Then, halve the current down to at least 100 microamperes and do it again. Then, set up the circuit and, using your 2N2222 transistor, measure the

Table 3-15. Data for Experiment 8, Step 1

Calculated Data				
I_c				
β	Take this as 50 throughout			
K_{v0}				
K_{v5}				
Measured Data				
I_c				
K_{v0}				
K_{v5}				

corresponding values of voltage amplification. Record your calculations and measurements in Table 3-15. Then plot the voltage gain against collector current for each case. Use Fig. 3-27 for your plots. What have you learned from this?

You almost certainly have found that the predicted and the measured curves are quite similar, as close as you can expect in view of the fact that you probably neglected the base-spreading resistance, taking its value to be zero. (If you used the approximate value, the results would be much closer.) The degenerative case would prove to be closer than the nondegenerative.

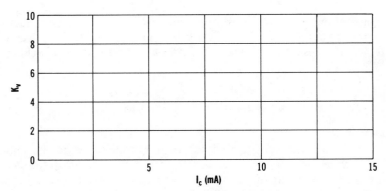

Fig. 3-27. Graph for Experiment 8.

Step 2.

Repeat your measurements, except this time, measure the signal voltage across the emitter resistance. As you do it, find out how much you can increase the input-signal voltage until there is distortion across the load resistance. How much net voltage did you have from base to emitter when the distortion began to show? Record your

Table 3-16. Data for Experiment 8, Step 2

I_c				
K_{ve}				
v_{bm}				
v_{em}				
Δv				

values in Table 3-16, where the subscript "bm" is used for the maximum signal to the base at just detectable distortion, and the subscript "em" the same value for the emitter. The Δv is the difference of these two values, and should be rather constant and less than 10 to 15 millivolts. Explain what you have found.

As the collector current is increased, the signal voltage which can be developed across the emitter resistance should increase and, as noted above, the difference between the base-signal voltage and the emitter-signal voltage will stay nearly constant. At small values of collector current, where value of the correction term is small, the effect of the emitter resistance is negligible. At what current does this change take place?

Step 3.

Next, you will want to find out what has happened to the input admittance for your amplifier as a result of the 5 ohms introduced into the emitter circuit. To do this, you can operate your amplifier with a resistance, in series with the signal path into the base, and measure the signal voltage across each end of it. Remember that we can operate our amplifier as a voltage amplifier even when it is configured as a current amplifier simply by measuring the input voltage at the base and ignoring the voltage lost across the series resistance. Then, we can read the total input signal at the input end of the series resistor. The net signal current is found by taking the difference of the two voltages and dividing by the series resistance. The input resistance for the transistor is the quotient of the voltage at the base divided by the signal current. Record your data in Table 3-17. What happens to the input admittance at several typical points?

Table 3-17. Data for Experiment 8, Step 3

v_s				
v_i				
R_s				
I_c				
i_b				
r_b				
y_i				

The effective value of the input admittance y_i is the reciprocal of the value of r_b. We find that the input admittance can be calculated from the equation:

$$Y_i = \frac{y_i}{1 + (y_i + y_f)R_e} \qquad \text{(Eq. 3-20)}$$

$$= \frac{\Lambda I_b}{1 + \Lambda(I_b + I_c)R_e}$$

where we have intentionally neglected the effect of the base-spreading resistance.

If you examine the curve for voltage gain as a function of the collector current, you will see that it rises rapidly up to about 5 mA of collector current, at which point it quickly levels. For larger values of current, it stays rather constant. This is exactly the condition needed for matching two amplifiers in a push-pull configuration. You can invert the collector-current axis by replotting it on the negative-horizontal axis and, then, combine the two curves by addition in a way which leads to the smallest irregularity in the transition region. If you set each transistor so that its static operating point is at this bias, you will have a combinaion showing minimium overall distortion (Fig. 3-28). For more information on this subject, the author suggests that you examine two of his other books.[1,2]

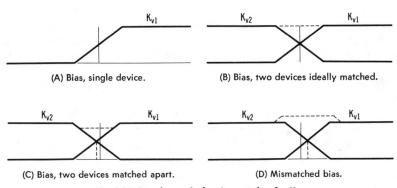

(A) Bias, single device.

(B) Bias, two devices ideally matched.

(C) Bias, two devices matched apart.

(D) Mismatched bias.

Fig. 3-28. Sample graph showing matches for K_v.

WHAT HAVE YOU LEARNED?

What does all of this mean? First, it means that both diodes and transistors are extremely nonlinear. An applied sine wave of over 5 to 10 millivolts amplitude, peak-to-peak, will be visibly distorted on a scope unless steps have been taken to compensate for doing this,

1. Pullen, K. A., *Conductance Design of Active Circuits,* (Rochelle Park: Hayden Book Co., Inc., 1959).
2. Pullen, K. A., *Handbook of Transistor Circuit Design,* (Englewood Cliffs: Prentice-Hall, Inc., 1962).

either using current control or using emitter-resistance feedback or degeneration. The use of current control leaves you subject to the inherent variations in current gain from device to device, and this means that you cannot reliably predict what the overall gain of your circuit will be. The use of emitter feedback, on the other hand, will linearize your stage and, at the same time, stabilize the voltage gain, so that you will be able to know what your circuit is going to do.

The second important lesson you should have learned is that you can predict what the forward-voltage gain of a transistor amplifier will be if you know the collector current and the values of the collector and emitter resistances. If you need a linear amplifier, you can include a properly chosen emitter resistance; if you want a nonlinear amplifier, you leave the emitter resistance out. You know that you can predict the amount of base-voltage variation that you will require (to be sure that your circuit will be relatively linear without an emitter resistance), and you can also tell how large a signal you must apply to get a desired magnitude of change in amplification from positive-to-negative peak-input voltage.

The third important lesson you should have learned is how to determine whether an amplifier will cause its predecessor to distort or not, and how to set the boundary line establishing limits on this. You can also establish the minimum value of current gain that you must have in a transistor in order to be sure that it will not load a previous amplifier severely. You should have found that the current level in the active devices control this nonlinear loading along with the output-impedance level in the interstage between the two transistors.

You should have further learned something about how the frequency response of a transistor amplifier varies with the way it is used. You should have found that the current-gain, or beta, mode is the poorest mode you can select from a frequency response point-of-view, and that the voltage-gain mode is substantially better in this respect. You should have discovered that the common-base configuration offers some advantages in specialized applications for this purpose.

The next chapter will help you to expand your understanding of these relationships by observing how pnp silicon transistors and both pnp and npn germanium transistors behave under similar conditions. At the same time, it will be appropriate to design and to test some simple amplifiers and to measure their characteristics and properties in more detail. You will find that if you make correction for the base-spreading resistance, your experiments and your calculations will agree rather well. In addition, you will design and test a variety of typical amplifiers using these devices, and will learn more about the critical parameters on which you have to depend for reliable design.

CHAPTER 4

Comparisons of Different Bipolar Transistors and Their Applications

The principal purpose of this chapter is to evaluate how well the characteristics that you observed for your small-signal npn silicon transistor are duplicated in a variety of other transistors. This will include pnp silicon transistors and both types of germanium bipolar transistors. This chapter will also discuss more about how all of them behave in typical circuits. Make a particularly careful note of those properties that appear to be common to all of these devices, whether they are power or small-signal transistors. This is extremely important as there is no other way that you can determine for certain what factors can be used as reliable parameters. These parameters must be used when making practical circuits. For the same reason, you will have guidance in what factors are the most important to you when attempting to explain the behavior of a circuit that is using a transistor.

OBJECTIVES

After studying this chapter and performing the experiments, you will understand more about the nonlinear characteristics of the pnp silicon transistor and both types of germanium transistors. You will have gained a better understanding of the following items during your study and measurements.

1. The static characteristics of three different transistor types.
2. The differences and the similarities of the various types.

3. The small-signal characteristics of the devices.
4. The high-injection physics of the devices.
5. The noise characteristics of the devices.
6. The parasitic parameters of bipolar transistors.
7. Device spreading resistances.
8. Bipolar transistors.
9. Darlington pairs.
10. The Miller effect.
11. The transistor amplifier as a feedback circuit.
12. How to set the collector supply voltage.
13. How to design audio amplifiers.
14. How to design transformer-coupled amplifiers.
15. How to design high-frequency tuned amplifiers.
16. How to design power amplifiers.
17. Amplifier linearization.
18. How to design oscillators.
19. Class-C amplifier characteristics.

You will find that all of these concepts are important to you as you develop a better understanding of the devices and how to use them.

DEFINITIONS

An understanding of the following definitions will be useful. A few of the most important from the last chapter are repeated.

pnp transistor—This is a three-layer, three-lead semiconductor device that contains two pn junctions in juxtaposition, with an electron-rich n-type layer between the two junctions. The n-type layer is extremely thin.

npn transistor—This is a three-layer, three-lead semiconductor device that contains two pn junctions in juxtaposition, with an electron-deficient p-type layer between the two junctions. The p-type layer is extremely thin.

doping—The process of introducing into an *intrinsic*, or neutral, semiconductor, a material which will cause it to be either electron rich or electron deficient. This process is vital in the construction of any semiconductor electron chip.

electron—The smallest readily used particle carrying a negative charge.

hole (*vacancy*)—The basic "positive carrier" which is the complement or dual of the electron. It acts as the carrier in p-type materials. It really represents a missing electron with a hole or vacancy in the spot where an electron belongs. It "moves" by having an electron combine with it, thus leaving a vacancy at some neighboring location.

silicon transistor—A transistor whose active regions are composed of crystalline silicon of the required purity and with the perfection of crystalline structure. The material has been doped to create the necessary junctions and other boundaries.

germanium transistor—A transistor whose active regions are composed of crystalline germanium of the required purity and with the perfection of crystalline structure. The material has been doped to create the necessary junctions and other boundaries.

heterojunction transistor—A transistor which is made by crystallizing two compatible semiconductor materials onto one another in such a way as to create the required pair of junctions.

gallium arsenide transistor—A transistor whose active regions are composed of single-crystal gallium arsenide of the required purity and perfection. Gallium is a valence-three element and arsenic a valence-five element. When they are properly mixed and processed, they can form a high-quality semiconductor material. Its active regions have been doped to create the necessary boundaries or junctions with the dopants, in this case, having valences of either two or six elements. Bipolar transistors made of gallium arsenide have proven difficult to manufacture, but field-effect transistors are made from it.

binary alloy semiconductors—These semiconductor materials are made by forming single-crystal compounds of either valence-three and valence-five elements, or valence-two and valence-six elements. As long as the crystal structure of the respective atoms are compatible, semiconductor materials can be made in this way.

ternary alloy semiconductors—These semiconductor materials are made by forming single-crystal compounds from combinations of either three- and five-valence elements or from two- and six-valence elements. One of the materials that comprises the compound consists of a single element of that valence, while the other is a combination of two elements of valence needed to match.

compound semiconductors—This is a semiconductor material that is formed from a compatible mixture of elements of an appropriate valence. The binary and ternary alloy semiconductors referenced previously are compound semiconductors. With all of these materials, the highest possible perfection in crystallization is required to assure that the compound will have a sufficiently small intrinsic conductivity.

totem pole—This term is used to refer to a group of transistors that are connected in series to perform some special function. Two transistors and a diode are typically connected in the output of a TTL gate circuit in this fashion.

diffusion velocity—The apparent velocity that is induced in a cloud of charges as a result of the influence field caused by the carrier distribution. The actual charges move more slowly than the diffusion velocity would indicate. It is an influence velocity, rather than a "particle" velocity.

drift velocity—The velocity induced in a carrier as a result of an actual electric field gradient. In many types of transistors, it is negligible compared to the diffusion velocity. Drift velocity only becomes significant with either very high collector voltages or with base regions where the doping decreases significantly from emitter to collector.

transit time—The actual time required for a single minority carrier to cross the base region of a bipolar transistor.

minority carrier lifetime—The average length of time that a minority carrier can survive when moving in the base region of a transistor. This lifetime is extremely important in establishing the beta for a transistor.

light-emitting diode (LED)—A semiconductor diode that emits light when properly biased.

beta—The beta for a transistor is the ratio of the current from the emitter (that reaches the collector) to that which is extracted via the base lead. It is roughly equal to the ratio of the minority carrier lifetime to the transit time.

diffusion capacitance—The apparent capacitance between the base lead of a transistor and its emitter that results from the charge distribution in the base region. It is much more than the static capacitance. (The two principal components of the input capacitance are the diffusion capacitance and the transition capacitance.)

neutralization—A process of balancing out a return signal at the input of an active device, thereby minimizing the feedback gain so that stable operation can be achieved. A bridge-type circuit is commonly used for the purpose.

unilateral network—This is a network capable of transferring an electrical or other type signal in one direction only. A signal applied at the output will not cause a response at the input. It is theoretically possible to unilateralize any active network. Neutralization is a simple and imperfect form of unilateralization.

tuned circuit—A circuit comprised of one element capable of "storing" potential energy, and of one element capable of "storing" kinetic energy. The two elements are so configured that an energy exchange occurs between them. The result is that a coherent sinusoidal oscillation can be produced with it.

tapped tuned circuit—A tuned circuit in which a tap of some sort exists on one or the other of the elements. With an LC circuit used in electronics, the tap may be on either the inductance or the capacitance.

maximum operating frequency—This is f_{max}, the frequency at which the output power from an amplifier just equals the input power. At higher frequencies, the output power normally is less than the power required to generate it.

transconductance efficiency—The ratio of the actual transconductance for an active device (like a transistor) to the theoretical value possible at the same output-current level. It can be obtained from the equation:

$$\kappa = \frac{y_t}{(q/kT)I_c} \qquad \text{(Eq. 4-1)}$$

where,

κ is kappa, the transconductance efficiency.

This parameter is important with all presently known solid-state active devices. Among other things, it is a measure of the power-handling ability of an active device.

crossover distortion—This is a form of distortion resulting from the imperfect matching of the amplification contours for different elements of a power amplifier. It is a form of distortion which is severe with small signals, but which may largely disappear with large signals.

transconductance-per-unit-current efficiency—See Transconductance Efficiency.

breakdown voltage BV_{ceo}—This is the maximum voltage that a transistor (with essentially an open base circuit) is designed to withstand without a voltage breakdown or failure. It may withstand more and many will. However, the margin of safety beyond this value is decreasing as processing is becoming better controlled.

breakdown voltage BV_{cer}—This is the voltage that the device is supposed to withstand, with a very small impedance between its base and its emitter. It can be as large as twice the magnitude of BV_{ceo}.

breakdown voltage BV_{ces}—This is the maximum voltage that a device is supposed to withstand, with a specified impedance between its base and its emitter. Its value is likely to be somewhat less than the value of BV_{cer}.

breakdown voltage BV_{cbo}—This is the maximum voltage from the collector to the base that a transistor is expected to withstand. The emitter is normally open-circuited for this test.

current gain—Also called β or h_{fe}. This is the nominal current gain for a transistor. Either the nominal value, or the maximum value, or the minimum value may be designated. With the latter two values, the symbol is supplemented by either the subscript (max) or (min). The more important of these is probably the minimum value.

dc current gain—Also called h_{FE}. This is defined as the ratio of the collector static current to the base static current. Again, the minimum value can be rather important. The subscript letters are capitalized to indicate that a dc value is given.

saturation voltage—Also called $V_{ce(sat)}$. This is the minimum value of voltage that you can get from collector to emitter at any value of collector current. No variation of the base voltage or base current can help to get a lower voltage.

corner frequency—A corner frequency is a frequency at which there is an abrupt change in the way some parameter behaves. For example, with an amplifier, there is a range of frequencies between which the voltage amplification of the stage is approximately constant. At each end of this range, there may be sharp breaks or points beyond which the voltage amplification changes rapidly with frequency. The change is typically by a factor of 2 to 1, or a multiple thereof for each octave of frequency change. The frequencies at which this change appears to occur are called corner frequencies.

noise corner frequency—A frequency that defines the point of change in the noise behavior of a device. On one side of such a frequency, the noise behavior follows some rule, such as being roughly uniform for white noise. On the other side, the noise may increase rapidly with a change in frequency.

white noise—A kind of noise whose energy distribution is such that the average energy-per-unit bandwidth is constant and is essentially independent of time. It is also called thermal noise. White noise is present in all known systems and is a function of absolute temperature. It is also sometimes called Johnson noise.

excess noise—This is the noise present in a system that is in excess of that due to white noise. It may be independent of frequency, in which case, it has the properties of white noise, or it may be a function of frequency, either increasing with an increasing frequency, or else increasing with a decreasing frequency.

lower noise corner frequency—This is a frequency below which the excess noise in a given device increases sharply as the frequency is reduced. The factor is written as f_{n1}. The nature of the increase depends on the source. It may be generated by surface or body leakage in a material or by traps and flaws in a semiconductor material.

upper noise corner frequency—This is a frequency above which the excess noise in a given device increases sharply as the frequency is increased. This noise is normally associated with the minority carrier flow through the base of a bipolar transistor. It is dependent on the fact that the current flow is granular in nature. The factor is written as f_{n2}. This frequency is defined in terms of the transit time for a minority carrier to cross the base region. Its value is roughly specified as the square root of the product of the alpha-cutoff and the beta-cutoff frequencies. The noise roughly doubles for each octave above this frequency.

MORE BASIC PARAMETERS

Most of the parameters that you measured and studied in Chapter 3 are ones common to all present-day junction transistors. It is now important to find out in what ways the various kinds of bipolar transistors actually differ. You can then make allowances for these differences. Based on the common phenomena that you have studied and measured in the previous chapter, and on the other ones that you will now study, you will learn how to approach the problem of selecting a desired circuit for use with an amplifier, and how to adjust it to behave as you wish it to act.

As you have observed in the definitions, there are basically two different ways that you can classify junction transistors, namely, by the polarity configuration and by the basic material used. Probably the most important basis is the polarity configuration, since germanium and materials other than silicon are relatively little used, either with junction devices or with arrays of junction devices. It is necessary to consider other materials at least briefly, however, as recent developments have been leading toward the introduction of compound semiconductor devices. There are even, on the far horizon, devices called heterojunction transistors that you will want to know about.

Regardless of the material from which they are constructed, most bipolar transistors are basically either pnp or npn devices. They consist of a heavily doped emitter junction, a lightly doped base layer (which is very thin), and a collector region which, with the possible exception of a very thin layer, is usually reasonably heavily doped to minimize its series resistance. Normally, the base region has opposite polarity to the emitter and collector, with the result that majority carriers in either the emitter or the collector are minority carriers in the base. (A "junction" in a heterojunction device can, under some conditions, have the same polarity doping in both the emitter and the base.) The base region is made extremely thin (as thin as possible) so that the ratio of the lifetime of the minority carriers to the transit time for crossing the base will be as large as possible. The doping level in the base is at least 100 times less than that in the emitter region.

The process of injecting carriers from the emitter into the base layer by way of a forward bias across the junction can upset the electrical equilibrium in the base layer very sharply, even almost violently. These carriers, when they enter the base region, become minority carriers because they are of opposite polarity to the carriers normally available in the base region. However, it happens that the small volume of semiconductor, that comprises the base region, cannot tolerate such a large number of minority carriers, and it draws in more majority carriers via the base lead, causing an increase in the number of both the majority *and* the minority carriers in the region. Removal of the forward bias leads to recombination and decay to equilibrium conditions. Under equilibrium conditions, the levels of minority and majority carriers in the two layers take the form:

$$p_p \times n_p = n_i^2 \qquad \text{(Eq. 4-2)}$$
$$n_n \times p_n = n_i^2 \qquad \text{(Eq. 4-3)}$$

where,

n$_i$ is the number of both the positive and negative carriers in an undoped or intrinsic semiconductor material.

When minority carriers are introduced, an equal number of additional majority carriers are drawn into the base region, leading to the inequalities:

$$(p_p + N) \times (n_p + N) > n_i^2 \qquad \text{(Eq. 4-4)}$$

$$(n_n + N) \times (p_n + N) > n_i^2 \qquad \text{(Eq. 4-5)}$$

where, the additive majority carrier term is taken as the same as for the kinds of minority carriers. Equation 4-4 applies for npn transistors, and Equation 4-5 for pnp transistors. The value of n$_i$ is dependent solely on the kind of semiconductor material involved. Values for n$_i$ for typical materials are listed in most basic books on transistor physics. The values for germanium and for silicon are approximately 6×10^{12} and 3×10^9, respectively. It is important to note that the value for silicon is much smaller than that for germanium. This tells you at least two important things:

1. The basic leakage current in silicon devices is much smaller than that in germanium.
2. The maximum operating temperature for silicon devices is theoretically substantially higher than that for germanium. Your own tests and measurements will verify this, and your examination of typical data sheets will further confirm that these facts are so.

The introduction of excess minority carriers, by drawing in additional majority carriers, causes a typical mass-action effect that is well known to chemists. It leads to an increase in the rate of recombination, and causes a sharp reduction in the current gain for the device. The point at which this becomes significant is closely related to the point at which both the small-signal and dc beta begin to drop as the device current is increased.

The influx of majority carriers into the base region has another important effect. The number of carriers in the outer base region is not altered, but it is markedly higher in the active base region. This leads to a voltage difference, the difference required to keep the carriers from rushing back out. This is a sort of contact potential, and it acts to either increase or to decrease the magnitude of the signal voltage that is applied to the base of the transistor. With an npn transistor, the polarity of this voltage is such that it, in effect, amplifies the applied signal voltage, and increases the apparent transconductance efficiency for your device. Under these conditions, the transconductance efficiency may be greater than unity and may, in fact, be as high as 1.5. With a pnp transistor, this voltage reduces the effective signal amplitude, and the transconductance efficiency in this instance may be as small as 0.6. It varies somewhat depending on the basic material used.

There is also a small flow of carriers from the collector into the base region, leading to a small number of minority carriers from this source. Since these carriers are not under the control of the base-to-emitter voltage, they do introduce a small reduction of transconductance efficiency. It will fortunately be negligible unless the base voltage and the collector voltage are approximately equal. The recombination of these carriers, which normally are swimming "upstream," is almost a certainty, and they also reduce the effective current transport efficiency from emitter to collector. Where beta is used in calculations, a correction may have to be made for this "reverse" flow, which can become significant when the collector-saturation condition is approached. The current gain from this flow is often called the "inverse beta."

IMPENDING DEVELOPMENTS

Now that the techniques of ion implantation are becoming rather well understood and are quite well controlled, we can expect that bipolar transistors based on the use of gallium arsenide may make their appearance, for applications where the high-diffusion velocity of electrons in this material can be helpful. Gallium arsenide is being used extensively in microwave field-effect transistors. The basic field-effect transistor is discussed in Chapter 5, along with some typical applications. The problem with this material has been that arsenic tends to leach out of the compound at diffusion temperatures. Gallium arsenide is by far the best high-frequency semiconductor presently available, with germanium next, and with silicon the poorest semiconductor. However, the other advantages of silicon have been sufficiently attractive that germanium has proven a less desirable compromise even for high-frequency applications. The leakage and temperature limitations for germanium are too severe.

The heterojunction is a curious affair. Its properties are not yet sufficiently well controlled that it has become a useful structure, but it is almost certain to prove valuable in at least specialized applications. It is possible to grow a compound semiconductor on, say, silicon if the lattice of the crystal structure matches adequately. Then, one has such a junction. In fact, it is possible to have a junction even if both layers have the same type of doping, as one layer will inject carriers into the other layer sufficiently heavily so that it will appear like a normal junction. This combination may prove to have reduced minority leakage current and, as a result, an abnormally high beta.

In short, there are intriguing possibilities with some of these developments.

DIFFUSION VELOCITY AND NOISE

The diffusion velocity for the minority carriers in the base region of a transistor does not represent the actual velocity of the charged particles, but corresponds more closely to a phase velocity, or a velocity of an electrical influence field. (In this sense, the group velocity of light corresponds to the particle velocity, and the wave velocity corresponds to the diffusion velocity.) It is interesting to note that for frequencies that are such that the transit time for a particular particle across the base region exceeds the radian period of the signal, the resultant noise figure for the device at that frequency will exceed the average noise level at lower frequencies. The frequency, at which this change in effect takes place, is sometimes called the "upper noise corner frequency." Above this frequency, nonuniformity of carrier flow of a statistical nature causes a degradation of the kind of uniformity of controlled flow which is encountered at lower frequencies, and leads to the introduction of an increased level of random noise.

There are two important frequencies in the noise characteristic spectrum for a transistor. One of these is the frequency just noted, and its approximate value may be determined from the equation:

$$f_{n2} = (f_a f_\beta)^{\frac{1}{2}} \qquad \text{(Eq. 4-6)}$$
$$= f_a / (\beta)^{\frac{1}{2}}$$

where it should be noted that the alpha frequency and the maximum frequency are close enough to being equal that either may be used for this calculation. (The maximum frequency is almost always within a half-order-of-magnitude of the alpha cutoff frequency.) This equation will define the upper noise corner frequency adequately for most applications. The second important noise corner frequency is sometimes called the "lower noise corner frequency," and it is the frequency below which there is excess noise. This noise is due to a variety of causes—internal imperfections, surface leakage paths, poor attachment of a deposited layer, etc. The excess noise below this corner frequency is, in part, responsible for the instability that is encountered in electronic feedback integrators and similar circuits. The quality of a given transistor can often be judged by measurement of this lower noise corner frequency (f_{n1}), as the lower its value, the better the overall properties of the device. Apparently the relation of f_{n1} to the early failure of transistors has not been studied adequately. This may, in part, be because the measurements of noise level and noise figure are some of the most difficult tasks faced by designers and technicians who are concerned with achieving the best possible operation of amplifiers. A matter of concern is the amount of noise per unit of bandwidth. Since the unit of bandwidth is a single cycle, and a measurement of this kind must, therefore, be made for a small but fixed number of cycles, it is difficult to get circuits that are suitable for providing for the measurement of "rms" noise in this width. As long as noise is uniformly distributed (white noise), as it normally is between f_{n1} and f_{n2}, the precise bandwidth used for making the measurement is not really critical, and the noise power is then defined in terms of the equation:

$$P_n = \frac{W_b}{BW}$$ (Eq. 4-7)

where,

P_n is the noise power per unit bandwidth,
W_b is the total power measured in the bandwidth BW.

The noise per unit bandwidth increases with decreasing frequency for frequencies that are less than f_{n1}, and increases with increasing frequency for frequencies that are above f_{n2}. As previously noted, the increase above f_{n2} is a consequence of the statistical degradation of uniformity of the current flow resulting from random processes, and is roughly proportional to frequency. (Statistical smoothing which places the noise outside the passband of the circuit occurs when the operating frequency is less than the corner frequency.) The increase above f_{n2} is roughly proportional to the frequency, as the apparent "granularity" of the current at high frequencies increases as the frequency is increased. The increase below f_{n1} is caused by a variety of effects, as has been previously noted, with the result that it is impossible to specify precisely the rate at which the noise increases with a decreased frequency. It should be noted, however, that it is a power law, with the power being between a square-root and a three-halves power. Your principal concern is that both of these turnover frequencies do exist, and that you, at least, have an idea what they do and what their causes are likely to be.

EFFECTS OF DEVICE PARASITICS

In the absence of a spreading resistance in any of the leads of a transistor, you have found (in Chapter 3) that the bipolar transistor behaves almost precisely like an ideal admittance device. Its input and forward admittances are linearly dependent on the current flow in the input or the output, respectively, with the constant of proportionality being (q/kT). When you placed resistances in series with any of the terminals of your transistor, you should have found that the properties of the device were changed substantially. You should have observed some of the consequences of these changes in your measurements. It is now important to learn how to detect devices having excessive values of spreading resistance so that you can avoid using them, thereby avoiding the consequences of their presence on the circuits you may be using or designing.

It is important, first, to consider the effect of internal shunt conductances before considering the spreading resistances, however. These admittances superficially can be lumped into the internal conductances for the bipolar transistor as long as certain important consequences are kept in mind. Needless to say, the consequences depend on whether the input or the output circuit is under consideration.

The ideal bipolar transistor should have a zero value of input conductance and a zero value of output conductance when voltages are applied so that the device is nominally "turned off." (Theoretically, of course, this does not occur.) It is, in fact, possible for the device to have a strictly conductance-type of "leak" between any pair of terminals, like between the collector and the emitter. It can also have a "constant-current" type of leak, and you will find both of these conditions with transistors you will test. The presence of a small constant-current leak may or may not introduce problems for you. You will have to decide that based on the problem at hand. If there is a significant amount of a conductance-type of leak, however, where the current increases roughly proportionally to the collector voltage over the range of use, the device should be discarded. For that reason, the first experiments in this chapter are directed toward showing you how to spot devices having this kind of deficiency. (This condition of a conductance leak *can* occur on the input, but it is much more likely to be found from collector to emitter.)

The experiments in Chapter 2 have already shown that one of the characteristics of a good diode is that the relation between the current change (through the diode) and the voltage change (across it) is that a current change of 2 to 1 results when a voltage change of approximately 18 millivolts occurs across the diode. For that reason, an estimate of the total spreading resistance in any circuit may be obtained by making the appropriate 2-to-1 change in the current and by determining the required voltage change across the appropriate junction. (This will give the *total* resistance on both sides of the junction.) Ideally, this test should be made across each of the two junctions in a bipolar transistor independently and, then, a test through the device as a whole should be performed. Fortunately, the internal emitter-spreading resistance is usually very small compared to the base-spreading resistance, with the result that a measurement of the input junction will, essentially, yield the base-spreading resistance alone. The emitter-spreading resistance is best determined by measuring the effective transconductance, as a function of the collector current, when the forward bias on the base is varied and, then, solving for r_e. Correction for the base-spreading resistance can be made with the help of the measured value, based on the input measurement. (The effect of the emitter-series resistance was demonstrated in Chapter 3. The use of the resulting technique for measuring r_e is straight-forward, and is applied in the experiments.)

Determination of the collector-spreading resistance is possibly the most difficult of the three measurements. It is best found by plotting the saturation curve for the transistor (the variation of the collector voltage with the collector current), with the base forward-biased to force the device into saturation. In the presence of significant collector-spreading resistance, the slope of the saturation curve will break sharply as the current of the device is increased, with the slope of the current at higher current levels indicating the approximate value of the spreading resistance. (Emitter-spreading resistance can contaminate this measurement, but its value can be determined from transconductance measurements.) The presence of collector-spreading resistance will *not* alter the measured transconductance under normal operation as will the emitter-spreading resistance. One must be careful in making this measurement to be sure that the base current is adequate to keep the device in saturation, as the curve normally rises sharply as the device leaves saturation.

STATIC CHARACTERISTIC CURVES AND THE BIPOLAR TEST SET

From the previous discussion, it is evident that several static tests are useful in the preliminary classification of transistors. The first test is to verify that there are two good diode junctions in the device. The basic circuit for this test is shown in Fig. 4-1. This circuit simply tests for the level of conductance. The LED diode placed across the

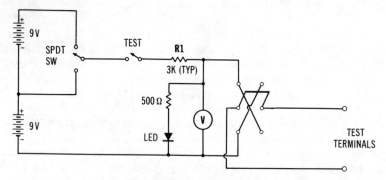

Fig. 4-1. Basic diode test circuit.

junction limits the voltage to a safe value in the reverse direction. (This can be important on very-high-frequency and microwave transistors.) The meter should be adjusted to read full scale when the diode lights. Once two junctions have been located, it is known that one represents the emitter-base junction and, the other, the base-collector junction. Which is which may not yet be known, but from the polarity of the conduction direction, it is possible to determine whether the device is an npn or a pnp transistor. Normally, only a current-gain test will identify which is the emitter and which is the collector.

The test configuration, as it stands, applies only a maximum of about 2.5 volts across the emitter-collector port because of the limiting effect of the LED diode. This is seldom an adequate leakage test. At least 9 volts, and possibly as much as 18 volts, should be applied for this test, with the voltage across the terminals measured with a series resistance in the circuit to permit the current flow to be estimated. High leakage leads to an abnormally low value of voltage in this test. A suggested form for this part of the test circuit is shown in Fig. 4-2. The series resistance of the meter should be adjusted to give a full-scale reading with the LED diode lit in the one case, and a full-scale reading without a transistor being tested in the second case.

The use of a test meter of this sort is highly recommended, even though it is not absolutely necessary. Points can be calibrated on the meter to indicate the voltage to be expected across either a silicon or a germanium diode. The resulting information can be used to identify not only whether the device is npn or pnp, but also whether it is a germanium or a silicon device. This arrangement is particularly useful in identifying Darlington pairs, which have *two* diode drops on the base-emitter junction, and only one on the base-collector junction.

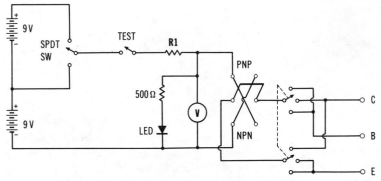

Fig. 4-2. A transistor stand-off voltage test circuit.

The introduction of a way to make a current-gain measurement is additionally useful. If the circuit is arranged with a resistor which can be connected from the base to the collector by pushing a button, and if the value of the resistance is properly chosen, it is possible to cause the transistor to be driven into saturation, at least weakly. Since the meter measures the voltage across the collector-emitter port when the test configuration is activated, a device with a low beta will have a high voltage across this port, as will a device with a high collector-spreading resistance, but with an adequate beta.

The bipolar test set will also detect field-effect devices, but it is not optimum for that purpose. Another configuration, which has proven satisfactory for that, will be described in the next chapter. FET devices, that are known to be good, can be tested on the bipolar tester, of course, to find out how they respond. The results can be used as guides for testing other FETs. However, all of the variety of FET devices must be tested before one can be sure of making reliable tests. Unfortunately, our semiconductor device-numbering system does not provide a way of distinguishing between bipolar and field-effect devices. Likewise, it does not enable one to differentiate dual diodes, SCRs, unijunction transistors, and other devices from either of these kinds of devices. Devices which appear not to be good as bipolar transistors could be any of the other kinds of devices that carry a "2N" code designation.

THE DARLINGTON PAIR

The Darlington compound has been mentioned earlier, and the fact that it has two diode drops in its input circuit was also mentioned. Its basic circuit is shown in Fig. 4-3. There are some precau-

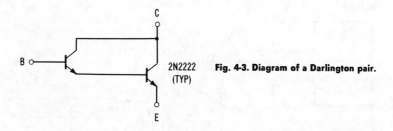

Fig. 4-3. Diagram of a Darlington pair.

tions which must be noted in its use in practical circuits but, when used properly, it is an extremely useful configuration. As you can see, it consists essentially of two transistors in "series," in that the emitter output of the first transistor is connected to the base input of the second. The two collectors are usually connected together, and the input is applied to the base of the first transistor. The output may be taken either from the common collector or from the emitter of the second device, or from both. This circuit is ideal for use with split-load phase dividers, as the total collector current of the pair and the emitter current of the output transistor are almost identical, usually within a very small fraction of a percent in most cases. The result is excellent balance.

In spite of its very high current gain, the Darlington pair is surprisingly stable. It does not tend to oscillation unless an excessive voltage gain is required from it. As long as the device is operated under conditions that do not lead to cutoff of the input transistor, few problems are likely to be encountered. Special precautions are required if cutoff is required, however, as then, an interstage return from the common emitter-base lead to the supply is an absolute necessity, and the current balance condition is lost. Often the collector-to-emitter leakage in the input stage is sufficient that cutoff of the second stage cannot be assured.

VOLTAGE GAIN AND DISSIPATION CONSIDERATIONS

A subject of primary importance, yet one which is largely ignored, is the control of power dissipation. The mean time between failure for all semiconductor devices is critically dependent on their operating temperatures, and the operating temperatures are dependent on first, the environmental temperature, and second, the total power dissipation per unit volume in their surroundings. It is critically important to understand what factors are the primary ones in limiting dissipation in order to understand what can be done to minimize it.

Two major factors place limitations on the amount of voltage gain which can be tolerated in a circuit. The first of these is the Miller effect capacitance, and the second is the voltage-gain limit which must be accepted, in order to minimize excess phase shift and possibly even parasitic oscillation. The fact that extremely large current gains can be realized with the Darlington-type circuits described earlier, shows that current gain per se is not a primary factor. It is easily shown that the primary problem is *voltage gain*.

Experience that has been accruing since early World War II shows clearly that the voltage gain per stage in multistage amplifiers must be limited. Typical gain values depend on the circuit configuration selected. When the carrier source (emitter, source, or cathode) is grounded, the maximum value is generally accepted to be a voltage gain of 10; when the control electrode is grounded, the voltage gain from input to output may be as high as 100, as a consequence of the very low input impedance of the current-source injection used.

Clearly, if one assumes that an input-signal voltage of 10 millivolts is accepted, the output voltage will be either 100 millivolts or 1 volt, depending on the configuration. As a result, one must ask the question, "Is that enough?" Fortunately, the answer is yes, as even a 200-millivolt change across the input junction of one of these transistors can generate over a 2000-times change in the output current. If the device just saturates with 1 milliampere of current, it is practically cut off when the base voltage has been reduced by 200 millivolts. This means less than a volt over the saturation voltage will be ample to generate the necessary output voltage needed to operate a following stage in the common-emitter configuration. (Even in the common-base configuration, only 5 to 10 volts is ample.) A good compromise value for many circuits is 1.5 to 2 volts.

Miller Effect

The Miller effect is the apparent increase of capacitance between the input and output resulting from the voltage gain developed in an active device. The apparent value of this capacitance is given by the equation:

$$C_a = C_{io}(1 - K_v) \qquad \text{(Eq. 4-8)}$$

where,

C_a is the equivalent input capacitance resulting from the voltage gain,
C_{io} is the effective capacitance between the base and the collector,
K_v is the effective voltage gain across the output junction.

The sign, or polarity, of the voltage gain can be important. With common-emitter amplifiers, it is negative, making the quantity in the parentheses in Equation 4-8 positive.

Clearly, from Equation 4-8, it is possible for the apparent value of the capacitance to be from ten to several hundred times the value of C_{io}. This can lead to substantial feedback, considerable excessive phase shift in the circuit and, in severe cases, actual oscillation. The presence of this capacitance converts the amplifier into a feedback amplifier, with the result that the correct general form for its voltage amplification takes the form:

$$K = \frac{K_v}{(1 - K_v K_f)} \qquad \text{(Eq. 4-9)}$$

where,

K is the overall resulting amplification,
K_f is the feedback amplification.

These are all *voltage* gains, *not* current gains, and the equation is the standard feedback amplifier equation used extensively in electronics, and particularly with the study of servomechanisms.

When it is vital that linear phase and high stability be maintained, it is an absolute necessity that the value of the denominator K_vK_f term given in Equation 4-9 have a value no larger than 0.05, or even less, if possible. Where phase linearity is not as important but freedom from oscillation is, the magnitude of this product should not exceed approximately 0.3 or 0.4 in magnitude. Otherwise, the damping factor will become too low, and a tendency of the circuit to ring or actually oscillate will begin to develop.

More on Voltage Gain

Based on these voltage-gain considerations, it is evident that there is little use in designing for more voltage gain in a circuit than is required to assure stability. If one makes calculations of the per stage gain in typical better quality fm and am radio receivers, one finds that typically there is one stage of amplification for each decade of voltage gain. (The range of voltage gains normally encountered will be between 5 and 20 per stage.) One is forced to conclude that for some reason the per stage allowed voltage amplification equation established during World War II is still valid.

Equation 3-3 (Chapter 3) has shown that the voltage gain for a typical transistor amplifier takes the form:

$$K_v = -(q/kT)I_cR_L$$

Since the product (I_cR_L) clearly is a voltage, it must be related to the supply voltage for the amplifier or $|V_{cc}|$. This relationship can be expressed by the relation:

$$I_cR_L = \eta\kappa|V_{cc}| \qquad \text{(Eq. 4-10)}$$

where the value of eta (η) is between 0.3 and 1.0. Then, Equation 3-3 can be rewritten in the form:

$$|V_{cc}| = |K_v|/(\eta\kappa q/kT) \qquad \text{(Eq. 4-11)}$$
$$\simeq |K_v|/20$$

(This assumes that the transistor is operating in the low-injection mode of operation, although the change due to high-injection is relatively small compared to other factors.)

It is evident from the above discussion that less than a volt of collector-supply voltage is absolutely necessary to provide the required amplification conditions. The next question, therefore, is, "Will that voltage provide enough signal drive to the next stage?"

Since we have already concluded that a voltage change across the base-emitter junction, of as little as 200 millivolts, can cause over a 2000-times change in the collector current, it is evident that a few hundred millivolts of change is more than what is needed. As a result, as little as 2 or 3 volts above saturation voltage is all that is required from the collector to the emitter to assure an ample signal output. Even with degeneration, less than 5 volts will usually suffice.

When the common-base configuration shown in Fig. 4-4 is used, the value of K_v can be higher than 10, in fact, up to as much as 100. This is, first, because of the very high input admittance (low-input impedance) of the emitter of the transistor and is shown by the equation:

$$r_e = (kT/q\ I_e) \qquad \text{(Eq. 4-12)}$$

Secondly, this is because a voltage step-down is required to match the collector to the following emitter. In this instance, voltage gains of 100 from the emitter to the collector may still lead only to an overall voltage gain from the emitter to the emitter of 10. The fact that base-spreading resistance can allow the base region to "rise" off ground does make the stage susceptible to parasitic oscillations, however. The common-base configuration should *always* be used when maximum power output (with reasonable stability) is essential, however, as otherwise, virtually all the odds oppose you.

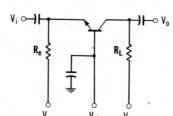

Fig. 4-4. An RC amplifier using a common-base transistor configuration.

HIGH-FREQUENCY OPERATION

When you are building rf circuits, it is important to select transistors whose f_{n2} is higher than your operating frequency, if at all possible, if low-noise operating conditions are important to you. Otherwise, you are potentially taking a noise figure penalty which you need not take. Admittedly, you will find statements that devices used in the frequency range above f_{n2} are "absolutely stable" or "will not oscillate." This does not guarantee that there is not excess phase shift, however. The network still is far from minimum phase throughout the region between f_{n2} and f_{max}.

The absolute maximum operating frequency for a transistor, in a form which can provide a useful function, is clearly that frequency at which the power gain is unity. For higher frequencies, you are in a losing situation for the most part; below it, you do gain at least somewhat. The frequency at this transition point is often called f_{max}. It is defined as the highest frequency at which a device in a unilateralized circuit can provide unity power gain. Also, it may be described as the maximum oscillation frequency when used in an idealized circuit. This frequency is near the alpha cutoff frequency, which is the frequency at which the common-base current gain drops to 0.707 of its low-frequency value.

Since the output impedance that can be used with the load for a common-base amplifier can be many times its input impedance, it is still possible to get a power gain at the alpha cutoff frequency. As a result, this frequency is less than f_{max}, although it may only be less by a factor of between 2 and 5.

AMPLIFIER DESIGNS

There are a variety of amplifiers which can be built based on transistors, and there is a wide range of applications for them, both in digital and in analog electronics. You examined and experimented with some basic transistor characteristics in Chapter 3 and made some simple amplifiers so, now, it is necessary to examine some of the specific kinds of important applications. In the experiments that follow, you will do some more work with the basic types and will find out how they are applied in practice.

The basic configuration of an amplifier based on the bipolar transistor (and the field-effect transistor) is the resistance-coupled amplifier. Except in analog integrated circuits, this amplifier has not been used as much as the transformer-coupled and the tuned-circuit-coupled amplifiers, primarily because of a lack of understanding of just how important they are. The reasons for that are based on the misconception that bipolar transistors are principally current-gain devices instead of transadmittance devices.

Resistance-Coupled Amplifiers

The basic resistance-coupled transistor amplifier is ideally suited to the amplification of voltages in the range from a few tens of microvolts to about 100 millivolts. At about the 100-mV level, correction for high nonlinearity in the device becomes necessary. In this range, the device can be used strictly as a voltage amplifier as long as the gain per stage is limited sufficiently so that the loading of a following stage does not significantly affect the output impedance of the signal source. Using the resistance-coupled amplifier as a voltage amplifier yields an amplifier having ample bandwidth for most ordinary applications. To use this configuration effectively, it is convenient to operate the collector circuits at a supply voltage between

1 and 5 volts, rather than the typical 10 volts usually used with a base circuit such as shown in Fig. 4-5.

It is possible to use resistance-coupled amplifiers with signal voltages above 100 millivolts, but several steps are required to assure that distortion levels are acceptable. (In this respect, the circuit really is not any worse than those of a transformer-coupled amplifier, but the distortion encountered with the latter apparently has been ignored because of its somewhat more elusive nature.) The first step is the introduction of emitter degeneration to limit the range of the effective transconductance variation for the active device. Typically, the value of the resistance required is 100 ohms or so for a device drawing 1 milliampere of collector current; the product of the current in milliamperes by the emitter resistance should be about 100. Then, the load resistance in the collector circuit is made large enough to provide a reasonable gain, but small enough so that the following transistor input will not load it severely. An amplifier with an emitter resistance of 100 ohms will require a collector resistance between 200 and 1000 ohms, depending on the exact function of the circuit and of the circuit configuration that follows.

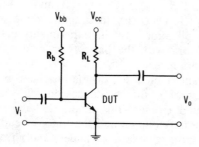

Fig. 4-5. Typical resistance-coupled transistor amplifier.

The important point to all this is that once you reach a signal level of about 100 millivolts, you may not want your voltage to build up until you have reached your final amplifier. The higher the signal level, the larger the amount of degeneration required to linearize the amplifier. It is better to use the amplifier chain, at this point, as true current amplifiers rather than as voltage amplifiers. (This is one reason why Darlington pairs have proven so satisfactory.) Each successive amplifier is designed to have a voltage gain of just about unity, but the current level in each amplifier is successively higher. The final amplifier in the chain can be driven as hard as desired, and the required power obtained from its collector circuit. However, if you wish, some other form of an output chain, such as a cascode pair, may be used.

The critical issue, then, is clearly that as long as the input signal to a resistance-coupled amplifier is less than 10 millivolts, it may and should be used as a voltage amplifier. The balance of the amplifiers, up to the final configuration, should be current amplifiers, with just enough degeneration to keep the distortion to the required limits.

Transformer-Coupled Amplifiers

The transformer-coupled amplifier was the first kind of amplifier that was built using electron tubes, and it also was the first useful amplifier to be built using bipolar transistors. Whereas the transformers used with tubes were usually connected in a way to boost the voltage for the succeeding amplifier, it is necessary to use the transformer as a step-down unit, when using transistors, in order that the required current drive for the succeeding stage will be available. In other words, the transformer typically functions as a *current transformer* in this type of operation (Fig. 4-6).

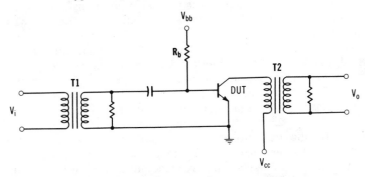

Fig. 4-6. A transformer-coupled transistor amplifier.

If the signal voltage is large enough to cause a significant change of the input admittance for the driven transistor, the transformer is being operated with a very definitely nonlinear load. When the driven transformer is forward biased, the load impedance on it decreases, and an increased current demand results. Likewise, when the driven transistor is reverse biased, the load impedance increases, and a decreased current demand results.

The driving transistor is required to provide the drive energy through the interstage transformer. In fact, the output ampere-turns must equal the input ampere-turns if one ignores the magnetizing ampere-turns. However, the current in the driver transistor increases and decreases, and it may do this in step with the demands of the driven transistor or out-of-step with its demands. In short, *polarization is very important* in a transformer-coupled transistor amplifier.

In a simple multistage amplifier, the current is strictly a pulsating dc, or a dc with a small ac signal superposed on it. In this situation, it is easily possible to saturate the core of the transformer and cause

165

it to be ineffective. Typically, a small air gap is usually left in the core of a transformer intended for this type of service. This air gap reduces the coupling coefficient and, thereby, reduces the overall bandwidth of the amplifier at the same time. On top of all these problems, one has to remember that transformers tend to distort waveforms unless they are reasonably heavily loaded.

In the transformer experiments described in Chapter 7, you will observe a number of the phenomena we are noting here. Perhaps you may think some of them are irrelevant. Unfortunately, they are quite relevant. When you have a transformer whose load is varying over its operating cycle, its frequency response is going to vary also. It may actually become resonant at a very high frequency on some low-frequency peaks! Needless to say, such a behavior is not very good for maintaining a high-quality operating condition in an amplifier. You may find it interesting to attempt to cause a transformer-coupled transistor amplifier to generate the intermittent resonant condition noted above.

Where compactness and modest intelligibility are all that is required, the use of transformer-coupled amplifiers can be very convenient, as long as precautions are taken to assure that a minimum number of problems are introduced by careless design. Clearly, the transformer must be used as an energy transfer unit, and the primary and secondary currents must be maximum at the same time. The loading on the transformer must be sufficient to assure that distortion due to the effects of the magnetizing current will not be a problem. And, above all, the frequency response must be adequate to suit the needs. Unfortunately, adequate inductance is also required to assure acceptable low-frequency response, and a high inductance and a small size just are not compatible. This can be helped to some extent by increasing the load current drawn from the transformer, thereby making the reflected input reactance at the primary smaller. However, that reduces the voltage gain and increases the dissipation in the circuits.

Before the discovery of the transistor, at least one transformer was used in most audio amplifiers. This was an output transformer. However, because of the extremely high transconductance of power transistors that is no longer necessary, and there are many excellent audio-amplifier circuits without a single iron-core device in them. Remember, however, that if a pm dynamic speaker is used, dc flowing in the voice coil can displace the coil from dead center and cause the speaker to distort due to nonsymmetrical motion.

Transformer-coupled amplifiers are commonly used with servo-mechanisms which operate at 400 Hz. (Some servomechanisms operate at frequencies as low as 60 Hz, and others operate at frequencies as high as a few thousand hertz, but 400 Hz is probably optimum for most applications.) With these applications, distortion is not as important, although balance and phase delay can be very important. The ability to isolate circuits, as can be done with transformers, can also be important. Care, nonetheless, has to be taken to ensure that waveform distortion is within reasonable limits and, also, that there is not an excessive amount of phase shift due to unbalance or other causes. These circuits are critically dependent on phase detectors for their proper operation, as the phase detector determines the direction of action for the servo. And unwanted phase shift can cause a displacement of the null and can introduce errors.

Tuned Amplifiers

Tuned amplifiers make use of one or more inductors and some associated resonating capacitors, as required, to cause the amplifier to pass only a relatively narrow band of frequencies. The inductors can be similar to the coils in a transformer (they usually are not), but the coupling between a set of such coils almost invariably will be much less than for a conventional transformer. The higher the coupling, the wider the bandwidth. The capacitance is used to resonate the "leakage" inductance (which theoretically is zero with unity coupling). The subject of the selection of the proper amount of coupling is beyond the mathematical scope of this book. You will note some of its effects in the experiments, however. The impedance level of the tuned circuit controls the voltage gain of the stage directly. A typical tuned-amplifier circuit is shown in Fig. 4-7.

As has been already noted, it is the overall stage voltage gain which is a primary factor in determining whether a circuit is likely to be unstable or not. With tuned amplifiers, the voltage-gain limitation is particularly important, because the variety of stray or parasitic capacitances and inductances in a circuit is such that if there is any possibility of regeneration (a near-oscillatory condition), it will

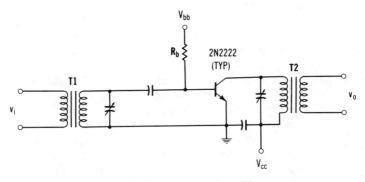

Fig. 4-7. Circuit for a tuned amplifier.

occur. Tuned transistor amplifiers operating in the common-emitter mode usually should not have voltage gains in excess of 10, and in the common-base mode of not over 100. As has been noted above, this is largely possible because of the extremely low input impedance of the emitter of a transistor. (It is approximately the reciprocal of the device transconductance at the operating point in question.)

Introducing a signal into the emitter of a transistor operating in the common-base mode can be a problem because of this extremely low value of impedance. There are several ways of providing the required signal. The simplest way is the use of a transistor operating in the common-emitter mode, with the emitter of the second transistor used as the collector load of the first. This configuration is referred to as a "cascode" configuration. A second mode of introducing the required signal is through the use of a tuned circuit, with the input of the second stage connected to a tap on either the inductance or the capacitance. With this configuration, the placement of the tap is critical. A third mode is using coupled inductors, one or more of which is tuned to the appropriate frequency.

A cascode amplifier can be used either as a tuned structure, with the tuned circuit in the collector lead of the second transistor. It is either coupled directly to the next amplifier or by way of some kind of a coupling circuit. However, it can be untuned if a broad-band amplifier is for some reason required. The impedances faced by both transistors can be controlled in a way which will assure that the amplifier is inherently stable and that it will have little excess phase if properly designed. Nonetheless, the tuned impedance or the untuned impedance should be so selected that a reasonable level of gain is demanded of the second transistor so that the following amplifier will not load it. Thus, any tendency toward instability will be fully suppressed.

With an amplifier using a tapped tuned circuit to provide a signal to a following common-base amplifier, the location of the tap probably will have to be determined by test. The reasons for this include not only the high loading effect of the emitter but, also, the relatively large capacitive loading and the fact that the coupling between the turns of the tuned circuit is relatively low, making it impossible to compute precisely where the tap should be placed. The effective overall impedance level that the tuned circuit should present to the collector must be chosen so as to limit the overall voltage gain to a safe value. The tap is then set to provide, under load, about 10% of that voltage to the next input.

When coupled circuits are used, it is usual to place the tuned circuit in the collector lead and the untuned link in the emitter lead. The principal reason for this is that a much larger inductance can be used if it is untuned. It is important to be sure that the input capacitance of the emitter does not have a severe effect by tuning the untuned link, however. The nominal reactance of the link should be $1/y_f$. Taking the effect of the input capacitance into account, the net reactance of the combination again should be that indicated by the equation:

$$|Z_u| = (y_f)^{-1} \qquad \text{(Eq. 4-13)}$$

where,

Z_u is the net effective magnitude of the input reactance, including both the inductance and the emitter itself.

The combination should have a net inductive component. Needless to say, this can be an extremely small inductance, and it may be unmanageably small with power amplifiers. Stripline and high-capacitance coaxial line may be required in some frequency ranges.

In addition to the use of tapped coils, it is possible to use what is sometimes called "tapped capacitors." They are used with the inductance in providing the impedance step-down. A tapped capacitor really consists of two capacitors in series. Usually the one having the larger value is adjacent to the return, so that only a small part of the overall voltage will appear at the tap. In an ideal situation, this combination can prove quite satisfactory but there are warnings that must be given with them. Unlike with the tapped inductor, you do *not* have a coupled magnetic field to boost the level of current that is available, with the result that the total current circulating in the circuit must exceed the amount of current to be withdrawn at the tap. This means that the Q of the tuned circuit must be substantially higher than the current transformation ratio required. Furthermore, this must be the case *under load*. In short, the tapped-coil approach is the better approach if it can be used. Further aspects of the coupling configurations will be described later in the experiments.

Oscillators

Oscillators are special forms of tuned amplifiers in which the output energy is returned to the input to provide for a continuously variable operating condition. The amount of energy returned at a frequency of oscillation must be sufficient to enable the active device to supply all of the losses in the tuned circuit. The phase of the returned energy must be such as to reinforce the natural vibratory currents in the tuned circuit. Any available energy in excess of that required to maintain oscillation may be withdrawn as a useful load (Fig. 4-8).

It is essential with oscillators that the energy returned from the output to the input at first be slightly more than what is required to provide losses. Further, it is necessary that as the oscillator starts to generate its signal, the amount of energy returned tapers off to just the required amount. Otherwise, erratic behavior can be expected. Where energy return is too great, either a blocking oscillator or a squegging may result. The first of these is indicated by a pulsing action within the audio or relatively low-frequency regime, accompanied with a very high rf noise content. The second is indicated by a switching or multivibrator type of action with a much lower high-frequency energy content. Blocking oscillator action is likely to be controlled, in part, by an LC tuned circuit (which may be based on parasitics), whereas the squegging oscillator is more likely to involve more than one active device and associated RC networks. Frequently the internal impedance of a power supply contributes to this latter kind of oscillation.

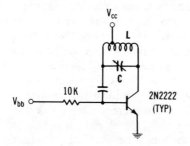

Fig. 4-8. A circuit for a tuned transistor oscillator.

There are many treatments of oscillators as linear network configurations which are relatively good but, unfortunately, all stable oscillators must be at least weakly nonlinear, or they cannot stabilize. The author's *Handbook of Transistor Circuit Design* gives a good treatment of the weakly nonlinear oscillator using transistors.[1] This treatment is both theoretical and practical in that it develops the nonlinear mechanics of oscillation in terms of the characteristics of the active devices being used. Also several books by Minorsky give excellent treatments of the underlying mathematics and describe in detail the phase plane approach commonly used in this development.

POWER AMPLIFIERS

Efficient power amplifiers must be based on circuits which draw little power in the quiescent state, but which can draw increasing amounts of power as the signal level increases. Essentially, two devices are required; both are almost turned off in the no-signal con-

---

1. Pullen, K. A., *Handbook of Transistor Circuit Design,* (Englewood Cliffs: Prentice-Hall, Inc., 1962).

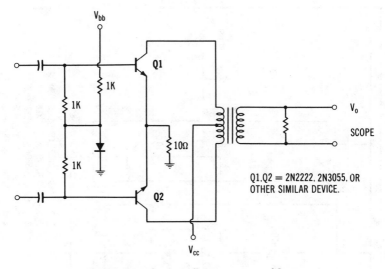

Fig. 4-9. Example of an efficient power amplifier.

dition. One device operates with one signal polarity, and the other with the other polarity. Figs. 4-9 through 4-12 illustrate some power amplifier circuits that give the efficiency needed. One can readily recognize, based on the high nonlinearity of the bipolar transistor, that a very large amount of distortion can be expected unless some means of linearization and a careful balance are achieved. Both of the devices are operated under large-signal conditions, so one *cannot*

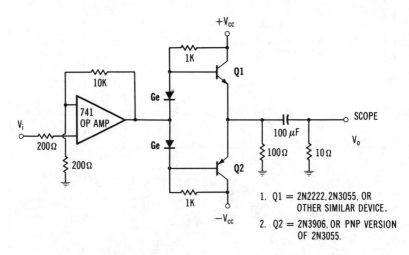

Fig. 4-10. Another efficient power amplifier.

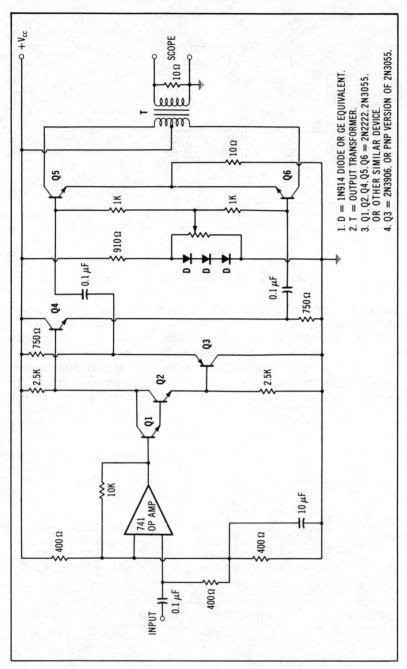

Fig. 4-11. The circuit diagram of an efficient power amplifier.

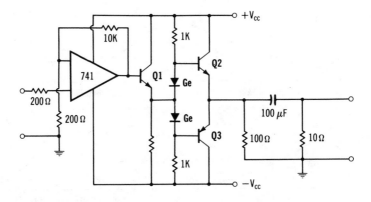

1. Q1,Q2 = 2N2222, 2N3055, OR OTHER SIMILAR DEVICE.

2. Q3 = 2N3906, OR PNP VERSION OF 2N3055.

Fig. 4-12. An example of an efficient power amplifier circuit.

count on beta being constant enough to minimize distortion even in the current mode of operation. In addition, in the current mode, the matching of current-gain levels would be very difficult, based on the nature of beta. Even if one gets an excellent large-signal operation, one is likely to find that, with weak signals, the waveform is likely to be badly distorted. This phenomenon is due to what is now known as "crossover distortion." This is a problem encountered about the quiescent operating point in particular. For further information on this factor, see the author's books *Conductance Design of Active Circuits*[2] and *Handbook of Transistor Circuit Design.*

The use of emitter degeneration and current balancing are probably the most important steps to take in minimizing this problem. The quiescent operating point must be set so that the currents assure that the sums of the effective amplifications of the two devices, as a team, match the effective amplification of either one by itself when the other is turned off. The consequences of improper adjustment of the bias point are shown in Fig. 4-13.

A considerable variety of output configurations have been introduced for use with transistors in power amplifiers. This has been possible for two reasons; first, because one can use an npn and a pnp transistor as a pair if they are reasonably well matched, and second, because the available transconductances for typical devices are high enough that with many of these configurations no output transformer at all is required. The transistors can easily drive a 10- or a 20-ohm speaker.

---

2. Pullen, K. A., *Conductance Design of Active Circuits,* (New York: John F. Rider Publisher, Inc., 1959).

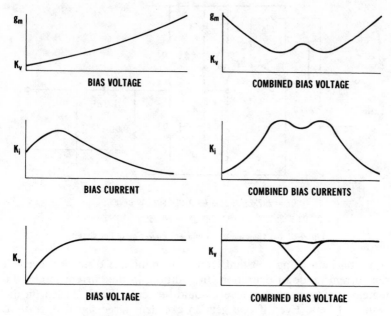

Fig. 4-13. Power amplifier bias-amplification relations.

In this connection, it is interesting to note that there are good reasons to use RC dividing networks prior to the final amplifier, when building multispeaker systems that use high-frequency, mid-range, and low-frequency woofers. This is because it can be shown that the transients introduced by the division network can be severe. They are more severe with LC dividing networks than with RC dividing networks. One thinks of music as being composed of a complex of sine waves. That is true to a point, but it really is an extremely complex array of waves which can be Fourier analyzed into distibutions of sine waves. The results of applying such a waveform on an LC divider network can really be quite unbelievable. The consequences of the use of an RC divider network are bad enough, but they are really much more acceptable.

You will find some suggestions for extra experiments included in Appendix E which will help you to determine for yourself what the consequences of some things are. They are best learned by doing!

EXPERIMENT 1
Bipolar Transistor Characteristics

In Chapter 3, you tested some 2N2222 silicon small-signal npn transistors rather extensively. In this experiment, you will evaluate the properties of silicon small-signal pnp transistors, and germanium npn and pnp small-signal transistors. You will want to make the same measurements on these devices as you made on the 2N2222 transistor in Experiments 1 and 3 in Chapter 3. Some transistors that you possibly can use are listed in Appendix C; some sources are given in Appendix D. You will want to pay particular attention to the comparative base voltages and the polarities required on each kind of these devices. Note the voltage required for a given level of current (for example, at 1 milliampere), and the voltage changes required to double and halve this current. You will want to plot the open-base collector-to-emitter voltage versus the collector-current contour for all of these devices. Then, you should vary the base-current value and repeat the curve. You will also want to measure the transconductances that are generated as a function of the collector currents to find out how it varies with each. You should also test to find the variation of transconductance at a fixed collector current, as the collector voltage is varied.

Step 1.

Set up your breadboarding socket for measuring a transistor. Connect the inner power stripes on either side to your ground reference, and connect the outer one on the left to the appropriate 12-volt supply. Then connect the outer one on the right side to the appropriate variable-voltage supply. (In the common-emitter configuration, both voltages are positive with npn transistors, and both are negative with pnp transistors.) Either a potentiometer or a decade resistance box is connected between the base supply and the base of the transistor under test. A small metering resistor may be placed between the collector supply and the collector. Remember, if you place it in the emitter circuit, you have to keep it very, very small to avoid emitter degeneration. If you do place such a resistor in the emitter circuit, be sure that the voltage across it is less than *2 or 3 millivolts*. The circuit diagram for this configuration is shown in Fig. 4-14. The

NOTE: CLOSING SWITCH SHOULD SATURATE TRANSISTOR. WITH SWITCH OPEN, THE LED GLOWS AND THE METER READS FULL SCALE.

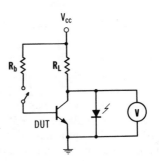

Fig. 4-14. Circuit diagram for testing transistors.

emitter resistance should have a value in ohms of less than three divided by the value of the emitter current (in milliamperes). You will want to meter your base current and your collector current, and still be able to meter the emitter current should the need arise. Only the emitter-current measurement will be concerned with the degeneration problem.

Step 2.

Connect your metering circuits to the solderless breadboard and wire up the circuit. Install the transistor and set the collector supply voltage to minimum. Check the polarities to be sure that they are correct, and disconnect the base lead so that the zero-base-current characteristic data can be taken. Adjust the zero reference of your differential voltmeter so that as the forward bias on the transistor is increased, the voltage change on the base can be measured. Then, connect power to the circuit. With the base circuit open, read the voltage and current on the collector (voltage with respect to the emitter). Record these data. Use a separate line in Table 4-1 for each different device that you measure. (Record the device type used in the margin of the book.)

Table 4-1. Data for Experiment 1, Step 2

V_c				
I_c				
V_c				
I_c				
V_c				
I_c				
V_c				
I_c				

Step 3.

Next, plot the zero-base-current contours for the npn transistors on the chart in Fig. 4-15, labeling each. Then, do the same with the pnp transistors on the chart in Fig. 4-16. These curves will show the relative leakage current in each device that you have tested. You will probably find that the leakage in the germanium devices is much higher than in the silicon devices. You may also find that there are some other differences between npn and pnp transistors. For example,

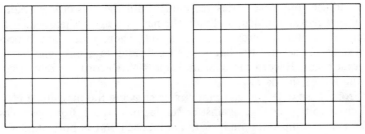

Fig. 4-15. Contours for npn transistors.

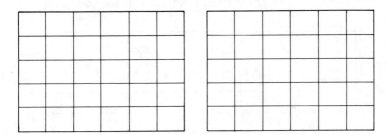

Fig. 4-16. Contours for pnp transistors.

it is apparently easier to get low leakage in silicon npn devices than in pnp devices.

Step 4.

Now, connect the base lead to the base supply voltage, and be sure that your differential voltmeter or your DVM is connected so that you can measure changes to a millivolt. Adjust the collector current in your transistor to 1 milliampere, and note the actual base-to-emitter voltage. Then return to differential operation, and "zero" if needed.

Table 4-2. Data for Experiment 1, Step 4

Times I_c				
ΔV_b				
Times I_c				
ΔV_b				
Times I_c				
ΔV_b				

Readjust your base series resistance until the collector current in your device has doubled, and note the base voltage change which has occurred. Enter it here: _____ mv. Then, repeat the process, first doubling several times, and then returning to 1 mA, and halving several times. Record the data in Table 4-2. Record the device code at the left of the data.

From these data you can plot curves showing the magnitude of the required voltage change against the collector current. For these plots, take 1 mA at the graph midpoint, and let each unit of horizontal scale represent a 2-to-1 change in current from its neighbor. Parameter I_{co} is 1 milliampere. Use graph given in Fig. 4-17A for your plots.

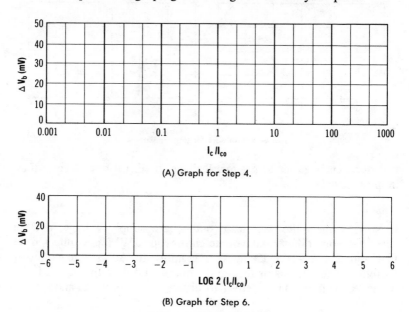

(A) Graph for Step 4.

(B) Graph for Step 6.

Fig. 4-17. Graphs for Experiment 1.

Step 5.

After you have collected data on one or more pnp silicon transistors in this way, compare your results with those that you obtained on the npn transistors. Explain your results.

Although the polarities of the voltages and currents for the two kinds of devices had to differ, and the initiation base voltages (at 1 mA collector current) differed, the differential values had magnitudes which were consistent within the limits to be expected for the various magnitudes for parasitic parameters.

Step 6.

Repeat the above tests with a germanium npn transistor, recording your data in Tables 4-1 and 4-2. (If you need more space, prepare additional tables on separate sheets of paper.) You need to compare the values of the base-to-emitter voltage (at your nominal 1 mA comparison condition) for this device with the corresponding values for the npn and pnp silicon devices (at the same current). Then, put this device through the complete set of tests discussed previously, and record your data. Use either the available table space or prepare additional space on a separate sheet of paper. Plot your curve for ΔV_b as a function of I_c in Fig. 4-17B.

Step 7.

Repeat Step 6, but use a germanium pnp transistor. Record your data as before, and discuss the results of these last two steps. Then,

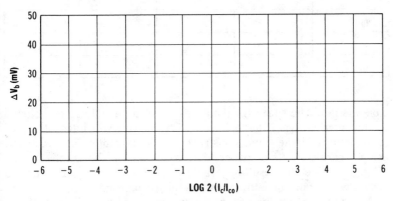

Fig. 4-18. Graph for Experiment 1, Step 7.

plot a curve of the change in base voltage as a function of collector current as you did in the preceding steps for this device. Use the graph in Fig. 4-18.

Step 8.

Now, plot the ratio of the dc beta (ratio of I_c to I_b) as a function of first I_b and, then, I_c. Make the measurements with 2 volts on the collector. Record your data in Table 4-3. Identify the device in the

Table 4-3. Data for Experiment 1, Step 8

I_c				
I_b				
V_c				
V_b				
I_c				
I_b				
V_c				
V_b				
I_c				
I_b				
V_c				
V_b				

page margin. The space for data on the collector voltage is included so that you can vary it too if you wish. So you will have plenty of room to record the data, construct additional tables, as necessary, using separate sheets of paper. Plot your data on the graphs of Fig. 4-19. Examine the resulting graphs carefully, as you will want to decide which current is the most useful for plotting.

Any preference between the two curves would have to be with the plot as a function of the base current, although historically, the curve has been given as a function of the collector current. The reason for this preference is that the distortion must be determined, based on a known waveform at the input (or base.) For this reason, in order to use the data conveniently, the data must be converted to a plot that is a function of base current. Actually, since the base current is a small difference of two currents, it is even better to use a plot in terms of base voltage. Then, you use transconductance and are not dependent in any way on small changes or differences.

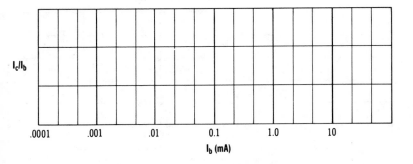

(A) Graph 1.

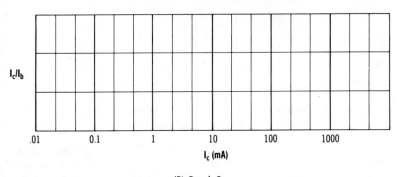

(B) Graph 2.

Fig. 4-19. Graphs for use with Step 8.

Step 9.

Now, you should simulate conductance-type leakage in some devices and examine the results. How do these results compare to those you have seen on actual devices? Can you observe degrading results? (Place a rather high resistance, such as 100,000 ohms, from the base to the emitter, and measure data like you did in Step 8. Record the data in Table 4-4. Compare the results. Do the same with a resistance from the collector to the emitter; use about 1000 ohms in this case. What is the effect on the dc beta for the leakage resistance in each place? Compare your data with that in Step 8. One can hardly say that the beta is well behaved when these leaks exist. There is not much doubt that we need to avoid devices having these characteristics. It is probably harder to detect a base-current leak than a collector-current leak.

Table 4-4. Data for Step 9

I_c				
I_b				
V_c				
V_b				
I_c				
I_b				
V_c				
V_b				
I_c				
I_b				
V_c				
V_b				

EXPERIMENT 2
Bipolar Transistor Transconductance and Current-Gain Characteristics

In Chapter Three, in Experiments 3 and 4, you measured the beta and the transconductance characteristics for an npn silicon transistor using a small sinusoidal signal. In this experiment, you will repeat those experiments using pnp silicon transistors, and npn and pnp germanium transistors.

Step 1.

Modify the wiring on your solderless breadboard so that you can observe the small-signal input sine wave from an oscillator both at the base and at the collector of your transistor with your scope, as illustrated in Fig. 4-20. As before, knowing the source resistance and the load resistance for your transistor, you can compute the beta and the transconductance. (In the measurement of the beta, you measure the source signal voltage and the base signal voltage. Then, using the known source resistance, calculate the source current.) Then, you can calculate both parameters. Record your data in Table 4-5. Use Equation 3-4 or 3-5 for your calculations for beta, and modify Equation 3-9 to read:

$$y_f = \frac{v_o}{(v_i R_L)} \qquad \text{(Eq. 4-14)}$$

Table 4-5. Data for Experiment 2, Step 1

I_c				
I_b				
V_c				
V_b				
v_s				
v_1				
v_o				
I_c				
I_b				
V_c				
V_b				
v_s				
v_1				
v_o				
I_c				
I_b				
V_c				
V_b				
v_s				
v_1				
v_o				

Next, compute your values of beta and dc beta. Also, compute the value of the transconductance as well as the ratio of the transconductance and the collector current at that transconductance. Plot the beta, dc beta, and transconductance-per-unit-current as a function of collector current, on the graphs shown in Figs. 4-21 through 4-23. Do the data show what you expected of them? Explain.

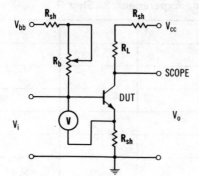

Fig. 4-20. Schematic for Experiment 2, Step 1.

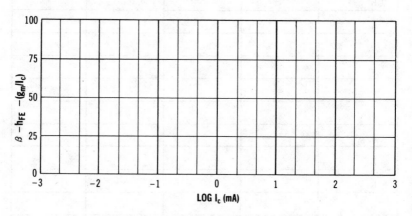

Fig. 4-21. Graph for Experiment 2, Step 1.

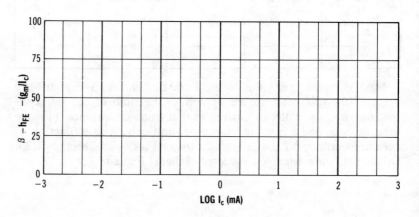

Fig. 4-22. Second graph for use with Step 1.

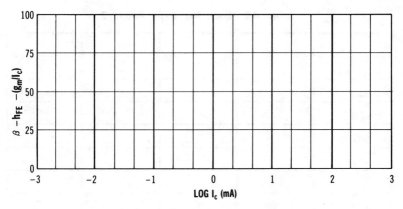

Fig. 4-23. Third graph for use with Step 1.

Your beta and dc beta curves will look very much like the ones that you obtained for the npn silicon devices. However, the vertical scales for each individual transistor will be different. On the other hand, the curves for transconductance efficiency will not only be similar, but those for npn devices will be nearly identical, and those for pnp devices will also be identical. You may be able to detect the point at which high injection becomes important. Those points will differ from code to code, but again, for any given code of device, they will be nearly identical. This shows that the transconductance and transconductance efficiency is the meaningful information.

Step 2.

You were not asked to tabulate the data for dc beta, beta, transconductance, and transconductance efficiency in Step 1, but you were asked to calculate them. Enter the resulting data now in Table 4-6. (Write in the data for each point as you go.)

EXPERIMENT 3
The Effects of Spreading Resistance

You have already simulated the effect of series base resistance, series emitter resistance, and series collector resistance on an npn silicon bipolar transistor in the experiments of Chapter 3. It will be helpful to try and find some more various kinds of sample devices that have some of these properties "built-in." The fastest and easiest way to make further tests is if you have access to a Tektronix curve tracer or a similar device. If not, you can set up a test circuit which you can conveniently use, although it is somewhat slower.

Table 4-6. Data for Experiment 2

beta				
dc beta				
g_m				
g_m/I_c				
beta				
dc beta				
g_m				
g_m/I_c				
beta				
dc beta				
g_m				
g_m/I_c				

Step 1.

If you have a curve tracer available, set it up first for npn transistors. Place a few npn silicon devices in one pile, and a few npn germanium devices in another. Then, you can plug them in one by one and test them. The following steps will guide you through the proper procedures. I might note that the older Tektronix 575 scope testers are better than the more recent ones in a very important way. With the older testers, you can display a set of constant base-current contours on a collector-voltage base-voltage field, which as we have already seen can be significantly more useful than the corresponding plot on a collector-current base-voltage field. (The latter only gives saturation characteristics that are of limited use in switching, whereas, the former enables you to find irregularities in the analog operating field, which is an area of considerably greater importance.) You need to know something about input impedance and transfer function in the analog area, but there is little to tell in the saturation area.

    a. First, adjust the collector power supply so that the peak voltage provided is at almost half of the rated value for your device. Then, you can set the value of the base series resistance so that the size of the base current step is not over 0.001 times the peak rated collector current. Next, insert a value of series collector resistance (R_L). Use a value so that at maximum current, the collector voltage will not exceed 50% to 60% of the peak voltage you have set for your supply. You are now ready to activate your transistor. Plug it in and do so.

b. When you power the circuit, you should get a set of curves that look like the set sketched in Fig. 3-9 in Chapter 3. If they look like Fig. 3-10, there is too much leakage current in the device. Mark the device and put it aside so you can identify it again. If the zero-bias contour follows along the horizontal axis exactly, and all of the other curves are reasonably horizontal (remember, common-emitter configuration), then it is safe to increase the collector supply voltage slowly to find where the zero-bias contour first begins to break away from the horizontal axis. (Increase the load resistance before increasing the voltage; doubling it should be fine.) The peak voltage applied to the transistor should never be greater than the voltage that is at the point just before where the curve first departs from the axis. If you notice that the higher base-current curves are curving as in Fig. 3-11, increase your load resistance more, as you may be getting into a trouble region.

c. Test a number of all types of transistors in this way. Keep each in a separate envelope with notes on its characteristics recorded on the outside of the envelope. (The collector supply you should use is described in Appendix B under the title "Collector Sweep-Voltage Tester.) Give a brief resume of what you discovered.

d. If you find a set of curves which behave somewhat like those shown in Fig. 4-24, you have found a device with either a high

Fig. 4-24. Curves showing presence of collector-spreading resistance.

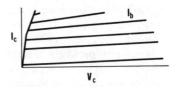

collector-spreading resistance or a high emitter-spreading resistance (or both). Take one of your best devices (one with the smallest apparent saturation voltage) and insert a pseudo-spreading resistance, first in the emitter side and, then, in the collector side. Observe how your curves behave, and explain your results.

If the emitter-spreading resistance is high, a measurement of the transconductance-per-unit-current efficiency will be substantially reduced. The approximate value of the effective resistance can be determined from the equation:

$$r_e = \frac{(y_f - y_{f'} - y_i y_{f'} r_{b'})}{(y_f y_{f'})} \qquad \text{(Eq. 4-15)}$$

where, under normal injection,
  y_f is 39 mhos per ampere,
  $y_{f'}$ is the apparent value (as measured by the techniques described above).

If $y_{f'}$ is practically equal to y_f, then both $r_{e'}$ and $r_{b'}$ are negligible. As long as your current level does not exceed about 10 to 20% of maximum, the effect of $r_{b'}$ is likely to be small, and the term involving it probably can be neglected. The effect of $r_{b'}$ on stage gain will likewise prove to be negligible compared to the effect of R_e or $r_{e'}$ (in any case where either of these is significant). Only a few ohms in the emitter return can cut the overall gain by a factor of as much as 5 or 10. In the absence of a significant emitter- or collector-spreading resistance, the voltage across the emitter-collector port under saturation conditions will be typically a fraction of a volt. (You can expect this with power devices, too.) A significant break in the slope of the saturation line is a warning of one of these conditions.

e. As the power dissipation in the chip approaches the rated value, the contours plotted on the curve tracer will open out and show hysteresis. (The trace back to zero voltage will follow a path at a higher current than the trace to maximum voltage.) Capacitive loading can also show this effect, but it will be present on all of the contours, not just those at maximum power input. There is typically a tendency for the contours to curve toward higher current under high-dissipation conditions. This phenomenon is typical of bipolar transistors, but is less prevalent with field-effect devices. It is one of the important reasons why these devices are prone to catastrophic failure under heavy loading, and is another important reason why the use of a minimum value of collector supply voltage consistent with required operation is vital.

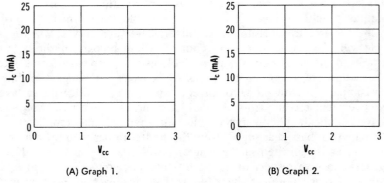

(A) Graph 1. (B) Graph 2.

Fig. 4-25. Graphs for Experiment 3, Step 1.

f. The search for spreading resistance should be made with a reduced collector supply voltage and with an increased sensitivity on your voltage-display axis on your scope. Sketches of the curves that you obtain should be plotted on the graphs given in Fig. 4-25, along with your interpretation of the cause—base-, emitter-, or collector-spreading resistance.

Step 2.

If you don't have access to a curve tracer, it is helpful to simulate one. Set up your solderless breadboard in accord with the circuit shown in Fig. 4-26. The collector sweep-voltage supply must be used

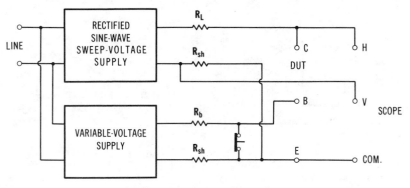

Fig. 4-26. Transistor sweep-voltage test circuit.

for operating your transistor in this configuration, and its voltage output should be controlled from a variable transformer. A push button is used to generate one of the typical sweep curves; when it is released, the zero-base-current contour will reappear. In short, this arrangement gives you a way of looking at sweep waveforms like those that are generated by a Tektronix tester but, one at a time, in comparison with the zero-current contour. You will want to take your current reading from the emitter return as, otherwise, you will have a difficult differencing problem in your current amplifier. Returning your base-bias supply from the emitter rather than the emitter-return, as is indicated in the diagram, will enable you to avoid degeneration with respect to the base. However, you still have to take your reference point at the emitter. That means that your current waveform for the collector current will be inverted. It is still wise to minimize the voltage drop across your current-metering shunt to minimize coupling problems. It is best to keep the maximum voltage across your shunt to less than 5 millivolts. As a result, you will probably want to insert a booster amplifier between the shunt and the scope input. An operational amplifier, based on the LM4250 IC, should be ideal if adjusted to boost the voltage to about a half volt. The circuit for such an amplifier is shown in Fig. 4-27. The advantage of this amplifier is that it can be powered from four 1.5-volt batteries and, if you use "D" cells, it may run for years without being turned off. You probably will find this amplifier useful in a variety of measuring problems, both with your vom or dvm, and with your scope.

As in the previous discussion, the value of the resistance chosen for R_L should be sufficient to keep the peak voltage (at high current) from exceeding half the supply peak voltage. Once again, the peak supply voltage, except when taking saturation curves, should be

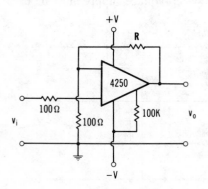

Fig. 4-27. Auxiliary amplifier using an LM4250 op amp.

NOTE: FOR $K_v = 100$, R = 10KΩ
FOR $K_v = 10$, R = 1KΩ

somewhat less than the voltage at which the zero-current bias contour leaves the axis. When you are examining the saturation characteristics, it should be less than 5 volts.

Step 3.

The purpose of the push button in Fig. 4-26 is to keep the transistor inactive until you are ready for it to function. The first test calls for R_b to be infinite (open-circuit). This causes the device to operate in the open-base configuration, and traces the normal zero-base-current curve you will see when using a curve tracer. As you decrease the value of R_b from infinite resistance, you will be tracing a series of curves like those traced with a conventional curve tracer.

To make a test, you press the push button momentarily to apply a signal to the transistor base. You adjust the value of R_b step by step to lower values, increasing the current of your device. In this way, you can observe a series of typical base-current contours such as might be viewed on a curve tracer. It is suggested that you test a series of transistors in this way, just as is called for in Step 1c. Keep notes on the behavior of each device, and identify each so you can find it again. Did you find any leakers? Are there any with high spreading resistance? Are there any which appeared to be thermally sensitive? Record your results here.

Step 4.

Take a sample of each kind of transistor that you have sorted out (npn, pnp, germanium, silicon, leakers, those with high spreading resistance, etc.), and run a set of curves on them. Record these data. You may wish to comment on this mode of testing.

Step 5.

Complete your evaluation of your test devices as described in Steps 1e and f. Remember that when you are measuring devices that have both relatively high voltage and relatively high current, you will be better off to use several short pulses than one long one. When you are examining the saturation region, keep your supply voltage to less than 4 or 5 volts peak. Since you are using a rectified sine-wave voltage, the power being dissipated in your transistor is somewhat less than the peak voltage might indicate. (You can change this supply to provide only one-half a wave if you wish to increase your ability to test at full power but at a reduced duty cycle.)

Step 6.

Since present-day Tektronic curve tracers no longer provide for input contours on the basis of base voltage and collector voltage, our next test is to examine swept input-curve characteristics. We will run this test both against the collector voltage and against collector current, so that you will be able to use either presentation that you like. However, we suspect that you will find that you prefer the voltage presentation. First, however, we need to devise a way of spreading out the base voltage contours so that we can see them clearly.

There are several techniques which can be used for this. If your scope has sufficient control range in the horizontal and vertical sweep positions, you may not need a means of reducing the static magnitude of voltage in this circuit. We would suggest, however, that you arrange to use a reference like the Intersil ICL8069 low-voltage reference. Use a multiturn potentiometer, with a bypass across its output, as a balancing source. These devices can be operated from a 9-volt transistor battery, and should prove to have more than adequate stability for the purpose. It is connected in series, with the base terminal of the transistor connected to the scope, and it is adjusted to balance out the desired level of static bias. Then, the curves can be spread out in a way that they are useful. (The older-model Tektronix 575 curve tracers can be modified to provide this same feature. The input of the

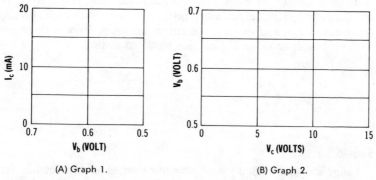

(A) Graph 1. (B) Graph 2.

Fig. 4-28. Input characteristic curves for a typical transistor.

amplifier that is used to meter the base voltage is balanced. It is easy to arrange for a balancing potentiometer in this lead that can apply either a positive or a negative balancing voltage. A polarity-changing switch is required, but source voltages, with adequate regulation, are available within the instrument.)

Once the voltage-offset arrangement is installed, the curves can be plotted either against the collector voltage or against the collector current. We suggest that you do it, and include a sample of each on one of the transistors that you tested. Plot your curves on the graphs given in Fig. 4-28.

Step 7.

Explain in your own words what you think the importance of this series of tests is in learning how to handle transistor amplifiers.

These tests have shown how you can, using available electronics equipment, learn most of the things that you need to know to solve rather difficult transistor evaluation problems. They have shown you how to locate most of the important characteristics and properties so that you can usually judge accurately as to what will happen when you put a specific device in a circuit. Furthermore, they will enable you to judge whether the run of production devices are likely to give you problems in most circuit configurations. These tests should have shown you how to locate commonly encountered deficiencies and defects so that you can avoid their consequences.

Step 8.

Repeat the previous tests with enough devices so that you can be reasonably confident that you have become familiar with the responses to the various tests of both npn and pnp transistors. This will also include both germanium and silicon transistors. You should, also, be reasonably confident that you can successfully apply what you have learned in making amplifiers.

EXPERIMENT 4
Let's Design an Amplifier!

This experiment can be performed using any of the devices you have tested so far. In fact, you do this experiment with at least one sample of each. You need to learn what this experiment can teach you in order to be sure that you have encountered all of the curious, but rather common, effects which occur with these devices. Then, when you see them again (as you undoubtedly will), you will recognize them and use them to your advantage.

The first amplifier as such that you will make will be a resistance-load type. (Most of the test circuits have been amplifiers of some kind.) You will need to try the amplifier in a voltage-gain configuration and, again, in a current-gain configuration so that you can verify all of the phenomena we have discussed. You will also want to put the circuit through some of its paces so that you will know what to do in case of trouble—if it distorts too much, if it loads your signal source too much, or any of the other things which might surprise you.

Before you tackle the experiment itself, though, it is best that a very important point is discussed. When, if one transistor amplifier is providing a signal to another, is the second amplifier being used as a voltage amplifier, and when is it being used as a current amplifier?

Is is simply a matter of the available signal current. If there is enough signal current being generated by the input amplifier to provide the required input signal to the second stage, without causing any significant effect on the characteristics of the input amplifier, the input amplifier is operating as a voltage source. Under these conditions, the effective source impedance that the input amplifier presents to the second amplifier is substantially less than the input impedance of the second amplifier. If, however, the effective source impedance of the input amplifier is significant or large compared to the input impedance of the second amplifier, the combination is functioning as a current amplifier. It is just that simple. The collector of the source or input amplifier acts as a current source as it stands. Unless the load impedance it sees is low enough to cause it to provide more than enough signal current to activate the following amplifier, the current-source element prevails and the amplifier is, in fact, a current amplifier. The low-voltage-gain amplifier demands a substantial signal current from the source transistor and, as a result, more than enough current will usually be available for the following amplifier. However, if there is not, the combination is a current amplifier; otherwise, it is a voltage amplifier.

Step 1.

Select an npn transistor that you have not yet used, and measure the base current required to cause 2 milliamperes of collector current to flow. Then, wire up the circuit shown in Fig. 4-29 on your solderless breadboard. In this test, you will vary your collector supply voltage and vary the output load resistance so that variations of bias can vary the full collector voltage with the current change from zero to 2 milliamperes. Then, you will introduce a small signal into the base. Use a variable-frequency sine wave of about 5 millivolts am-

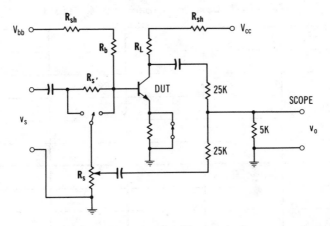

Fig. 4-29. Transistor amplifier test circuit.

plitude. For the various settings of current, find the 3 dB frequency at which the voltage gain decreases as the frequency is increased. (It may vary somewhat with the collector current.) At each point, before you begin your frequency variations, you will want to estimate the approximate effective input resistance in the signal path. Use a resistance in series with the signal source. Record your data in Table 4-7.

Step 2.

Select a group of collector voltages for use in your tests; for example, 12 volts, 6 volts, and 3 volts. Select a series of load resistances for use with these tests, setting the maximum value by the equation:

$$Z_{Lmax} = \frac{|V_{cc}|}{2|I_{cmax}|} \qquad \text{(Eq. 4-15)}$$

A good series of impedances to use would be the maximum value, and both a half and a quarter of the maximum and, possibly, an eighth of it. When the voltage gain approaches unity (near full current), you have reached a small enough value. In each case, measure the input impedance by use of the comparative source voltage and the base voltage and the known series-input resistance, in terms of the equation:

$$y_i = \frac{(v_s - v_b)}{v_b R_s} \qquad \text{(Eq. 4-16)}$$

This may be checked by using a potentiometer for R_s and adjusting its value so that v_b is half of v_s. Then the product of R_s and y_i must be unity. Record your data along with the voltage gains in Tables 4-8 and 4-9.

Table 4-7. Data for Experiment 4, Step 1

I_c				
V_c				
ω				
v_s				
v_b				
v_o				
ω				
v_s				
v_b				
v_o				
ω				
v_s				
v_b				
v_o				
ω				
v_s				
v_b				
v_o				

Table 4-8. Data for Experiment 4, Step 2

V_{cc}				
R_s				
R_L				
V_c				
I_c				
I_b				
v_s				
v_b				

v_o				
K_v				
R_s				
R_L				
V_c				
I_c				
I_b				
v_s				
v_b				
v_o				
K_v				

Table 4-9. More Data for Step 2

R_s				
R_L				
V_c				
I_c				
I_b				
v_s				
v_b				
v_o				
K_v				
R_s				
R_L				
V_c				
I_c				
I_b				
v_s				
v_b				
v_o				
K_v				

You should clearly see now how the amplification varies with bias with one of these amplifiers, and why it is necessary to know that transistors really are not linear. It the next step, you will make some further calculations with these data. It is important that your value of v_b be small enough to avoid distortion.

Step 3.

As has been repeatedly noted, it is necessary that all of your signal-voltage waveforms be essentially the same. Otherwise, the analysis becomes too complex to handle in any way but the "current-gain" approximation approach. Since you cannot really define any of the voltages or currents unless they are sinusoidal or in some other identifiable pattern, this is not really a limitation.

You have enough data in the tables of Step 2 to calculate the small-signal current gain, the dc current gain, the voltage gain and the transconductance. The equation you require for the current gain takes the form:

$$K_i = \frac{|v_o| R_s}{|v_s| R_L} \qquad (Eq. \ 4\text{-}17)$$

Table 4-10. Data for Step 3

V_{cc}				
K_i				
K_v				
y_f				
K_i				
K_v				
y_f				
K_i				
K_v				
y_f				
K_i				
K_v				
y_f				

In a similar way, the forward conductance or transconductance can be determined in terms of the equation:

$$y_f = \frac{|v_o|}{|v_b|R_L} \qquad \text{(Eq. 4-18)}$$

The magnitude of the voltage gain can be obtained by multiplying through by R_L. Record your data on the current and voltage gain and the forward conductance in Table 4-10, entering each value in proper order in the proper column.

Step 4.

What gain do you want from your amplifier? Clearly, the total output voltage available from your amplifier is a function of both the signal current generated in the transistor and the load impedance. If your load impedance is controlled by R_L rather than the input impedance of whatever it excites, then you can in effect control the input characteristics of the following stage and the stage will operate as a voltage amplifier. If, however, the value of R_L equals or exceeds the value of the input impedance of the following stage, then the behavior of this amplifier becomes totally dependent on the characteristics of that following stage, and you will have a current amplifier that has an unknown gain. For practical purposes, you can say that the current from one amplifier is fed directly into the input of the following amplifier, and the circuit voltage depends on the impedance of the following amplifier. In short, current amplifiers are not quite as desirable as one might hope.

To verify this, take one of the amplifier configurations with one of the larger load resistances that you used in Steps 2 and 3, and capacitively couple a load resistance to it, returning the back end of the resistor to negative supply. Select values of resistance that are four times, twice, equal to, one half, and one quarter of the nominal load resistance, and observe the effect on the voltage gain. Select typical

Table 4-11. Data for Step 4

R_L				
R_L'				
K_v				
K_v'				
R_L				
R_L'				
K_v				
K_v'				

points and enter the data in Table 4-11. (The resistance $R_{L'}$ is added load resistance.)

In Table 4-11, the unprimed values are the values without the additional load, whereas the primed values apply for the additional resistance and the voltage gain with it in place. Discuss your results.

As long as the additional resistance has a large value when compared to the load itself, relatively little change is noted. However, when the load is equal to or less than the load resistance in the collector lead, then the voltage gain degrades badly.

As a supplement to this experiment, you should find it interesting to couple two amplifiers under severe load conditions. First, use a pair of npn transistors and, then, an npn transistor followed by a pnp transistor. In this instance, allow the amplification in the first amplifier to be of sufficient value that it can drive the second amplifier into its nonlinear region. Record the results that you observed.

Step 5.

Next, it would be of interest to try and introduce some emitter degeneration. If the circuit is using 2 milliamperes maximum collector current, 10 or 12 ohms of resistance is suggested. Once again, measure the current, voltage, and the data for the voltage gain of the circuit. Tabulate your results in Tables 4-12 and 4-13, and explain the reasons for the results you get.

Table 4-12. Data for Experiment 4, Step 5

V_{cc}				
R_s				
R_L				
V_c				
I_c				

I_b				
v_s				
v_b				
v_o				
K_v				
K_i				
R_s				
R_L				
V_c				
I_c				
I_b				
v_s				
v_b				
v_o				
K_v				
K_i				

Table 4-13. More Data for Step 5

R_s				
R_L				
V_c				
I_c				
I_b				
v_s				
v_b				
v_o				
K_v				
K_i				
R_s				
R_L				

V_c				
I_c				
I_b				
v_s				
v_b				
v_o				
K_v				
K_i				

Once again, the signal voltage developed across the emitter resistor decreases the effective input signal from the base to the emitter, and the signal voltage developed across the linear resistor reduces the effective nonlinearity inherent in the transistor. The input signal voltage v_s, as a consequence, can be increased in magnitude, and the overall linearity is significantly better than previously, either in the current- or the voltage-gain mode of operation.

Step 6.

It is important to know how the mode of operation affects the frequency response of your amplifier. For this purpose, select a series of test conditions to find out how the mode of operation and how the capacitance (from collector to base) affect the overall frequency characteristics of the amplifier. For this purpose, choose different overall voltage or current gains, with and without emitter degeneration. For your convenience, recall that the effective input-output (or Miller) capacitance as a function of the voltage gain is given by the equation:

$$C_{eff} = (1 + |K_v|)C_{bc} \qquad \text{(Eq. 4-19)}$$

where,

C_{bc} is the base-to-collector capacitance,
C_{eff} is the effective value of the capacitance.

You should note that whether your amplifier operates in the current-gain mode or the voltage-gain mode, correction for this capacitance *must* be made in terms of the actual base-to-collector voltage gain, which is negative in sign. (That is the reason for the absolute-value bars shown in Equation 4-19; in a strict sense the sign should be negative, and the absolute-value bars could be omitted.) Effectively, this capacitance is in parallel with the input for your transistor, and for that reason must be added to the input admittance as a loss element. Enter your data on the revised design in Tables 4-14 and 4-15.

Table 4-14. Data for Experiment 4, Step 6

V_{cc}				
R_s				
R_L				
V_c				
I_c				
f				
v_s				
v_b				
v_e				
v_o				
K_v				
K_1				
R_s				
R_L				
V_c				
I_c				
f				
v_s				
v_b				
v_e				
v_o				
K_v				
K_1				

Table 4-15. More Data for Step 6

R_s				
R_L				
V_c				

I_c				
f				
v_s				
v_b				
v_e				
v_o				
K_v				
K_1				
R_s				
R_L				
V_c				
I_c				
f				
v_s				
v_b				
v_e				
v_o				
K_v				
K_1				

What have you learned about amplifiers and the effect of degeneration on their frequency response? In these tables, the value of "f" should be the corner frequency, or 3 dB, in each case.

You, of course, can add more capacitance from the base to the collector and see what happens. There should be plenty of room in the tables to record additional data. We suggest that initially you select frequencies that are about a decade apart until you find where the corner frequency is. You can then use smaller increments until you reach the approximate 3 dB point. You might find it interesting to try a load resistance of about 1500 ohms with a 10-volt supply, and a load resistance of about 220 ohms with a 4-volt supply. You can get an estimate of the value of C_{bc} by using the equation:

$$C_{bc} = (1/2\pi)(1/f_1 - 1/f_2)\{(1-K_1)R_{L1} - (1-K_2)R_{L2}\}$$
(Eq. 4-20)

where, f_1 and f_2 are the 3 dB corner frequencies at the appropriate 3 dB points for the corresponding gains and load resistances.

Step 7.

Repeat the above tests with a pnp transistor to see if you can find any other things of interest. Watch in particular for parallels and for differences. Record your data in Tables 4-14 and 4-15. Summarize your conclusions here.

You can expect to get results with the pnp transistor that are very similar to those obtained with the npn transistor. The corner frequency for a completely equivalent transistor would be perhaps 30% lower than for the npn transistor, but you really cannot tell equivalences that well. The most obvious difference may be the difference in behavior of the transconductance efficiency at high collector currents.

Step 8.

Repeat the above experiments using some power transistors and some rf power transistors. Discuss and explain how the results of your tests compare with the measurements you make. Use separate sheets of paper to set up a data table for your results, selecting the parameters you wish to use. Include your discussion on the paper with your data table.

EXPERIMENT 5
Transformer-Coupled Amplifiers

The purpose of this experiment is to determine how transistors and transformers interact in transformer-coupled amplifiers. For this purpose, you will need to set up a two-transistor amplifier, with the input amplifier capable of being operated as either a current signal-source or a voltage signal-source device. The interstage must be so arranged that the input polarity to the base of the second amplifier can be reversed. The output from the second transistor will be taken from a load resistance in its collector circuit, as shown in Fig. 4-30. Your scope will be very useful in this experiment, as you will need to be able to examine signals at several points in the circuit. A rotary switch will prove to be very helpful in giving you access to the source generator, and to the input and output points on both amplifier stages.

Step 1.

Wire up your circuit on a solderless breadboard in accord with the circuit diagram given in Fig. 4-30. Your scope observation points are identified by capital letters—A, B, C, etc. Then, mark your signal polarities on the reversing switch used to reverse the polarity of the signal into the base of the second transistor. Mark the polarity (which will bias the transistor base in the forward direction when the current is maximum in the primary) as direct and, then, mark the other position as reverse so that you can tell which is which. You can deter-

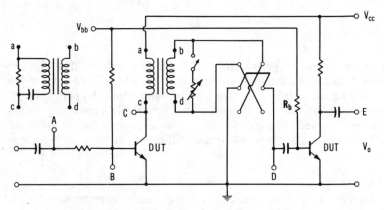

Fig. 4-30. Circuit diagram for transistor amplifier.

mine this status by using the transformer input signal as the source for the horizontal input to your scope. Find out which is the terminal on the secondary to ground, in order to have a negative output when the signal input is positive. If, instead of using both pnp or npn transistors, you use one of each, then, both should be positive at the same time.

Once again, it is helpful to be able to drive the input transistor in either the current or the voltage mode. The switch at the input will enable you to make that change and once the potentiometer is calibrated, it can help you to estimate the effective input impedance of the input transistor. However, you cannot really use this to help you get the current gain of the input amplifier, as its load is a transistor whose input is highly nonlinear. Since your transformer has an input impedance that is totally dependent on the input impedance of the base of the second transistor, the nonlinearity is moved back into the collector circuit. This makes determination of the signal current there difficult. You can get an estimate of the overall current gain and the voltage gain for the two stages, however, as you have a known load resistance in the output stage. A possible interstage transformer for this experiment is the Archer number 273-1378, obtainable at your neighborhood Radio Shack. However, any equivalent transformer will be satisfactory.

Step 2.

First, set up the amplifier as a voltage amplifier. Then, introduce the requisite amount of series input resistance needed to cut the *overall* voltage gain in half. Then, return the circuit to a voltage source, and increase the input amplitude to the point where you just get a noticeable distortion in the output. Switch over to the current-gain mode, and readjust the signal input level until you again get a corresponding amount of distortion. Has the output level changed any? Explain what you have observed.

At low frequency, you probably found that you could get more undistorted output in the current-gain mode, but your frequency response was much more limited. Also, the magnitude of the output varies much more from transistor to transistor than when you operate the circuit in the voltage-gain mode.

Step 3.

Now, reverse the polarity of your signal into the second transistor. This causes the second transistor to draw more current as the first is providing less, and vice versa. Vary the amplitude of the signal into the amplifier, and compare the waveform you get now with that from Step 2. Do you find the amplifier sensitive to polarity distortion-wise? Note down the output voltage levels at which distortion is noted with both configurations. Also, run a frequency-response curve for the amplifier using both configurations. (For these tests, the collector voltage, under static conditions, should be about one quarter of the supply voltage for the stage.) The frequency-response curve should show an overall gain as a function of the frequency for the input stage.This should occur both when operating as a current amplifier and when operating as a voltage amplifier with the secondary of the transformer both unloaded and, then, with different values of resistive load. See the curves shown in Fig. 4-31. A separate run

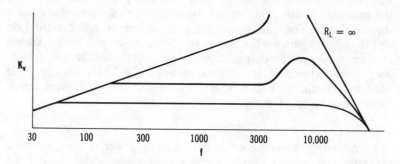

Fig. 4-31. The load effect on the transformer frequency response.

Table 4-16. Data for Experiment 5, Step 3

V_{s1}				
V_{b1}				
V_{o2}				
Sw(+/−)				
f				
R_{s1}				
R_{L1}				
R_{L2}				
V_{o2max}				
V_{s1}				
V_{b1}				
V_{o2}				
Sw				
f				
R_{s1}				
R_{L1}				
R_{L2}				
V_{o2max}				

should be made for each polarity of the inverting switch. You should also determine the signal level that causes the distortion to become just visible for each frequency, load, and polarity value. Record your data in Tables 4-16 and 4-17, and plot a curve of amplification vs. frequency. Also plot a curve of maximum undistorted voltage vs. frequency for each set of operating conditions. Adjust R_{b1} for 1 mA of collector current in the input transistor and R_{b2} for a little over a milliampere, possibly 1.25 mA. Do you find the distortion-level polarity sensitive at any frequency? If so, circle them in the table with a colored pencil line. Tabulate your data in Tables 4-16 and 4-17.

Table 4-17. More Data for Step 3

V_{s1}				
V_{b1}				
V_{o2}				
Sw				
f				
R_{s1}				
R_{L1}				
R_{L2}				
V_{o2max}				
V_{s1}				
V_{b1}				
V_{o2}				
Sw				
f				
R_{s1}				
R_{L1}				
R_{L2}				
V_{o2max}				

Plot your frequency-response and maximum-signal graphs on the drawings given in Fig. 4-32. Enter a discussion of your results here.

Since signal output ampere-turns on a transformer must equal the signal input ampere-turns (plus magnetizing ampere-turns, which should be negligible for good results), one can expect that the signal amplitude at which difficulties are encountered will be substantially smaller when the output load-current requirement increases and the input source current decreases. This effect probably will be much more noticeable with microtransformers than it will be with larger transformers. Careful testing is required to show these effects clearly.

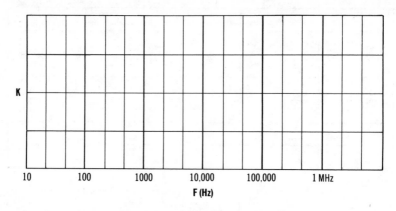

(A) Gain.

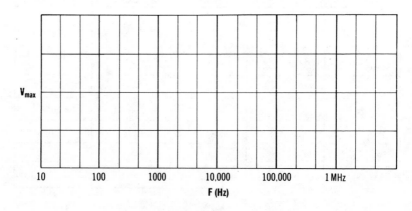

(B) Amplitude.

Fig. 4-32. Gain and amplitude graphs.

Step 4.

You may find it interesting to repeat this experiment using various combinations of transistors, such as an npn transistor driving a pnp transistor through a transformer, an npn transistor driving an npn transistor, a pnp transistor driving a pnp transistor, or a pnp transistor driving an npn transistor. Describe and explain the results that you obtained. Select the data that should be recorded and enter them in a table that you construct on a separate sheet of paper.

You probably discovered that the combination of polarities which gives a high primary current (at the same time that you have a high secondary current) will give you the best overall operation and the best waveform. The kind of device that you select will make little difference otherwise.

Step 5.

Now, you should evaluate the effect of the secondary load resistance on the transformer, and its effect on the overall amplifier operation. You collected some data in Step 3 on this factor. You should take more data at this time. You will want to revise the circuit given in Fig. 4-30 so that you can use the dc decoupling circuit shown there to isolate the dc component of the collector current from the primary of the transformer. Transformers work much better with a minimum of net dc in both the primary and the secondary! Repeat the frequency-response tests and the distortion tests, and see if matters are improved. Just be sure that your coupling capacitors are large enough so that they do not limit the frequency response on the low-frequency end. Select the data you need to record. Record them on a separate sheet of paper along with the data you previously recorded in Step 4. Discuss what you have learned.

Loading the transformer when the increase in load current is matched by an increase in source current and when the direct current is otherwise balanced out by decoupling will cause the transformer to "think" it has zero dc, and will lead to the best overall operational balance. The only way that the loading on the transformer can actually be balanced is through the use of emitter degeneration, which will reduce the loading and at the same time make it much better balanced. The amount of degeneration required is surprisingly small.

EXPERIMENT 6
The Principles of Higher-Efficiency Amplifiers

There are two main kinds of higher-efficiency amplifiers you are likely to be concerned with, although there is a very extensive array of new types, based on transistors, which have been developed. The two types that will be considered here are the push-pull audio power amplifier and the class-C rf amplifier. Both of these types were first used with electron tubes. The same principles apply here, in some cases in somewhat modified forms, with many of the corresponding transistorized types.

The basic push-pull amplifier uses one active device to develop power half the time, and another device to develop it the rest of the time. The important advantage in operating an amplifier in this manner is that one can minimize the amount of power being drawn from such an amplifier when there is no signal or very little signal. However, it can draw substantial amounts of power when a strong signal is encountered. If one attempts to use a single transistor in this manner, it is first necessary to ensure that the device is properly linearized and, second, it is necessary to have a substantial amount of power flowing into the device so that a current flow can be obtained in either direction (an increase or a decrease). On the other hand, when a pair of devices are used, first one may draw current and then the other. As long as the direction of current flow for one appears to be the reverse of the other, distortion-free results can be obtained. Linear operation is still required, however, and, as a result, linearization is of prime importance in either case.

There are two ways in which the data for a push-pull amplifier can be plotted, and each has its advantages. The first, and commonest, way is to plot the output vs. the input voltage (or currents), and the second way is to plot either the voltage or current gain against the appropriate input variable. The results are interesting enough that you should try both methods. (Whether you use input voltage or input current, of course, depends on how you have designed your earlier stages; in doing this, you must not forget that a careless shift from one mode of operation to the other can lead to large amounts of distortion.)

Step 1.

Assume that you have two matched transistors, both having identical beta curves (small-signal) like in Fig. 4-13. *Notice that I have given you the curve of beta as a function of "base" current, not col-*

lector current. This curve is usually plotted as a function of collector current and, as a result, the process to be described is made more difficult if not impossible to do outside of the laboratory. That is how these amplifiers are usually designed.

It is helpful to have a curve of dc beta as a function of the base current in addition to the curve for small-signal beta when using current-gain design techniques. Typically, you have only one curve, and that one is almost always plotted as a function of the collector current. To make the conversion from a curve plotted as a function of collector current, it is necessary first to generate both the dc beta and the small-signal beta curves as a function of the collector current. Then, rescale both using the dc beta curve onto a new base-current abscissa. (It is not easy and, therefore, will not be elaborated here.) The alternative is knob twiddling in the laboratory unless you use voltage-gain techniques.

When you use the voltage-gain basis for design, this problem is simplified considerably. The fact that transconductance is a known linear function of the output current, and that the overall transconductance can be limited to a desired value using emitter degeneration, gives you the tools you need to design on this basis. You simply have to provide a voltage-type signal source to your amplifier.

Step 2.

You need to convert a sine-wave signal source of signal input into a sine-wave voltage and current at the output of your amplifier whether you are operating current source or voltage source. First, the current-source problem shall be examined. It is necessary to find out how to get the required sine-wave signal output for both small and large signals at the input. To do this, it is convenient to measure the small-signal amplification as a function of either current or voltage displacement from a selected static operating point, and to select that point in a way that will assure the best possible linearity of gain. As a current amplifier, it is important to find how the current amplification varies as a function of current displacement from the selected static point. It will also be convenient to know the collector current at each trial point.

To find the small-signal current gain for any value of base and collector current, you use the technique we have described for measuring the beta at each selected trial operating point. Record both the base and collector currents at each point. The ratio of the currents, again, will give you the dc beta should you need it.

To collect the required data, record the small-signal data that you used in measuring the current gain, and the base and collector currents, using equal increments of base current. (The fact that you are actually measuring the small-signal current gain makes the equality of the increments somewhat less critical, as interpolation to adjust values of small-signal gain can be done more precisely than an interpolation that is based on static current values.) These values should be obtained at roughly five or more equal increments up to the maximum bias deviation to be used. Then, plot your curve of current gain against bias displacement, and you will have a curve of current gain as a function of input signal.

Ideally, each device should turn off abruptly when its device pair turns on abruptly. Practically, this is not possible, so that the next best choice is to seek a region of uniformly rising amplification and a second region of uniform amplification. Then, the static operating point can be selected at the "halfway" point of the rising amplification region, and, as a result, matching two devices that have the same characteristics will yield very nearly constant amplification at all signal levels.

Record in Table 4-18 your data that you will use to seek the required matching point. You will need to take data quite closely together to seek out the matching point and, then, when you have that

Table 4-18. Data for Experiment 6, Step 2

I_b				
I_c				
V_b				
v_s				
v_b				
R_s				
v_o				
R_L				
I_b				
I_c				
V_b				
v_s				
v_b				
R_s				
v_o				
R_L				

point, record your data at increments of base current that are roughly equal. (With the voltage amplifier circuit, you do the same, but base your decisions on voltages and voltage amplifications.) Once again, you will want data on I_b and I_c, and you will need signal-voltage data at v_s, v_b, and v_o. You may wish to have data on the static base voltage V_b, as well, as it will help you to make decisions on the voltage configuration when you are ready to try a voltage-design procedure. Based on your data, select the point you think may be best for use as a static operating point. Explain your reasons for your choice.

Plot your data on the graph given in Fig. 4-33. You probably will have some trouble selecting an adequate operating point. What you, in effect, are trying to do is to find the best value of base current to achieve a uniform amplification sum, point-by-point. Since the current gain decreases, but does not approach zero as the base current

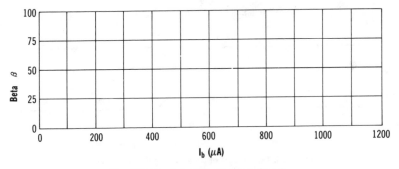

Fig. 4-33. Graph for Experiment 6, Step 2.

is decreased, getting a proper match is difficult at best. There is nothing you can do about the variation of current gain with base current except to apply overall feedback as a means of correction of the problem.

As an exercise to reinforce this in your mind, try plotting the mirror image of the curve of beta vs. I_b. See if you can obtain curves like those shown in Fig. 4-34. Plot your curves on the graph given in Fig. 4-35. Then, sum the curve and its image using different values of base current as the static operating point. Explain the results of your attempts.

We could not find a really good match point, except for zero base current.

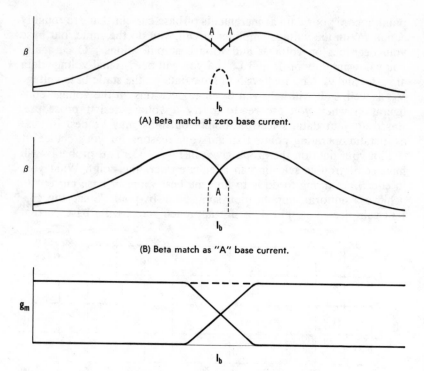

(A) Beta match at zero base current.

(B) Beta match as "A" base current.

(C) Transconductance matching with degeneration.

Fig. 4-34. Typical plot of beta against the base current.

Step 3.

Repeat the above process, first based on voltage gain and no emitter degeneration and, then, with degeneration. With the latter, select your emitter resistance to cut the amplification in half at your selected quiescent operating point. Can you control distortion and operate satisfactorily in this way? Can you use a common-emitter resistance

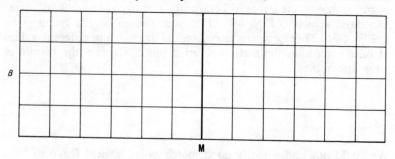

Fig. 4-35. Mirror image graph for Experiment 6.

for the two transistors in doing this? Explain your views, and then try it.

First, take your data in the usual way. When you try the push-pull configuration, adjust the collector currents that are to be balanced by trimming the base-current-control resistances. Record your data in Tables 4-19 and 4-20.

Table 4-19. Data for Experiment 6, Step 3

I_c				
R_L				
V_b				
V_s				
V_o				
K_v				
I_c				
R_L				
V_b				
V_s				
V_o				
K_v				

Plot a graph of effective g_m vs. V_b, both with and without the emitter degenerative resistance. Use the graph given in Fig. 4-36. You will notice that, in order to get a smoothly controlled behavior, you need to operate in the voltage-gain mode with a properly selected emitter resistance. The emitter resistance can be common to the two stages. However, you can have most of it common with a small additional independent degenerative resistance added to each transistor to give a little independent stabilization. A design can be developed based on this kind of a configuration that will have a rather small crossover distortion.

Step 4.

If you wish to evaluate the properties of a transistor based on existing current-gain vs. collector-current curves, it will be necessary

Table 4-20. More Data for Step 3

I_c				
R_L				
R_e				
V_b				
v_s				
v_o				
I_c				
R_L				
R_e				
V_b				
v_s				
v_o				
K_v				

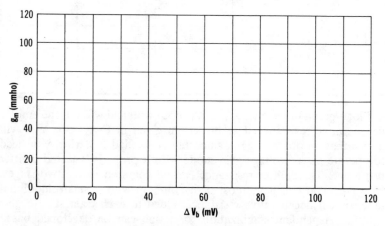

Fig. 4-36. Graph for Experiment 6, Step 3.

to determine the data. To determine the data, it will be extremely helpful to generate the same two kinds of curves that the author has provided to show the dc and small-signal beta as a function of the base current. Explain how you might do this.

The mechanics of converting a curve of dc beta as a function of the collector current into a dc beta curve as a function of the base current is simply to divide each abscissa current by the corresponding value of dc beta to get the corresponding base current. Then, replot the data on a uniform base-current scale. Generating the appropriate small-signal beta curve to get a match involves finding the slope of the dc beta curve. The slope of the dc beta curve must be found point-by-point along its length from zero current and then, the resulting small-signal contour is plotted point-by-point. This is seldom if ever done, even by the experts in the field. If you start out with a plot of the small-signal beta as a function of collector current, your first task will be to try to get the corresponding contour of dc beta. From that curve, you can rescale your curve in terms of the base current, as is required. Then the small-signal beta curve can be replotted in terms of the base current on a point-by-point basis. This too is seldom if ever done, even by the experts in the field. Theoretically, it is possible, but practically, no.

Step 5.

Finding the optimum match point for a pair of transistors that are operating in the current-gain mode is almost impossible analytically. One has to select a trial operating point and, then, set up the amplifier and test it. In making the test, one must arrange to vary the operating points for the pair of matched transistors in the exact same manner as they would be varied by the applied signal. Then, small-signal measurements of the current gain must be made under these conditions. How do you think you might do this?

You have to set up a circuit that will enable you to introduce a small given amplitude of current-test signal, and arrange it so that you can make appropriate static values of current change at the base(s) of the two transistors. The signal will have one polarity in the transistor that has a positive-bias current change, and an opposite polarity in the transistor that has the negative-bias current change. Thus, the total signal-current amplitude in the output circuit is monitored as the bias currents are changed. The static bias value is adjusted so that the over-all sum is as constant as possible. If this adjustment is correctly made, both the crossover distortion and the large-signal distortion can be minimized.

Step 6.

Assume that you choose to operate your amplifier as a voltage-gain unit. Also, assume that you have a 200-millivolt signal either side of the quiescent point that you desire to amplify. First, using *no* feedback, plot what your amplification vs. input voltage curve will look like. (Use a small-value load resistor so that you will not have to concern yourself about the value of the collector supply voltage. Also, use 5 volts.) Make your calculations in increments of 20 milli-volts from a starting point of 100-microamperes collector current. Make a table of your results (use Table 4-21), and plot them as best you can on the graph given in Fig. 4-37. In this table, v_s is taken as the input signal voltage, to distinguish it from the static voltage at the base. You can plot either the effective transconductance or the voltage gain in your graph.

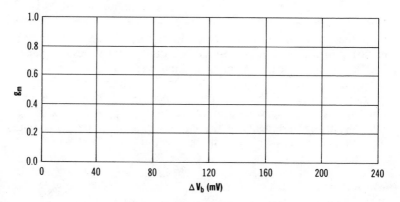

Fig. 4-37. Graph for Step 6.

Clearly, the results are extremely nonlinear, something we have already repeatedly observed with solid-state devices. What do you think you should do about this dilemma? Or, do you prefer to use a current-gain approach, attempting to match betas? (Remember that you will need to match values throughout the range of operation

Table 4-21. Data for Experiment 6, Step 6

I_c 100 μamp				
V_b				
v_s				
v_o				
R_L				
I_c				
V_b				
v_s				
v_o				
R_L				

and will, also, need to be sure that the variation over the range is what you want.) Explain your options here.

Neither the matched-beta approach nor the nondegenerated voltage-gain approach appears to be acceptable. It is suggested that degeneration be introduced. Select a suitable value based on the already described technique, and include it in the common-emitter return. Then, record the resulting data, adjusting the static operating point as has already been described. Record your data in Table 4-22.

The subscripts 1 and 2 in Table 4-22 refer to the two separate devices. The absence of a subscript indicates a combined value. The subscript "s" on the small-signal voltage input indicates the value for either side at the base for the devices. If you have adjusted the quiescent point correctly, you will find the sums of K_{v1} and K_{v2}, or K_v, surprisingly uniform over the operating range.

Step 7.

This is a calculation back-up step for what you have just done. Calculate the voltage gain using a range of emitter-current values from 100 microamperes to at least ten milliamperes for the emitter-resistance value you have selected. Then, repeat the calculation using other values of emitter resistance. Some typical suggested values are

Table 4-22. Second Set of Data for Step 6

I_{c1}				
I_{c2}				
V_{b1}				
V_{b2}				
v_s				
v_{o1}				
v_{o2}				
K_{v1}				
K_{v2}				
K_v				
R_L				
I_{c1}				
I_{c2}				
V_{b1}				
V_{b2}				
v_s				
v_{o1}				
v_{o2}				
K_{v1}				
K_{v2}				
K_v				
R_L				

1, 3, 10, 30, 100, 300, and 1000 ohms. You may wish to go back and try some in-between values after you have finished. Tabulate your results in Table 4-23. The equation for calculating the approximate effective transconductance is:

$$y_f = \frac{\Lambda I_c}{(1 + \Lambda I_e R_e)} \qquad \text{(Eq. 4-21)}$$

where,

$\Lambda = (q/kT)$ and has the value of 39 mhos-per-ampere.

For convenience, the emitter and collector currents may be taken to be approximately equal. Record your data in Table 4-23.

Table 4-23. Data for Step 7

I_c				
R_e				
y_t				
I_c				
R_e				
y_t				
I_c				
R_e				
y_t				
I_c				
R_e				
y_t				
I_c				
R_e				
y_t				
I_c				
R_e				
y_t				
I_c				
R_e				
y_t				

As always, the voltage gain is the negative of the product of y_t and the load impedance. Ideally, you should select a resistance that will degrade the value of y_t to half its undegenerated value at the quiescent point. At 1 milliampere, this works out to be 26 ohms. The voltage across the resistance is again 26 millivolts.

Step 8.

Assume that you have two transistors, each of which carries the operation half of the time. Vary the quiescent point about the one that is defined by solving the denominator of Equation 4-21 for the unity value of the added term. Select the resistance for the match point at 200 microamperes and use that as the initial quiescent point. Further, assume that the effect of the added term is linear, from zero current through half the limit value at 200 microamperes, and it reaches full value at 400 microamperes. Then, using the graph in Fig. 4-38, plot how it actually varies against the curve for y_f. Next,

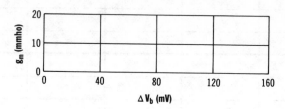

Fig. 4-38. Graph for Step 8.

using very small increments about the selected quiescent point, find the point which appears to give you the best overall constancy of voltage amplification. Describe what you have learned from this.

We find that there is no perfect match point, but the suggested one is quite close to the best one. You may find a point which is somewhat better rather near this point, but it is doubtful whether much effort should be made in finding a more precise match point. The variation is small enough that a small amount of overall feedback will be effective and completely adequate.

EXPERIMENT 7
Testing a High-Efficiency Amplifier

You will want to observe examples of these amplifiers in action to see just how they behave in practice. Essentially, what you need to do is to set up a push-pull current-gain and a push-pull transconductance amplifier and find just how the appropriate gain factor varies with the required current or voltage bias variation. Then, a gain-summing about a series of match quiescent points will give the

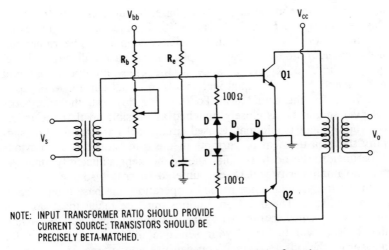

Fig. 4-39. Current-gain amplifier measurement configuration.

required amplification variation. Finally, you will want to set up a full amplifier (including both halves) and measure the gain as a function of the bias excursion to see if your theoretical data are confirmed in practice.

In the current-gain mode of operation, it is necessary to measure the gain as a function of the base current for the single stage, and, then, make a similar measurement on a pair of stages whose base currents are varied in opposite directions (down to a nominal zero, of course, for one). The test signals should be signal currents that

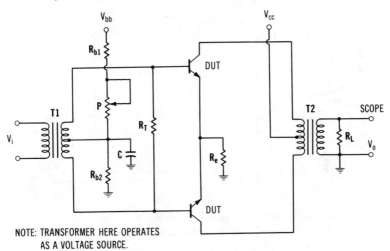

Fig. 4-40. Voltage-gain amplifier measurement configuration.

are equal in magnitude and out-of-phase. Suitable circuits for doing this are shown in Figs. 4-39 and 4-40. Balance is of extreme importance with this kind of high-efficiency amplifier. The rather unusual configuration of a base supply with a push-pull circuit is required—first, to achieve balance of the quiescent operating point and, second, to obtain equal increments in signal current in *opposite directions* for the actual test. To start for this test, the transistors selected should be as closely matched as possible so that the quiescent adjustment is minimal. Closed-circuit jacks are required to simplify the measurement of currents in the transistors. As in previous experiments, the ac test signal should be kept sufficiently small so that no nonlinearity will be encountered from this source.

In the transconductance mode of operation, the base input signals are all voltages. The usual precautions must be taken to assure that linearity is achieved, otherwise, the measurements could be invalidated. You will be using your sensitive differential voltmeter for measuring the offset voltage. You can use your scope or a sensitive ac meter for measuring the small-signal components. (The scope is the better device.) Suggested circuits for a "single-ended" emulation and a full-emulation are shown in Figs. 4-41 and 4-42, respectively.

Step 1.

Wire up the single-ended emulation of the current-gain circuit on your solderless breadboard. Set your quiescent point for a modest amount of current (not over 10% of the maximum that you expect

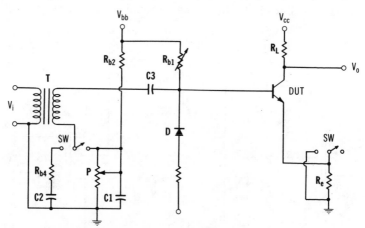

NOTE: SWITCH IS TURNED TO THE RIGHT FOR A VOLTAGE AMPLIFIER;
SWITCH IS TURNED TO THE LEFT FOR A CURRENT AMPLIFIER.

Fig. 4-41. Single-ended emulation of a high-efficiency power amplifier.

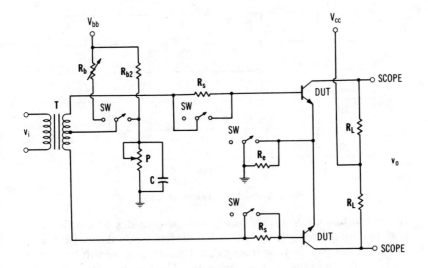

NOTE: SWITCH IS TURNED TO RIGHT FOR A VOLTAGE AMPLIFIER;
SWITCH IS TURNED TO LEFT FOR A CURRENT AMPLIFIER.

Fig. 4-42. Push-pull emulation of a high-efficiency power amplifier.

to draw). Apply a signal current from an audio generator, and adjust its amplitude so that the transistor signal-input voltage does not exceed about 10 millivolts. Measure the output voltage across the load resistance, either directly or by use of an isolation transformer, and plot the results. Shift your bias current step by step in both directions so that you can get a full picture of the behavior of the stage. Record your data in Table 4-24 and plot your data for the best operating combination on the graph in Fig. 4-43.

Step 2

From the previous data, determine how the small-signal amplification of your amplifier varies with bias, and plot out how it will vary with a push-pull combination. Then, select a pair of transistors that are reasonably well matched, and set up and measure the push-pull configuration. How do your results conform with what you predicted? Can you find any way to improve linearity in the neighborhood of the quiescent point? Explain.

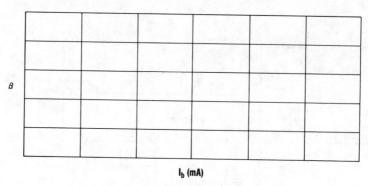

β

I_b (mA)

Fig. 4-43. Graph for Experiment 7, Step 1.

It is important to note that this technique provides a sensitive way of determining probable distortion, and it will easily detect variations in the current gain in your transistors. Therefore, there will be a good bit of variation in the gain you measure. The best way to control the

Table 4-24. Data for Experiment 7, Step 1

I_c				
I_b				
v_s				
R_s				
i_s				
v_o				
R_L				
K_1				
I_c				
I_b				
v_s				
R_s				
i_s				
v_o				
R_L				
K_1				

linearity is in the overall feedback; with a current amplifier, you probably will need a substantial amount. The introduction of emitter-return resistance will have a relatively small effect in this configuration.

Step 3.

Repeat Steps 1 and 2. Only this time, operate the transistors as voltage amplifiers. Voltage sources are now needed for the signals, and the bias will be achieved by shifting the base voltage instead of changing the base current. The initial step is to set the bias voltage to the chosen quiescent point for the devices. This is done by setting the collector currents at the chosen values. A possible bias circuit for doing this is shown in Fig. 4-44. Again, the test should be run

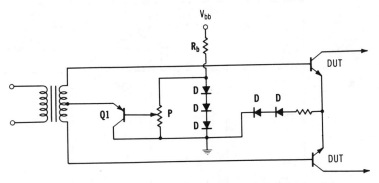

Fig. 4-44. A bias circuit for Experiment 7, Step 3.

first with a single transistor and the operating characteristics tabulated. Then, conditions for achieving a proper match can be selected, a pair of transistors can be wired up on your solderless breadboard, and their operating conditions can be adjusted for proper values of voltages and currents. The same basic test, for voltage gain, may be made to find out how well balanced the amplifier can be and what the actual variation of amplification with the operating point is. The data can be tabulated and the overall amplification variation with operating point determined. The net bias should be varied to help find the best quiescent point. The data can then be recorded in Table 4-25.

Your test circuit has been set up to provide proper bias displacement for both the signal and the static bias. It does not prevent a dc unbalance in the output transformer, however. The turns ratio of the transformer is not too critical, as long as the value of R_L is small enough that the transformer is properly loaded. When you set up an amplifier without emitter degeneration, you will find that your

Table 4-25. Data for Experiment 7, Step 3

I_{cs}				
I_{c1}				
I_{c2}				
V_{b1}				
V_{b2}				
v_s				
v_{o1}				
v_{o2}				
R_L				
R_e				
K_v				
I_{cs}				
I_{c1}				
I_{c2}				
V_{b1}				
V_{b2}				
v_s				
v_{o1}				
v_{o2}				
R_L				
R_e				
K_v				

amplification curve looks like a parabola or a catenary curve. When emitter-degeneration resistance of the proper value is included, however, you will find that you can get good linearity. If you have not already confirmed this with practical circuits, you should do so now. Note any special effects you may have found.

The distortion that you noted from improper adjustment of the quiescent bias is now called "crossover distortion."

EXPERIMENT 8
Tuned Transistor Amplifiers

Tuned amplifiers using transistors have important similarities and, also, important differences to transformer-coupled amplifiers. You have already discovered that loading the secondary is an ideal method if you want to broad-band a transformer-coupled amplifier. You reduce the overall voltage gain this way and you also make the response much more uniform.

However, with a tuned amplifier, you may not wish to broad-band your circuit beyond a certain point. This point represents the optimum passband. How do you adjust your gain to a safe level?

It turns out that the frequency of oscillation for a tuned circuit is defined in terms of the equation:

$$f = (2\pi\sqrt{LC})^{-1} \qquad \text{(Eq. 4-22)}$$

and the impedance of the same tuned circuit is defined in terms of the equation:

$$Z = \frac{L}{CR} \qquad \text{(Eq. 4-23)}$$
$$= QX_L$$
$$= QX_C$$

where,
  Q is the approximate ratio of the inductive or capacitive reactance to the total effective loss resistance for the tuned circuit.

This means that the impedance of the tuned circuit is proportional to both the inductance and to the Q of the circuit. You can, therefore, reduce the impedance *either* by reducing the Q-factor or by reducing the inductance (and correspondingly increasing the capacitance to keep the LC product constant). Or, you can reduce both.

If you do not wish to reduce the Q, therefore, you reduce the inductance. However, the impedance levels at which you must operate these tuned amplifiers are almost unbelievably small and, often, tuned circuits having low enough impedances for transistor amplifiers are difficult to design. This results in a dilemma.

There are different approaches to solving this problem. The first is to tap the coil near the supply end and use a larger inductance, as shown in Fig. 4-45A. The inductance from the tap to V_{cc}, then, should be that value that is required to provide the impedance level needed to furnish the chosen safe gain. The second alternative is to use a capacitive tap on the tuned circuit. In using this circuit, however, it is necessary that the ratio of the capacitances be so selected that the quotient will be significantly smaller than the loaded Q for the tuned circuit. The reason for this is that there is no way for the capacitor totem-pole to increase the current that is available to the larger capacitor, with the result that the circulating current due to the Q of the tuned circuit must be larger than the load current demanded at the tap (see Fig. 4-45B).

A third method of doing this can take either of two forms. These are illustrated in Figs. 4-45C and 4-45D. In this instance, an untuned inductance is used in the collector circuit. It is magnetically coupled to the tuned circuit. Frequently, optimum operation requires the use of a second coupling coil to provide the signal to the input of the following stage. This configuration, shown in Fig. 4-45D, is

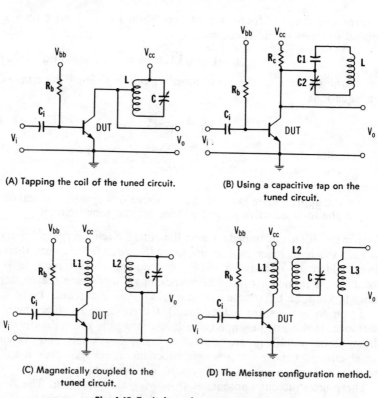

(A) Tapping the coil of the tuned circuit.

(B) Using a capacitive tap on the tuned circuit.

(C) Magnetically coupled to the tuned circuit.

(D) The Meissner configuration method.

Fig. 4-45. Typical tuned transistor amplifiers.

sometimes called the Meissner configuration after its inventor. Since there is no Q build-up in the coupling coil(s) with these configurations, it is much easier to achieve the low impedance levels required for input and output. The reactances of these coils should be nominally equal to the required impedance level. Magnetic coupling makes it possible to obtain the required current levels without any difficulties.

Step 1.

Select an operating frequency within the range of both your scope and your signal-generating equipment for this test. A convenient frequency is 1 MHz. Select a transistor whose f_{max} is above 200 MHz. Determine the component sizes that are required to provide a voltage gain of about 10 to 20 (take $Q = 100$) for a device with a current of 1 milliampere. Enter your calculations here:

The equation to use is $2\pi fL = R_L/Q = 1/2\pi fC$, where $R_L = 10/.04 = 250$ ohms. Our calculations indicate from this that an inductance of 0.4 microhenries and a capacitance of 0.06 microfarads is required. An inductor can be designed from the equation:

$$n = \left(\frac{L}{Fd}\right)^{\frac{1}{2}}$$ (Eq. 4-24)

where,

n is the number of turns,
L is the inductance in microhenries,
d is the coil diameter in inches,
F is a form factor.

The factor F is equal to 0.018 for coils having a winding length equal to their diameter, and 0.010 for coils having a diameter half the length. Based on these equations, you would need a coil of eight turns of a 5/16-inch form with the coil being 5/16 inch long. The wire that is chosen should be of such a diameter that about sixteen turns can be wound in the nominal length, and the turns are then spaced to fill the 5/16-inch coil length. Either 24 or 26 gauge wire should be satisfactory.

Step 2.

Assemble a transistorized amplifier using a 0.05-μF ceramic capacitor with your inductance for your tuned circuit. Observe the output of the amplifier as you vary the input frequency from about 0.5 MHz to 2 MHz. You should cross the resonant frequency. Apply an input signal of about 10 millivolts, and you should get about 100- to 200-millivolts output. (Be sure that you have adjusted the collector current to 1 milliampere.) If you have access to a digital counter, measure the center frequency of the amplifier and the frequencies for the minus three decibel (3dB) points (0.7 times peak voltage); then determine the effective Q of your tuned circuit from the equation:

$$Q = \frac{f}{\Delta f} \qquad \text{(Eq. 4-25)}$$

where,

f is the peak frequency,

Δf is the frequency difference between the two −3dB points.

The voltage gain you obtain should be approximately one tenth of the Q that you calculate. Record your data.

f = _____. K_v = _____.

Δf = _____.   Q = _____.

Do your data check?

Step 3.

Next, you may set up a Meissner-type circuit, with an untuned coupling link into your tuned circuit. The inductor you will need for the collector circuit should have about ten times the number of turns for the tuned circuit itself. An output link may also be made which has about three times the number of turns on the tuned inductor. Your circuit should look like the one shown in Fig. 4-45D.

The reason for the large number of turns on the untuned link is that you will get optimum results if the coils are designed to have a net reactance equal to the nominal impedance you wish to place in the circuit. Under these conditions, you can use minimum coupling for the desired result, and you will get an optimum operating condition. If shunt capacitance increases the effective impedance, the inductance value chosen should be reduced so that the net reactance conforms with this condition (which can be verified analytically). With this circuit, the collector link has 250 ohms of inductive reactance, and the output link has about 25 ohms.

Step 4.

Replace the tuned circuit with the appropriate coupling link, and connect the smaller link to your scope. Couple the three coils together, as indicated in Fig. 4-45D, with the collector link next to the tuned circuit next to the output link. You will find that up to a certain point that as the coupling between the coils is tightened, the output observed on your scope will increase and then it hardly changes. The point at which the rate of increase becomes small represents the critical coupling; beyond that point the Q of the tuned circuit is adversely affected. How can you demonstrate this? Explain, and then demonstrate it.

As long as coupling is below the critical level, changing the coupling will change the amplitude observed at peak frequency, but the shape of the curve observed as you vary the frequency will change rather little. Once you reach critical coupling, however, the peak frequency does not change much, but the shape of the curve as you vary the frequency will change significantly. You will find two peaks on either side of the original peak if the coupling is far enough beyond the critical point.

Step 5.

Fasten the output link to the tuned coil at approximately its critical coupling point, and vary the coupling between the pair and the collector link. Using your scope, plot the response curve for the amplifier as a function of the distance between the collector link and the tuned circuit. Also, find the 3-dB points at each setting. (There will be some shift of peak frequency as the coupling is varied, so you will have to retrim your frequency each time you readjust the coupling.) Estimate the peak frequency and the 3-dB bandwidth at each setting, and plot your data. When you are operating in the overcoupled condition, find the 3-dB points with reference to the midpoint frequency, not the peaks. Plot your data in Table 4-26, and explain what you have learned. In Table 4-26, D is the minimum axial distance, Q is the effective Q as defined before, u/c/o designates the state of coupling (either under, critical, or over), v_o is the maximum response (or the center response with overcoupling), and v_{op} the approximate peak response when overcoupled. The graph illus-

Table 4-26. Data for Experiment 8, Step 5

D				
f_0				
Δf				
Q				
u/c/o				
v_0				
v_{op}				

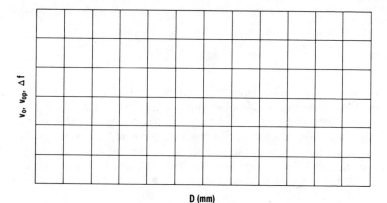

D (mm)

Fig. 4-46. Graph for Experiment 8, Step 5.

trated in Fig. 4-46 should show the values of v_0, v_{op}, and Δf as they vary with D. (Set your own scales on this graph.)

Step 6.

Now, interchange the input and output links after both are secured in the critical coupling position. (You do not want them to move until you are ready.) Then, vary the position of your new input coupling link (the lower-inductance one), and observe how the combination responds. Measure the same data as taken in Step 5 and record them in Table 4-27. Explain what you have observed.

Table 4-27. Data for Step 6

D				
f_o				
Δf				
Q				
u/c/o				
$\dot{v}_o$				
v_{op}				

Using a different color of ink, plot your data on the same chart as you used in Step 5 (the graph given in Fig. 4-46). Explain what you have observed.

You probably found that you had to tighten your coupling somewhat to get critical coupling, and the response curve for the system was perhaps broadened somewhat because of the mismatch. This optimum exists, but it is broad. Had you doubled the current in your transistor, you might not have had to change the setting. If, on the other hand, you had increased the current by a factor of 5 or 10, you might have found the behavior somewhat unstable. The way you can detect this is through tuning across the resonance curve. The amplitude observed on your scope would not vary smoothly but would appear to be somewhat "jumpy." Vary your current level and see if you can detect this condition.

EXPERIMENT 9
Further Discussion of Tuned Transistor Amplifiers

Usually only a single tuned circuit, without any coupling links, is used with a tuned transistor amplifier. With the almost unbelievably small inductance that is required at 1 MHz, and the very large associated capacitance (even with 1 milliampere of collector current), you can recognize that there are going to be problems as you try to go to higher frequencies and/or higher levels of current in your transistor. What can be done about this? The simplest answer to this problem, other than the coupling link used in Experiment 8, is to use a tapped tuned circuit. This requires either putting a tap on the coil or using two capacitors in series to achieve an equivalent result.

With the tapped-coil configuration, the inductance to the tap point has to be the value that you have calculated for the simple tuned circuit, whereas, the total inductance can be tuned by a smaller and better behaved capacitor. (See the discussion on capacitor-resonant frequency in Step 7.) It is much easier to select a suitable smaller capacitor for the purpose, as all capacitors do have a self-resonant frequency. It is likewise easier to make an inductor having a suitable Q if it can be somewhat larger than necessary so that it has a reasonable amount of inductance. In a sense, you get the best of both the high-Q circuit and the coupling-link approach without any associated mechanical problems.

The inductance of a coil is roughly proportional to the square of the number of turns in it. You can set your tap point by using Equation 4-24 as a starting point. Then, double or triple the number of turns and decrease the capacitance value by either a factor of 4 or 9 as a trial method. Adjust the inductance or the capacitance as required. When putting a tap on the enameled wire used for the coil, it is best if you remove a little bit of enamel at the appropriate point and, then, neatly tin the bare spot. You will need to solder on a separate piece of wire or a decoupling capacitor to make your lead, rather than trying to make a loop in the wire of the coil itself, which would have an adverse effect on the Q of the coil.

Step 1.

Temporarily remove the tuned coil you used in Experiment 8, and place a tap on its midpoint. You will be connecting various values of load resistance from this point to one end of the coil to observe the effect. Start with a 1000-ohm ¼-watt carbon (or carbon-film) resistor.

Step 2.

Set up the circuit as it was arranged in Experiment 8, with the coupling links in their original places. Measure the output voltage level, the center frequency, and the Q for the circuit. You will need these data to see what the loading does to the tuned circuit. Record your data in Table 4-28.

Step 3.

Connect the 1000-ohm resistor from the tap to one end of the tuned coil, and tabulate the same data as in Step 2. The peak frequency may be shifted slightly, and the Q may be decreased a little. Also, the overall amplification may be reduced. Then, methodically repeat your tests with successively smaller values of resistance, taking the appropriate data at each step. We suggest using resistance values of 700 ohms, 500 ohms, 330 ohms, 250 ohms, etc. until you find a substantial reduction in the Q of the coil. Record your data in Table 4-28.

Table 4-28. Data for Experiment 9, Steps 2 and 3

f_0				
Δf				
Q				
v_i				
v_0				
K				

Step 4.

Repeat the above test, placing the resistance across the full coil, and record your data. What is different now?

Table 4-29. Data for Step 4

f_0				
Δf				
Q				
v_i				
v_0				
K				

Step 5.

Plot the data you have obtained on the graph given in Fig. 4-47. Plot the overall gain as a function of load resistance, and the Q and peak frequency as a function of the load resistance. Use different colors of ink when plotting the data for Steps 3 and 4.

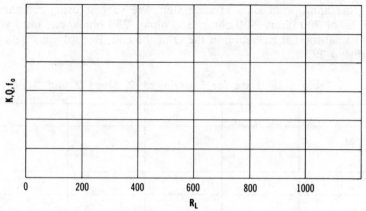

Fig. 4-47. Graph for plotting data from Tables 4-28 and 4-29.

Step 6.

Examine all of the data you have taken and write a short review of what you have learned about tuned amplifiers so far.

In the above steps, you found that you could set both the gain of an rf amplifier and its bandwidth, independently. In fact, you found that circuit-impedance levels are usually too high. You discovered that when the amplifier stage gains are too high, instability can result, and that this can be detected by erratic stage tuning. This is a result of excessive feedback. You also learned that tapped-coil tuned circuits could help to alleviate the problem better than capacitive taps could, and that untuned links could also be used. You also found ways of determining the effective impedance of tuned circuits.

Step 7.

To find the self-resonant frequency for a capacitor, you want to short the leads as directly as possible and solder the joint. Then, using either a dip meter (Fig. 4-48) or a signal generator and a detector circuit, find the lowest frequency at which the circuit draws

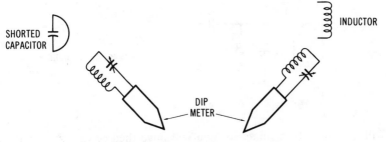

Fig. 4-48. Measuring a capacitor-resonant frequency.

energy. That is the lowest self-resonant frequency. Try it with several ceramic and mica capacitors. The resonant frequency of an inductor is measured similarly but *without* a short circuit.

EXPERIMENT 10
The Class-C Amplifier

The class-C amplifier is more efficient power-wise than the audio power amplifier because it operates under conditions where its non-linearity can be "washed out" with the help of a tuned circuit. This circuit is used largely in radio transmitters where either the amplitude of the carrier is constant (fm, for example), or where it can be "modulated" in a linear fashion. This amplifier is normally biased so that without an rf input it draws practically no current. Typically, it might be biased to 200 to 300 millivolts below peak current. The circuit is designed so that every time peak current is drawn, the device voltage drops to approximately saturation voltage, and a large amount of energy is stored in the magnetic field of the tuned circuit. The purpose here is to simply observe the operation of this amplifier and to learn more about its behavior.

Step 1.

Set up your tuned rf amplifier on your solderless breadboard again, and arrange to apply about ½ volt, peak-to-peak, on its input from your signal generator. The stage can be operated in the common-base configuration (Fig. 4-49), as that will make it easier to assure stability. Use your variable-voltage supply for the collector, and vary it to find the voltage giving the highest efficiency. (As you increase the collector voltage from a very small value, the power output will at first steadily increase but, at some point, the increase in output

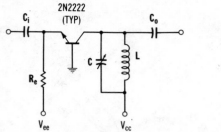

Fig. 4-49. An rf amplifier.

will no longer be significant. You will have then passed the point of maximum efficiency.) Describe what you observe.

Step 2.

After you find the "knee" of the power curve, note down the collector voltage at which it is obtained. Also note what the collector waveform is if your scope can display it. (This is best observed with an *extremely small current-metering resistance in the emitter circuit.*) What do you find?

Step 3.

Then, increase the collector voltage by about 25%, and loosen the coupling to the load (or increase the value of the load resistance). What happened? Determine the loaded Q for the output circuit by the frequency-variation method. Has it changed? Then, determine the overall voltage gain of the stage and its effective impedance. Your scope will help with these tests. Vary the load resistance to halve the output voltage. (This rf loading can be directly on the collector, but dc decoupled. Because of the short pulses of current the transistor passes, this load is, in effect, directly on the tuned circuit.) Remember, also, that a carbon resistor is not an ideal load so the method can only give you a fair indication of what is going on. Discuss what you have observed and explain. In particular, did you observe any evidence of instability?

As you allow the impedance of the tuned circuit to increase, you are running the risk of oscillation or instability. As you lower the impedance, you reduce the power output available and also stabilize the circuit. The onset of instability is indicated by jumps in the output as either the frequency or the tuning is varied.

The previous steps can be repeated using higher values of collector voltage and a higher load impedance. Repeat the process until you see evidence of potential instability. Determine the approximate value of the voltage gain when this condition appears. Describe what you have learned.

Step 4.

If your scope is sufficiently sensitive so that you can get a reasonable deflection on the horizontal axis from the input signal to your amplifier, you can repeat Step 3 while watching the phase variations as the signal source is tuned across the operating range of the amplifier. (This is again a Lissajous figure measurement.) This test may be carried out for each of the above conditions. The vertical input of the scope is connected to the output as before. You will find it a little difficult to estimate the exact phase shift at a series of different frequencies, but see if you can work out how to determine the sine of the angle between the voltages (in terms of the intercept and the peak value for the ellipse along one or the other axis). Then, plot the phase variation as a function of the frequency for the various conditions. Describe how the phase variation (as a function of frequency) behaves as the amplification of the circuit is increased. How would you test your circuit if phase stability were of prime concern to you?

You will find that the Lissajous figure measurement is a much more sensitive indicator of the presence of phase problems than is the tuning-variation method described in earlier experiments. The incipient instability will be much more noticeable in this kind of a presentation.

EXPERIMENT 11
The Class-C Frequency Multiplier

Class-C amplifiers are commonly used as frequency multipliers and, when properly designed and adjusted, they are very effective. The purpose of this experiment is to learn a little about what are some of the critical features of a frequency multiplier.

Step 1.

Change the input frequency to the amplifier used for Experiment 10 to half of its previous value. Place a *very small resistance* in series with the emitter of the transistor so that you can monitor the current flow of the device as well as the output voltage across the output tuned-tank circuit. Be sure that the maximum signal voltage drop across this resistor does not exceed about 5 millivolts. How much output signal amplitude do you get when your source-signal frequency is half the previous value? (Tune your frequency source to the peak value; remember that your apparent output frequency may be altered as a result of changes in the feedback behavior.) Describe your results.

After you have found how the circuit behaves at the same input amplitude as was used in Experiment 9, reduce the drive by about 25% and observe how your output changes. Repeat this test several times and record the data in Table 4-30. Plot a curve of output against input on the graph in Fig. 4-50. (You will find it interesting to re-

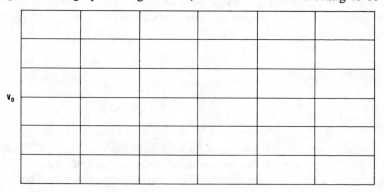

Fig. 4-50. Graph for Experiment 11, Step 1.

peat Experiment 9 while varying the input amplitude in the same way to discover how the amplitude of the output signal varies, when the amplifier is used as a straight-through amplifier with no change in frequency.) Then, the effect of drive level on the effectiveness of a frequency doubler will be noticeable. Discuss what you have observed, and explain any differences you have noted. Then repeat the experiment, increasing the drive from the original value. Explain your results.

Table 4-30. Data for Experiment 11, Step 1

V_I				
V_O				
V_I				
V_O				
V_I				
V_O				
V_I				
V_O				

You will find that the power output drops much more rapidly with a decrease of drive in a multiplier than it does in an amplifier that has the input and output operating at the same frequency.

Step 2.

Now, reduce the input frequency to a third of the output frequency, and repeat the previous experiment. Describe and explain what you observe. Be sure that you both reduce the level of drive and increase it from the original starting level. Tabulate your data in Table 4-31 and plot the results in Fig. 4-51. Then, explain your results.

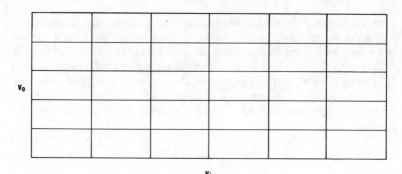

V_i

Fig. 4-51. Graph for Step 2.

Table 4-31. Data for Step 2

V_1				
V_o				
V_1				
V_o				
V_1				
V_o				

As a tripler, the output amplitude will be even more sensitive to input amplitude and other operating conditions.

Step 3.

Once more, reduce the input frequency to a quarter of the desired output frequency, and repeat the experiment while varying the input

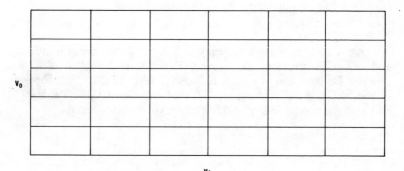

V_i

Fig. 4-52. Graph for Step 3.

drive. Explain your results, and plot them on the graph form given in Fig. 4-52.

Table 4-32. Data for Step 3

V_i				
V_o				
V_i				
V_o				
V_i				
V_o				

Step 4.

What have the above tests told you about the sensitivity of the frequency multiplier to the amplitude of its input signal? How does this sensitivity vary as the amount of multiplication is changed? Explain your results in terms of what you know about the characteristics of transistors?

You saw earlier that the approximate amount of second-harmonic distortion, as a proportion to the fundamental, was a quarter of the difference in the amplifications at the two extremes divided by the sum of the amplifications. This means that distortion is a prime requirement for harmonic multiplication. We can say that the time during which the collector-current pulse flows *must* be less than a half-cycle at the harmonic frequency. This means that the current pulse must be less than a quarter of a cycle of the input frequency for doubling, a sixth for tripling, and so forth. The transistor must turn on, conduct, and turn off again within this part of the cycle of the fundamental. On the sine wave waveform of Fig. 4-53, these time durations are indicated for doubling, tripling, and quadrupling. Clearly, the required magnitude of the input-signal voltage must increase rapidly as the multiplication ratio increases.

Why is this requirement so? If the current pulse lasts longer than

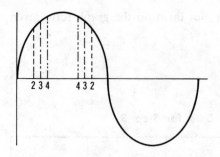

Fig. 4-53. Sine-wave conduction angles.

NOTE: THE NUMBERS INDICATE THE LIMITS
OF CONDUCTION FOR THE
CORRESPONDING HARMONIC.

a half-cycle at the output frequency, it damps and loads the tuned circuit and reduces its efficiency. Since the wave between pulses is generated by the near sinusoidal decay of the signal waveform between the current pulses, this loading is *very* critical.

SUMMARY

In this section, you have made a variety of different kinds of amplifiers. You have also put them through their paces and observed how they behaved as their operating parameters were varied. You observed the effect of emitter degeneration in giving you control over the behavior of your amplifier, a control which is rather more difficult to obtain with a current amplifier than it is with a voltage amplifier. You have seen how distortion can be either controlled or introduced as required. You have seen how transformer-coupled amplifiers behave, and you have seen how audio-power amplifiers can be constructed using them. You have seen how tuned amplifiers behave, and some of the interesting things you can do with them. And, you have learned a great deal about testing all of these circuits, and how to determine the amount of distortion present in them.

You should now be able to design almost any kind of an amplifier that you wish that is based on bipolar transistors. You should plan some additional circuits and design them based on what you have learned. You will find some additional interesting circuits studied in the next chapter. Whereas, the study and design in the next chapter are directed principally toward the application of field-effect transistors, most of these circuits can also be used with bipolar devices. You should try adapting some of them for bipolar devices, remembering, however, that bipolar devices are extremely nonlinear in practice, whereas, field-effect devices, although theoretically as nonlinear, in practice, are normally much less nonlinear. You will observe this as you test the devices.

CHAPTER 5

The Field-Effect Transistor

In this chapter, the goal is to develop an understanding of how the field-effect transistor functions. You will also acquire an understanding of the constructive ways in which it can augment bipolar transistors in the solution of difficult circuit design problems. Much of what you have learned so far can be applied directly to circuits involving field-effect devices, so the efforts here will be directed to the development of a clear understanding of what the devices are and how they behave. You will also learn how to modify the techniques you already know to create successful circuits using these devices.

OBJECTIVES

After you have studied this chapter and performed the experiments, you will understand the basic characteristics of field-effect transistors and how they differ from bipolar devices. You will have verified the following properties and concepts in your study and experiments:

1. The physical relationship of the source, gate and drain in the field-effect transistor (FET).
2. How the flow of current is controlled in these devices.
3. What the depletion region is in a FET.
4. What the channel region is in a FET.
5. What the Debye region is in a FET.
6. How the Debye region is important.
7. How depletion-mode FETs work.
8. How enhancement-mode FETs work.
9. How to test FET devices.

10. How to measure FET characteristics.
11. How to design FET amplifier circuits.
12. What the diffusion mode of operation is in FETs.
13. When to choose bipolar devices and when to choose FET devices.

As usual, however, we will start with an extensive list of definitions which will be important in the pages ahead.

DEFINITIONS

You will find the following definitions to be important in your understanding of FET devices and their mode of application.

field-effect transistor (FET)—This is a device constructed on a small bar or chip of semiconductor material. The chip has been doped with either n- or p-type material, with the density of the doping adjusted to suit its intended purpose. It contains a source (used to introduce carriers), a drain (used to collect carriers), and a gate region that is located between the other two (used to control carrier flow).

diode field-effect transistor (DiFET)—This is a FET in which the channel widening and constriction is achieved by the application of diode junctions in the gate positions. The gate diodes serve to widen and narrow the width of the channel, thus controlling the flow of current from source to drain.

junction field-effect transistor (JFET)—This is another name for the diode field-effect transistor. Bias on the diode(s) controls the flow of current in the channel through which the current flows.

metal-oxide-semiconductor field-effect transistor (MOSFET)—A field-effect transistor in which the control of the current flow in the channel is obtained through the use of capacitors that are placed along the channel, in the position normally occupied by the junction diodes in JFET units (Fig. 5-1).

metal-insulator-semiconductor field-effect transistor (MISFET)—Another name, possibly more correct semantically, for the MOSFET device.

insulated-gate field-effect transistor (IGFET)—Another name for the MOSFET device.

source—The point of origin of carriers that flow through the field-effect transistor. It corresponds to the emitter in junction transistors and to the cathode in electron tubes.

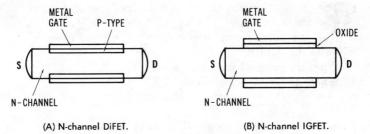

(A) N-channel DiFET.　　　　(B) N-channel IGFET.

Fig. 5-1. Junction and MOS field-effect transistors.

gate—The electrode that controls the flow of carriers in the channel of the field-effect transistor.

channel—The weakly conducting semiconductor material through which charge carriers can be caused to flow from the source to the drain under the control of the gate electrode.

drain—The electrode which receives or collects the carriers that are released by the source.

depletion region—A segment of the channel under the gate in which the field applied to the gate locks the carriers that are present and keeps them from moving. Variation of the extent of this region is what makes possible the variation of the magnitude of current flow in the device.

Debye region—The boundary between the channel in which the current flows in a FET and the depletion region in which carriers are bound by the field. This region is extremely thin, on the order of a few tens of Angstroms thick. It has a voltage increment across it of about 0.026 volt at room temperature.

enhancement mode—A condition of operation of a FET device in which it is necessary to forward bias the gate in order to cause current to flow. When the gate is left open circuited with this kind of a device, the current flow through it is almost indetectable.

depletion mode—A condition of operation of a FET device in which it is necessary to place a reverse bias on the gate to reduce the current flow to the drain to a negligible amount. With this kind of a device, a forward bias will increase the drain current beyond that encountered with zero gate bias.

forward bias—A voltage applied from gate to source whose polarity is the same as the polarity from drain to source.

reverse bias—A voltage applied from gate to source whose polarity differs from that applied from drain to source.

VMOS FET—A field-effect transistor in which the channel is located at right angles to the surface of the chip, typically on the side of a "vee" groove cut into the surface of the chip. This configuration permits the production of very short and very precisely controlled channels. A variant of this structure is described in U. S. Patent 3,274,462.

fan out—This is a measure of the number of separate devices or configurations of devices which can be controlled by one active device or circuit. The term is mostly used with digital circuitry, but it has some importance with analog circuits as well, as it defines the number of "equivalent transistors" one can load on another driver transistor without degrading its operation beyond specified limits.

transconductance-per-unit-current efficiency—This is the ratio of the actual transconductance, as measured at a given value of current to the product:

$$(qI/kT),$$

where,

    q is the charge on the electron,
    k is known as Boltzman's constant,
    I is the device current,
    T is the absolute temperature.

transconductance efficiency—This is a shortened expression for transconductance-per-unit-current efficiency.

n-channel device—A FET device constructed on a semiconductor material which has been doped to provide an excess of electrons through the use of an electron-rich doping material.

p-channel device—A FET device constructed on a semiconductor material which has been doped to provide a deficiency of electrons through the use of an electron-deficient doping material.

CONSTRUCTION AND OPERATION OF A FET

The field-effect transistor is usually constructed on a segment or chip semiconductor material that is either nonconducting or weakly conducting. In contrast to the construction of bipolar transistors, the entire segment (including the source and drain contacts) are all doped alike, with the result that there are no diodes in the path of the current flow. (The VMOS device limits channel thickness by having a thin region that is similar to the base of a bipolar transistor, but the actual current flow is through a true channel.) The gate itself may be doped differently, as it is with the DiFET devices. The result is that the FET device is strictly a majority-carrier device and, theoretically, is not subject to the minority-carrier limitations encountered with bipolar devices.

Control of the amount of current flow through the path from the source to the drain, called the *channel,* is obtained by placing an electrical field against the channel which immobilizes the carriers in a region adjacent to it, thereby preventing them from moving and being replaced by additional carriers from the source. The carriers passing through the channel are collected by the *drain,* which serves the same function as the collector does in a bipolar transistor. The side contacts, called *gates,* are typically either diode-type contacts or capacitive contacts. They serve to lock some charges in the segment in place, making it more or less difficult to draw charges from the source to the drain (Fig. 5-2). The voltage applied to the gate(s) can either enhance the flow of charge by releasing carriers, or it can reduce the number by binding some in place. The result is that under high-current conditions, the carriers are allowed to pass through a wide channel (as sketched in Fig. 5-3), and under low-current conditions, through only a narrow channel.

The *source* is designed so that it can introduce large numbers of charges into one end of the channel. The sketches shown in Figs. 5-2 and 5-3 would make the channel appear to be rather long but,

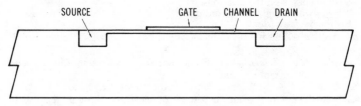

Fig. 5-2. Structure of a field-effect transistor.

actually, it is made as short as it can be and still achieve the required function and withstand the applied voltage. (Channels can be as short as fractions of a micrometer in microwave devices.) The channel must be short for several reasons: first, in order to introduce a minimum of channel resistance to the point at which the field control is effective; second, to enable the device to operate at an adequately high frequency; and, third, to minimize the amount of resistance in the channel between the point of control and the source.

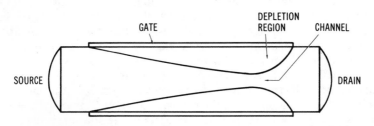

Fig. 5-3. Channel current conditions in a FET device.

As noted previously, the gate controls the flow of current in the channel by binding carriers to the side of the channel under the gate surface. The region in which the charges are immobilized in this way is called the *depletion region*. Its size and shape are dependent on the total field from the source to the drain and, also, on the voltage from the source to the gate. The general form of the depletion region is somewhat as sketched in Fig. 5-3. The space between the edges of the depletion regions is what determines how much current can flow through the gap.

The extremely thin boundary region between each depletion region and the channel through which the carriers flow from source to drain has some important properties. It is known as the "Debye region." It is at most a few tens of Angstroms thick. There is a "potential jump" across this region of (kT/q), its magnitude being about 26 millivolts at room temperature. Normally, the field resulting from the presence of this potential jump is less than that existing in the depletion region (which keeps the carriers bound in place). When the channel is nearly closed, however, the depletion fields on the two sides of the channel become rather small and, then, the potential jump can become the controlling field. Under these conditions, the FET device behaves very much like a bipolar transistor that has an infinite beta. The resulting mode of operation is a true diffusion mode, and the device will show a transconductance-per-unit-current value which is nearly the same as that for the bipolar transistor. This diffusion mode of operation may exist over as wide a current range as

100,000 to 1, typically with values that are from fractions of a nano-ampere to tens of microamperes.

As the channel widens out, a considerable amount of charge can flow through the center of the channel without being influenced by the control field that is exerted by the Debye region. Then, the trans-conductance-per-unit-current value of the FET device is markedly lower than that encountered with a bipolar transistor. This wider-channel condition is the one that was considered by Shockley in his original paper on transistors. The effect of the depletion field leads to the power-law relation between voltage and current for these devices, and makes them freer of the high-order components of distortion which plague exponential devices like bipolar transistors. This condition explains the reason that these devices are used extensively both as mixers and in the front-ends of receivers.

When these transistors are operating in the diffusion mode, the amount of drain voltage required to activate them is just a few milli-volts. As the amount of drain current increases, however, one observes something akin to the "saturation" region observed with bipolar transistors. As the drain voltage is increased, the current increases up to a point, and then it levels off. Above this point where it levels off, the FET device is said to be "pinched off," and the current through the device no longer varies significantly with the applied drain voltage. One consequence of this is that the devices can be used as variable resistors or as constant-current sources. As the gate voltage is varied, the drain current varies, but it does not vary significantly with drain voltage.

The amount of gate voltage required to induce changes in drain current through a FET device can vary over quite a wide range. The JFET devices were the first FET devices which could be made relatively reliably, so that the predicted kinds of control could be documented. As long as the doping and length of the channel are properly controlled, it is possible to get a stable and reliable operation of these devices. The insulated-gate devices proved to be a more complex problem, however. Drifts in the insulator layer led to different charge distributions under the gate and changed the amount of channel current. The consequence was that until that problem was controlled, stable devices could not exist, and the devices remained laboratory curiosities. These insulated-gate devices can now be made with sufficient stability and high enough yields that even integrated circuits (containing thousands of these devices) are made routinely.

The principal advantage of FET devices is that the input impedances on their gates can be from as much as a few megohms to thousands of megohms. You will find that they make excellent "emitter followers" where it is essential to repeat a voltage without loading the source, yet at the same time be able to operate a driven circuit adequately. That is the main reason they are commonly used as input devices in scope amplifiers and in similar instrumentation amplifiers. They can also be adjusted to repeat a voltage with nearly zero offset from input to output.

Some of the insulated-gate devices use plain silicon dioxide as their insulating layer, and some of them use more complex insulating structures. The problem here is that the insulating layer must have both precisely controlled thickness and extremely precisely controlled deposition characteristics. Otherwise, one encounters drift both with time and with temperature. A wide variation in the amount of voltage may be required from device to device to get given operating conditions. With an *enhancement-mode* device, for example, which is turned off until some forward bias is applied, it is ideally desirable that the initiation of the turn-on bias require from a few tenths of a volt to at most one or two volts. The turn-on point should be consistent from device to device and the current should be stable with time as well. As long as the source-to-gate voltage does not change, the drain current should not change, whether the applied voltage is in the form of a pulse that is a fraction of a microsecond in duration or whether it continues for hours. Insulated-gate FET devices probably will replace junction FETs in many applications now that the stabilization problems have been largely solved. Needless to say, similar stabilization is required of *depletion-mode* devices if they are to operate satisfactorily in scope-input circuits.

There are various combinations of insulators which have been used with IGFET devices. The commonest is ordinary silicon dioxide (sand), but it must be extremely pure and be crystallized with neither traps nor impurities nor defects in it. Combinations of silicon dioxide and silicon nitride are used on some of the most critical devices, as the drift seems to be much better controlled when a layer of each is used. However, both of the layers must be extremely pure and perfectly crystallized even then.

The layer of insulation used in IGFET devices must of necessity be extremely thin if the control voltages are to be kept to a volt or so. The consequence of this is that the electric field in the insulator can exceed 100,000 volts per centimeter, and the insulating layers are very vulnerable to static electricity.

One of the standing stories is for someone to take a pocket calculator into a repair shop, saying that something had happened to it. The owner had just walked across a rug, and when he picked it up, a spark jumped to one of the keys. At that point, the repairman remarked, "All right. I know what's wrong with it. It will cost $35 to repair."

A spark had destroyed some of the integrated circuits in spite of the protective diodes built into it. These diodes (or low-voltage zener diodes) are often placed on the chip adjacent to the external leads. They help a great deal, but electrostatic fields can still kill them if a strong spark gets into the circuits.

Most of the field-effect transistors that one finds are of the n-channel type. This is because it is much easier to achieve ohmic contacts at the source and drain in a segment of n-type semiconductor material and still have proper channel behavior as well. The n-channel device requires a positive supply voltage on its drain just as an npn transistor does on its collector. The n-channel FET can be either an *enhancement-mode* device, or a *depletion-mode* device. This is a consequence of the fact that it is quite easy to establish an excess of electrons in the channel region with these devices. At the same time, it is possible to have them conduct with at least modest amounts of reverse bias. Many of the scope applications of FET devices are based on depletion-mode n-channel devices. You will find an application experiment based on this mode and type of great interest.

The p-channel device requires a negative voltage applied to its drain. The current through it is increased by making the gate more negative with respect to the source. Most of the p-channel devices are *enhancement-mode* devices, in that the gate must be forward biased with respect to the source (biased toward the drain), in order for even small amounts of drain current to flow.

ENHANCEMENT AND DEPLETION MODES

One particularly significant feature of all of these FET devices is that their drain currents do *not* normally increase or decrease with gate bias anywhere near as rapidly as you have observed with the bipolar transistor. This is because you will probably not normally operate one of them at as low a value of current as is required to demonstrate a sufficiently high transconductance-efficiency mode of operation. It is usually possible to distinguish between present-day FET and bipolar devices in this way.

This does *not* mean, however, that it is very difficult to make FET devices show a much higher value of transconductance efficiency than the values ordinarily found. Achievement of such higher values is dependent partly on the basic design of the device, in that a channel, whose conductivity varies in a properly chosen fashion from its center to its outer edges, can lead to substantial increases in this efficiency factor. (This is most likely to be observed with VMOS FET devices.) The limit on the transconductance efficiency of the JFET device is unity, but with some IGFET devices it may be found to be a maximum of about one half of unity.

Silicon is the principal material used for the manufacture of FET devices at the present time. Germanium has not proven satisfactory because of its high leakage-current level and its low peak-temperature rating. Gallium arsenide has made significant inroads into the high-frequency device market, however, as it has the highest electron-diffusion velocity of the presently available materials. The GaAs devices are substantially harder to make, and possibly the fall-out during production is higher, so they are largely used for microwave amplifiers and similar applications.

Field-effect transistors are said to be in the "pinch-off" mode of operation when they are functional as amplifiers. Their so-called "linear mode" is really a saturation-type of mode in that there is little or no control available by way of variations of gate voltage. The pinch-off mode, therefore, really should be called the amplifier mode of operation. Because of the nearly total independence of drain current on the variations of drain voltage, the device can be used as a voltage-controlled current source in the same way that a pentode tube can. They are extremely useful when it is necessary to achieve a precisely controlled level of current that is nearly independent of the applied voltage.

The length of the gate region along the axis of the FET channel controls its maximum operating frequency and its maximum drain voltage. The diffusion velocity of the majority carriers determines the minimum length of time for control action to take place under the gate, and contributes to the determination of the maximum operating frequency. The diffusion time along the effective gate length is therefore a critical parameter. Device gain at signal frequency can be expected to be degraded if the transit time past the gate exceeds roughly a quarter of a cycle. As with bipolar transistors, the maximum operating frequency is that frequency at which the device in an optimum circuit configuration is just able to develop a power gain of unity.

One of the major problems that has been encountered with FET devices has been the problem of getting short enough gate regions without causing device failure. Two recent developments have pointed to almost ideal solutions to this problem. What basically is done is to start building a bipolar transistor and then make it work like a FET device. In order to do this, one grows either a mesa or a planar transistor, and then etches a suitable array of V-grooves in the surface of the structure. If then, the surafce of the groove is covered by a very thin layer of silicon dioxide and/or silicon nitride, one can deposit a gate on top of the insulator and, lo, one has an enhancement-mode device. It looks like a bipolar transistor without the usual base connection but with a capacitive electrode in its place. These so-called VMOS field-effect transistors are proving to have an extremely high potential both as elements for complex large-scale integrated circuits and also as power transistors. They tend to have higher transconductance efficiencies than ordinary FET devices, but not such high values that their power limitation is as severe as with bipolar devices. You will find it very desirable to use your special microammeter/milliammeter (or DVM on millivolt ranges) with its very low voltage drop with all of these devices, because any of them may prove to be operating under conditions where an emitter or source degeneration can occur without your realizing it.

The development of successful power FET devices, like the VMOS devices, has been a goal almost ever since the first FET devices were successfully made, since it was almost immediately evident that these devices were less likely to "self-destruct" when overloaded. (Drain current decreases with increased temperature, but collector current increases, forming hot-spots and generating failures.) Another important feature of the VMOS devices is the fact noted previously that their transconductance efficiencies can be higher than for ordinary FETs, but not as high as for a bipolar transistor. As a result, their power-handling capacity is significantly better than that for bipolar devices. Interestingly enough, it is easily shown that power handling ability has an inverse relation to transconductance efficiency (see Appendix A).

TESTING FIELD-EFFECT TRANSISTORS

The testing of FET devices can be done with some transistor testers, and it also can be done with some tube testers if the supply voltages are small enough. Electrically, they are quite similar in operation to either bipolar transistors or pentode electron tubes. Because of their reduced transconductance efficiency in the normal operating range, and because of the different diode configuration, they can be differentiated from bipolar transistors fairly easily. It is useful, for that reason, to design a circuit to test them.

Since the 2N codes used with many solid-state devices are useless to help in classification, it is probably convenient to sort these devices, as with bipolar transistors, into several classes, such as n-channel enhancement, n-channel depletion, p-channel enhancement and, possibly, p-channel depletion mode of operation. (Devices in the latter class are very rare at this time.) If FETs are tested as a function of frequency in order to make an approximate classification based an the frequency response, it may also be helpful. Further separation of the devices into junction and insulated-gate devices may also prove helpful. The insulated-gate devices can only conduct through the channel, if at all; whereas, the junction devices should disclose at least one diode. Testing an assortment of known good devices is the best way to find out what classification rules to use when testing with an ohmmeter. Depletion-mode devices should show conductivity through the channel and the presence of diodes with respect to the gate.

Field-effect devices have proven superior to bipolar devices in many respects in complex integrated circuits. Probably one of the more important reasons is the extremely small amount of control power required to cause them to operate. (Both the input control power and the output supply power can be extraordinarily small.) Very high values of fan-out (the number of device inputs a single device can control) often can be achieved because all the input current required by the load stages is for capacitance charging. Also, the input capacitance of FETs is orders-of-magnitude less than that for bipolar transistors. The consequence of fan-out loads will be primarily one of limiting the maximum switching frequency for a circuit, rather than a direct limitation from fan-out per se. Transconductance efficiency can enter into the maximum switching rate for these circuits, since the frequency figure-of-merit takes the general form:

$$f_{max} = F(g_m/C) \qquad \text{(Eq. 5-1)}$$
$$= F(\kappa \Lambda I_c/C)$$

where,

F is a function of undefined form, possibly linear,
g_m is transconductance,
C is the device input capacitance.

With bipolar transistors, which have kappas of approximately unity, the capacitances may be large enough for FET devices to eliminate the kappa as an advantage. Kappas for typical FET devices under normal operating conditions are in the range from 0.03 to 0.2. The input capacitance for bipolar transistors is high because it is generated by variations of minority-carrier density over exceedingly short distances.

The handling of all kinds of FET devices, particularly IGFET devices as noted previously, must be done with extreme care, particularly in winter and in areas where the humidity is low. The use of any kind of rugs or cloths which are water-repellent and high resistant is not advisable, as the static electrical charges that can build up can produce voltages of thousands or tens of thousands of volts, and can easily destroy the oxide layer. It is common practice to wear special clothing and grounding straps under the kinds of low-humidity conditions noted. The utmost precautions are taken in production areas where these kinds of devices are handled. Special conducting shoes and conducting floors are also required. Working in a "clean room," as is required when working with electron devices and with systems designed for high reliability, just is not enough.

Once you have familiarized yourself with the characteristics of the FET devices by measuring their parameters, as described in the following experiments, you will want to build and test some circuits that use them. After you get used to the transconductance approach, you will find that the conversion from bipolar transistors to field-effect transistors is relatively simple and straightforward. *All* of these devices are really transconductance controlled. You will, therefore, be asked to perform some experiments which are very similar to the ones you have already performed in Chapters 3 and 4. However, you will find that there are distinct differences in both the basic characteristics and in circuit applications.

EXPERIMENT 1
Measurement of FET Characteristics

You have two important concerns with respect to testing FET devices. The first of these is the need to configure a circuit that will enable you to measure what you need to know about these devices. This includes all types, both junction and insulated-gate FETs, n-channel and p-channel FETs, and depletion- and enhancement-mode types. The second need is a simple configuration that will permit you to both test the devices and, also, to classify them so that you will know what you have. The first of these circuit configurations can be based on your bipolar-transistor analyzer circuit. However, the second requires a special unit like a bipolar transistor tester. Circuit diagrams for some special testers are included in Appendix B. You will want to familiarize yourself with both the analyzer circuit and its operation, as well as the simple tester circuit, as at different times you may need to make a detailed curve analysis or may possibly need to make a routine check for correct functioning. Once you have a known good device, you can use the previously described procedures to establish the proper values of resistance and bias for the circuit.

Step 1.

Wire your solderless breadboard as shown in Fig. 5-4. Once again, a regulated multiple-voltage power supply will be useful but, in order to get the most effective results, you will have to make some minor variations in the way you apply it. As with the testing of bipolar transistors, you first start by plotting constant-bias contours. This time, the drain voltage (with respect to the source) is used for the horizontal axis and the drain current is used for the vertical axis. (This is the equivalent of what you did with your bipolar transistors.) Instead of using input (gate) current, however, you will want to plot in terms of input or gate voltage. Were it not for the temperature sensitivity of the base voltage on the bipolar transistor, it might

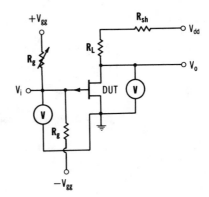

Fig. 5-4. A circuit for testing
FET devices.

really be better to use base voltage rather than base current with
them also. (When you use contours of constant base current, as is
normally done, it is very helpful to have an additional set of constant-current contours as a function of base voltage and collector
voltage to complete the set of information you will need for an effective design.) Fortunately, you can avoid the problem by the use
of transconductance efficiency and $(q/kT)I_c$ with both FET devices
and bipolar devices.

A drain-supply circuit configuration is wired similar to the circuit
you used with bipolar devices. It will enable you to get a single-polar-

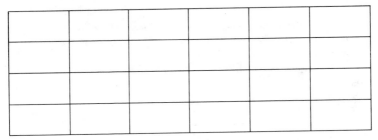

Fig. 5-5. Waveform plot.

ity swept supply voltage from your transformer that varies from
essentially zero to the available peak voltage. DO NOT FILTER
THE OUTPUT, as you need the variation at twice the line-frequency
rate. Examine the waveform with your scope to be sure that it is
correct. Draw a picture of it on the graph in Fig. 5-5. It should look
very much like the curves in Fig. 5-6. This varying voltage is applied
to your FET device through the usual load-protective resistance. (It
should drop the peak drain voltage in half at maximum current. This
resistance normally goes by the name of "load resistance.") You

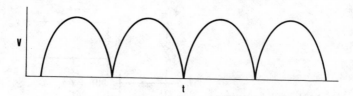

Fig. 5-6. Typical waveforms for a swept-output voltage supply.

will also be applying it on the horizontal axis of your scope, from drain to source, as indicated.

Your gate-bias circuit differs from the base-bias circuit used with bipolar transistors. First, you need to be able to forward bias or reverse bias a device under test and, second, you need a bias voltage instead of a bias current. A suggested circuit for this is shown in Fig. 5-7. Your problem here is that you must set the bias just on the verge of conduction for your device and, then, you must provide for step increases in the bias, each step being of a known size. A digital voltmeter is almost ideal for this measurement. Resistor R1 is adjusted to set the zero-current point, and R2 is used to establish the overall range of bias. Then, R3 can be used to set the actual bias. The push button applies it to the device, and the series resistance R4 is used to permit the introduction of an ac signal on the gate. The proper steps are:

1. Set R3 near zero.
2. Set R1 for a slight drain current (50 μA typical).
3. Set R3 to maximum and R2 to the desired maximum drain current.
4. Draw contours for different settings of R3.

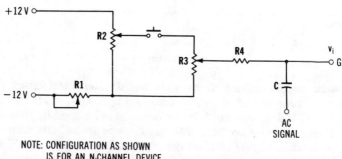

NOTE: CONFIGURATION AS SHOWN
IS FOR AN N-CHANNEL DEVICE.
FOR A P-CHANNEL DEVICE,
INTERCHANGE +12V AND −12V.

Fig. 5-7. Bias circuit for a field-effect transistor.

Step 2.

Since you are using a separate drain supply for this experiment, you could bridge your gate-bias circuit across your two variable-voltage supplies if you prefer. If you wish to do that, simply bridge the bias circuit from the plus variable point to the minus variable point, and adjust each to give the range of control you require. (Adjust the one to the verge of conduction, and the other to limit the current to the desired peak value. The negative supply sets the zero for the n-channel device, and the positive supply sets the zero for the p-channel device. The rest of the circuit is as shown in Fig. 5-8. In locating the zero drain-current point, choose R3 so that it introduces an extremely small voltage increment and, then, adjust the appropriate return supply so that a very small increment in drain current from zero is observed. With enhancement-mode devices, R1 may be returned to ground. With depletion-mode devices, it must be wired as shown in Fig. 5-7.

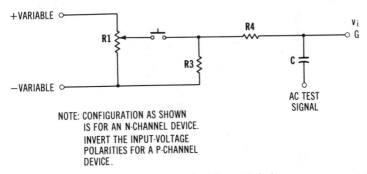

Fig. 5-8. Alternate bias circuit for a FET device.

Step 3.

Keeping the voltage drop across the current-metering circuit in the source lead extremely small is not always as critical here as it was with bipolar transistors, but since you may get into the high-transconductance-efficiency range, it should be kept to less than 20 millivolts. A simple way of checking this is to place a milliammeter or microammeter in the drain circuit and see if there is any significant drain-current change as you short the metering resistor in the source lead. If you have a problem, there will be a significant increase in the drain current when you do. Then, you must reduce the metering resistance and increase the sensitivity of your scope using a booster amplifier with a stage gain of ten or one hundred at the metering point. An LM4250 operational amplifier using a positive and negative 3-volt supply in the circuit as shown in Fig. 4-27 is ideal for this.

Step 4.

Now, take a series of operating contours at a convenient set of values of gate voltage. Select these values so that you can get about a 20 to 25% increase in drain current with each increment after the first. The values of voltage selected should be convenient ones, for example, 0.1, 0.2, 0.5, 1.0 volt, etc. You will have to use your judgment on this. Record your data in Table 5-1 and plot the curves on the graph shown in Fig. 5-9. If you prefer and can do it, take Polaroid pictures of the curves shown on your scope. Be sure you identify the device you are using so you can make further tests on it later. Also, remember that before you finish your tests you will want to test both types of n-channel and both types of p-channel devices. An n-channel depletion unit is a good device to use as a starting point.

Step 5.

Select another FET device, one having an opposite channel polarity, and repeat the tests you made with your first device. Again, plot your curves and identify them with the device so that you can coordinate them in later experiments. (Use a separate sheet of graph paper to plot your curves.) Plot similar curves for other devices that you may test. Record your data in Table 5-2.

Table 5-1. Data for Experiment 1, Step 4

V_g				
V_d				
I_d				
V_g				
V_d				
I_d				
V_g				
V_d				
I_d				
V_g				
V_d				
I_d				

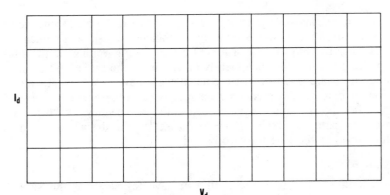

I_d

V_d

Fig. 5-9. Graph for Experiment 1, Step 4.

Table 5-2. Data for Experiment 1, Step 5

V_g				
V_d				
I_d				
V_g				
V_d				
I_d				
V_g				
V_d				
I_d				
V_g				
V_d				
I_d				

Examine the data you recorded in these steps. Describe and explain them carefully in the following space. Do you have any evidence that these devices have reduced transconductance efficiencies? Are they as nonlinear as the bipolar transistors and diodes you tested?

Since more bias-voltage change is required to make a 2:1 change in drain current, the transconductance efficiency must be significantly less than for bipolar transistors. Further, the grouping of the bias contours is such that they clearly do not have the same kind of non-linearity characteristics that bipolar devices have. The forward direction of bias on the n-channel devices is positive, whereas, forward bias for the p-channel devices is negative. Some devices "turn on" with the gate bias reversed (with respect to the drain), and others turn on with the gate forward biased.

Step 6.

Continue this procedure until you have tested all of the various classes of devices that have been discussed. Be sure to include the various kinds of JFETs and, also, the various kinds of IGFETs. Describe and explain what you have found, plotting graphs on at least some of the devices. Either record the data in the preceding tables or on tables you have constructed on separate sheets of paper. Then prepare a condensed table which identifies the properties of the different devices you have tested. Use Table 5-3. Remember that the value of kappa can be approximated by dividing 0.018 by the gate-voltage change that is required to cause a 2:1 change in drain current.

Table 5-3. Characteristic Data for Step 6

Device	R_{in}	κ	Depletion Type?	Enhancement Type?
N-channel junction				
P-channel junction				
N-channel insulated gate				
P-channel insulated gate				

EXPERIMENT 2
Transconductance Values for Typical FETs

As has been shown repeatedly (starting in Chapter 3), you have to establish an operating environment for your transistor. However, in the final analysis, its operation is dependent on its small-signal behavior. This is equally true of switching circuits, as they will *not* switch unless the voltage gain throughout the active region is sufficient to assure transition. (This requires a transconductance over the operating range that is adequate to assure a loop gain in excess of unity.) You will need to modify your circuit so that you can introduce a small alternating signal in the gate circuit. You can then observe the resulting alternating signal that is generated in the drain circuit. You should start just as you did when you were applying a signal voltage on the base of the bipolar transistor—adjust the input signal voltage so that there is a good sine-wave output that has almost no distortion. You can probably apply at least 100 millivolts this time instead of the 5-millivolt limit that you used with the bipolar device. The circuit connections you need to make are shown in Fig. 5-10. Use one of the FETs that you have already tested. Set your bias for a 2-mA drain current. Here again, the component values depend on your device so experiment carefully. Your experience with bipolar devices can be a useful guide.

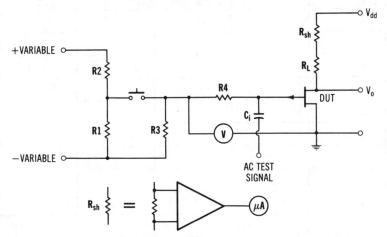

Fig. 5-10. Circuit for small-signal testing of a FET device.

Step 1.

Adapt your circuit and install your transistor. Then, adjust your input-signal voltage and measure it so that you know how much it is. Now, based on the value of load resistance you have cho'en, calculate the transconductance for your device. (A good value of load resistance is one giving a voltage gain for the stage between 2 and 5.) If the voltage gain should exceed 10 to 20, the load resistance is too large for your purpose and should be reduced. Vary the gate bias and record the static and small-signal input voltage and output voltage. Read the output current at each point, and calculate the voltage gain, the transconductance, and the transconductance-per-unit-current at each point. Then, vary the drain-supply voltage to see what effect that has on your output current and transconductance. Record your data in Table 5-4, and plot contours of constant value of the transconductance as well as contours of constant bias on a graph having drain voltage on the abscissa and drain current on the ordinate. After you have done these, discuss your results.

Table 5-4. Data for Experiment 2, Step 1

V_g				
V_d				
I_d				
v_s				
v_o				
K_v				
g_m				
κ				
V_g				
V_d				
I_d				
v_s				
v_o				
K_v				
g_m				
κ				
V_g				
V_d				
I_d				
v_s				
v_o				
K_v				
g_m				
κ				

Step 2.

Substitute one of the other FET devices that you have tested and have curves for, and repeat the processes of Step 1. In the process, trace out the paths of several convenient values of transconductance

Table 5-5. Data for Experiment 2, Step 2

V_g				
V_d				
I_d				
V_s				
V_o				
K_v				
g_m				
κ				
V_g				
V_d				
I_d				
V_s				
V_o				
K_v				
g_m				
κ				
V_g				
V_d				
I_d				
V_s				
V_o				
K_v				
g_m				
κ				

on your curve sheet for the device and label them in accordance with the value of transconductance. Values such as 500, 1000, 2000, 4000 micromhos, etc. should prove suitable. Select values for the device that will give you a convenient set to work with. If any significant similarities and differences are noted in testing the device, record them as well as the data. Record the data in Table 5-5.

The positions of the constant contours for g_m may be superimposed on the bias contours for the corresponding transistor used in Step 1. What are your observations?

One thing you probably will notice is that the transconductance contours for any given type or code number of device will be almost identical in position on the graph as a function of drain current, whereas, the gate voltage contours will vary from one device of a given code number to the next. This can only further confirm our conclusion that the important parameter is the transconductance. The value of kappa may vary, but by a surprisingly small amount, *except* when you go from one code number to another. Then, the difference may be significant, as the mechanical structure of the channel can differ very substantially from a device of one code to another of a different code.

Step 3.

Repeat Step 2 for various other devices that you have tested and have curves for. In each case, add dashed lines for the contours of selected values of transconductance. Record your data in Tables 5-6 and 5-7. Also, draw a separate curve of transconductance as a function of the drain current at a selected drain voltage, for example,

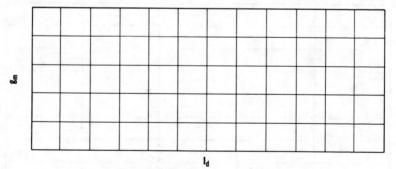

Fig. 5-11. Graph of transconductance as a function of drain current.

a voltage between 2 and 5 volts. Use the graph given in Fig. 5-11 for your curves. Discuss and explain what the above results tell you.

Table 5-6. Data for Experiment 2, Step 3

V_g				
V_d				
I_d				
v_s				
v_o				
K_v				
g_m				
κ				
V_g				
V_d				
I_d				
v_s				
v_o				
K_v				
g_m				
κ				

Step 4.

Calculate the values of transconductance efficiency at different values of drain current for each of your FET devices. Plot the curves showing how this parameter varies with drain current. What does this tell you about your devices? Are these devices linear, as nonlinear as diodes and bipolar transistors, or somewhere in between? Explain your reasoning, and plot your transconductance efficiency curves on the graph in Fig. 5-12.

If the devices are linear, the transconductance will be constant as a function of drain current. We suspect that you did not find this to

Table 5-7. More Data for Step 3

V_g				
V_d				
I_d				
v_s				
v_o				
K_v				
g_m				
κ				
V_g				
V_d				
I_d				
v_s				
v_o				
K_v				
g_m				
κ				

be the case with most of the operating ranges of the devices you tested. (The presence of such constancy is indicative of an emitter or a source degeneration resistance.) If the device obeys the solid-state diode relation, the transconductance efficiency will be either

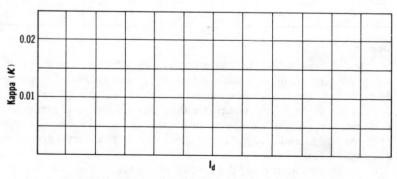

Fig. 5-12. Transconductance efficiency curves for Experiment 2, Step 4.

a constant or unity, the overall transconductance (in mhos) in the latter case being approximately 39 times the current (in amperes).

If the device obeys a square-law relation of current to voltage, the transconductance will be proportional to the square root of drain current, and the transconductance efficiency will be inversely proportional to the square root of drain current. It can further be shown that the following additional possible relations might exist depending on the relation of current to voltage.

I	V^2	$V^{3/2}$	V^3
g_m	$I^{1/2}$	$I^{1/3}$	$I^{2/3}$
g_m/I	$I^{-1/2}$	$I^{-2/3}$	$I^{-1/3}$

Other relations can be developed. You should plot some of these relations on the graph in Fig. 5-13, and see if the transconductance efficiency relation seems to match any of them. You can work out some additional relations and try them also if you wish. At very low values of drain current, you will find that the device does in fact seem to obey the diode equation. At higher values of current, however, is seems to obey a power law, but not too precisely.

EXPERIMENT 3
Design of an Amplifier Circuit

How do you use all this information to decide what values to use for the components in a circuit? To show this, let us take a set of typical curves for a hypothetical FET device like those in Fig. 5-14 and go through the steps. The same steps apply to bipolar transistors,

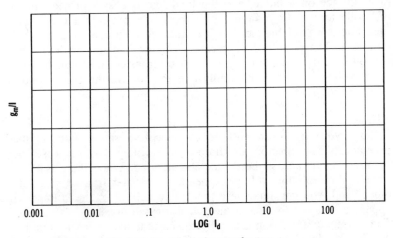

Fig. 5-13. Second set of curves for Step 4.

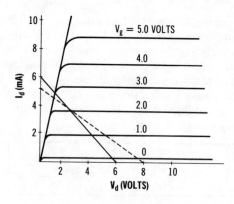

Fig. 5-14. Typical field-effect transistor curves

with the main difference being that with bipolar transistors you may have to work with a projection of your load line onto the base-voltage, collector-voltage field from the collector-voltage, collector-current field. (The sample load line, the dashed line, is for 8 volts and 1600 ohms.)

Since output current is a function of at least two variables (V_g and V_d with a field-effect device, and V_b and V_c with a bipolar device), the operation of your device is defined by a surface. It is possible to express the relations for the field-effect device in terms of the two voltages and the output current but, with the bipolar device, it is necessary to use the base current as a plotting parameter, making the second set of curves necessary. On a set of curves such as you are using, you are really looking at the projection of a set of contours on a surface on a plane, and the load line is also a projection. On the collector or drain family of curves, the projection of a linear load line is a straight line, but it cannot be a straight line in a general case. The points of intersection of the load line with the bias contours on the collector family must be transferred at constant collector voltage to the input family and marked on the corresponding base-current lines.

Step 1.

Since your transistor will be connected in series with its load resistance, the collector- or drain-supply voltage, as the case may be, must divide across the device and its load. When there is a zero current flow, all the voltage appears across the transistor; on the other hand, you can imagine a point at which all of the voltage will appear across the load, and the transistor will appear to be shorted. The correct point along the collector-current axis is then found by dividing V_{cc} by the load resistance. The two resulting points may be connected with a straight line. In the absence of inductance or loading effects, this will show how the current and voltage vary for the transistor. If your transistor is a bipolar unit, you will need to transfer the intersections of the load line with your base-current contours at constant collector voltage to the corresponding point on the V_b/V_c plot. Then, you can draw in the input projection for this load line and, also, read the nominal base voltage. (Note that the exact position of any I_b contour for a bipolar device may vary with respect to base voltage as a function of temperature, but the spacings between the successive contours does not vary.)

Our situation with the FET device is simpler because we do not have to use base current as an intermediary, but can work directly with the gate voltage. The curves on a hypothetical FET device are shown in Fig. 5-14, along with a load line (solid line) for a device operating with a drain supply of 6 volts. The theoretical maximum drain current is 6 milliamperes, defining the position of the load line as can be seen. (The load resistance is 1000 ohms.) At each intersection of this load line with a gate voltage line, you will want to read an approximate transconductance by estimating between the line positions. If you then multiply this transconductance by your load impedance, you have the voltage gain at that point. Do this for each of your points, and insert them in Table 5-8.

Table 5-8. Data for Experiment 3, Step 1

I_d				
g_m				
K_v				
I_d				
g_m				
K_v				
I_d				
g_m				
K_v				

Once you have the voltage gain as a function of the various gate bias values, you can calculate distortion by using the equation:

$$\% \ D = \frac{25(K_p - K_n)}{(K_p + {}_n)} \qquad \text{(Eq. 5-2)}$$

where,
>subscript p indicates the most positive value,
>subscript n is the most negative value.

(This equation assumes that the variation of amplification is linear. More complex cases are beyond the scope of this book.)

If the distortion is excessive, or the amplification appears to be excessive, then source degeneration may be added. The voltage amplification equation then takes the form:

$$K_v = \frac{- g_m R_L}{1 + (g_i + g_m) R_s} \qquad \text{(Eq. 5-3)}$$

In each case, the transconductance g_m depends on the product of the transconductance efficiency, the Fermi parameter (q/kT), and the output direct current. With an FET device, the value of g_i is negligible. With a bipolar device, it is roughly g_m/β, which often can be neglected. Equation 5-3 applies to both bipolar and field-effect transistors.

Once you have these data, you can decide how you wish to operate your device, and you simply adjust the resistance values in the circuit to give you the correct value of output current. Now select a value of operating current and a value of drain current for the FET. A value of 2 milliamperes should be acceptable. If you have already taken curves on the device, read the value of transconductance at that current. To start, assume a drain-supply voltage of 6 volts. From the known device transconductance, select a value of R_s which will give a $g_m R_s$ product of approximately four at the selected current. Determine the voltage which will develop across this resistor, and be sure that the drain-supply voltage is about fifteen times this voltage (rather than the preselected 6 volts). Then make static and small-signal measurements on the transistor if you have not already done so. It is assumed that you are using a transistor you have already measured and therefore, it is assumed that the required data and curves are already available from an earlier experiment. Record your data on the transistor amplifier as it will be configured with a source resistance, and with a load resistance that is ten times the source resistance. Record your data in Table 5-9. These values are to be taken along the load line. You have room to try several other values of drain-supply voltage, so we suggest that you try that too.

Step 2.

Try some other values of supply voltage and, also, other values of source and load resistance, and repeat the measurements taken in Step 1. Record the data in Table 5-10. In doing this, you should find

Table 5-9. Additional Data for Step 1

$V_{dd} = $ _____ volts; $R_L = $ _____ ohms; $R_s = $ _____ ohms				
V_g				
V_s				
V_d				
I_d				
v_i				
v_s				
v_o				
K_v				
$g_{m(eff)}$				

the average amplification and the percent distortion from Equations 5-2 and 5-4.

$$K_{ave} = 0.5(K_p + K_n) \qquad \text{(Eq. 5-4)}$$

where,

the p subscript indicates the most positive value,
the n subscript indicates the most negative value.

This equation *assumes* that the variation of amplification with bias is approximately linear. The sign in Equation 5-2 is of no concern as it only indicates something about the phase of the distortion. These equations can be used to obtain an idea of how the amplifier will behave with reasonably large signals. If the distortion equation indicates substantial amounts of distortion, it means that either source or emitter degeneration is desirable to ease the demands on the overall feedback which may be applied to the system (Equation 5-3).

Step 3.

Repeat the process with devices that you have curves for, using both FET and bipolar devices. Determine the distortion and voltage gain as a function of the input parameter in both cases, and enter the information in Table 5-11. (In this chapter, it is necessary to use the subscript "s" for the *source element* on the FET device instead of the signal source as has been done in previous chapters.) Also, observe the results on your scope for several sample devices, using both an output-time configuration and an output-input configuration as you have learned to do. Examine your data and the

Table 5-10. Data for Step 2

V_{dd}				
R_L				
R_s				
V_g				
V_s				
V_d				
I_d				
v_i				
v_s				
v_o				
K_v				

Table 5-11. Data for Step 3

V_{dd}				
R_L				
R_s				
V_g				
V_s				
V_d				
I_d				
v_i				
v_s				
v_o				
K_v				
%D				

scope presentations carefully, and explain what you have learned. (You now know most of the basic science that is needed for the designing of ordinary transistor amplifiers. You can probably apply it to some electron-tube applications as well.

Comparison of voltage changes and the actual small-signal amplification data shows quickly that the best way to appreciate how an amplifier will behave is in the examination of small-signal amplification data. Just as the use of dc beta hides the nonlinearity of small-signal beta, use of voltage changes hides the nonlinearity of actual amplifiers. The FET device, as the bipolar, is nonlinear, but it might be called a relatively well-behaved nonlinearity in that it is essentially monotonic, or always is in the same direction. Beta variation is not. As you have seen, the Shockley mode of operation of FETs is much less transcendental than is the bipolar transistor, and for that reason will not need as much feedback for linearization as is required with the bipolar device.

Step 4.

You are starting with a signal that has an amplitude of 1 millivolt ac, at about 1 kHz. You wish to amplify it to 1 volt, using two amplifier stages. First, design a bipolar amplifier, calculating the probable distortion, and introducing the required degeneration, etc., based on what you learned in Chapters 3 and 4. Then repeat this design, making a circuit using field-effect transistors. After you get the circuits wired, test them with your signal source and scope. Describe the similarities and differences. Which circuit would you choose if you had to make a choice of type?

The factors that make the difference in your selection are the factors we have been examining—input impedance, output impedance, linearity, and the required voltage and power levels. If minimum loading with fairly easy linearization are of prime importance, your selection will be FET devices. On the other hand, if the lowest possible output impedance is more important, then you will select a power bipolar device. If survival under difficult conditions is critical, you would probably select FET devices, possibly using VMOS-type output devices. (VMOS-type devices are discussed briefly in a later experiment.) If an absolute minimum of power is the prime consideration, then you probably would use some configuration that is based on insulated-gate FET devices.

Step 5.

Take the two amplifiers that you just designed (the FET unit and the bipolar unit), and wire them up and apply power. Vary your collector voltage and your drain voltage and find out how critical the value of these voltages really are to the proper operation of the circuit. You will need to make some calculations of the voltage gains based on your device curves. Then, measure to see how the results check with your calculations. Explain what you find.

You probably found out that as long as you had enough voltage so that the last active device did not either saturate or cut off, you could use that voltage in the circuit just as well as any higher voltage. There was little if any decrease in distortion once the circuit operation did not crowd either of these regions. Since this will also minimize circuit and device dissipation, this is usually a desirable way to go.

EXPERIMENT 4
Very-Low-Current Characteristics of FET Devices

Possibly one of the most interesting operating areas for FET devices is in the low-current area. It is in this region that one has an opportunity to find a diffusion mode of operation. As has been pointed out previously, this mode of operation is a result of the field, due to the potential jump across the Debye region becoming the controlling field for the channel. But operation in this region requires measurement of very small voltages, well under 25 millivolts. These need to be made with reasonable precision. It also requires detection of 2-to-1 changes in drain current at very small values of current, and possibly with very small voltage differences (as small as 5 millivolts) at the same time. It is best to assume that it is necessary to meter the current (with a voltage drop of at most 5 millivolts) to be sure that the voltage gain and other operations of the circuit will not be altered significantly as a consequence of circuit impedances.

Step 1.

Set up the required high-sensitivity microammeter (if you do not already have one) using either a National LM4250 or some other quality operational amplifier that is set to have a voltage gain of about 100. This amplifier can then be used with a milliammeter having a full-scale sensitivity of 1 milliampere or less (100 microamperes is probably optimum) and a series resistance to make it into a 500-millivolt full-scale voltmeter. This will give you a millivoltmeter of the required sensitivity. Then, using current-metering shunt resistances with values of 5000, 2000, 1000, 500, 200, 100, 50, 20, and 10 ohms, you will have full-scale sensitivities of from 1 microampere to 500 microamperes. You should use 1% resistors so that your 2:1 current changes will be reasonably precise.

Step 2.

Once the meter in Step 1 is checked out, you should wire the circuit shown in Fig. 5-15 using a typical FET device. Start by reducing the current in your device from 500 microamperes in several graduated steps, cutting it in half each time and measuring the gate-

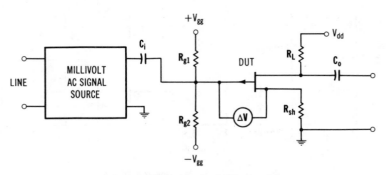

Fig. 5-15. Diffusion mode of FET operation.

voltage change for each step. Continue reducing the current until you get to approximately 0.25 microamperes. Record your data in Table 5-12, and plot a curve of the amount of millivolt change in the gate voltage for each halving of the device current. Describe and explain what you have found.

Table 5-12. Data for Experiment 4, Step 2

Initial I_d				
Δv_g				
Initial I_d				
Δv_g				

Step 3.

How much farther do you think that you can decrease the current, and still get less than 40 to 50 millivolts change for each halving of the current? Adjust your shunts so that you can go to the low nano-ampere range. Devices can be found with 2:1 current changes for less than 20-millivolt gate-voltage change over a current range as great as 100,000 to 1. It is common to find it over a range of 3000 to 1. Measure the data and plot a curve in Fig. 5-16 into the nanoampere

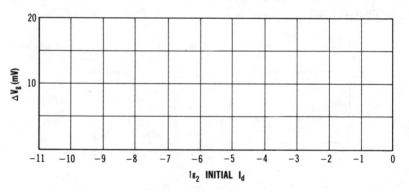

Fig. 5-16. Graph for Experiment 4, Step 3.

range, noting the point at which the required voltage change once again starts increasing. Plot your data in Table 5-13. How do you account for this? Explain.

Leakage paths are a common cause of reduced transconductance efficiency at very low values of current. It should be repeated here that some kinds of insulated-gate devices limit toward half the typical

Table 5-13. Data for Experiment 4, Step 3

Initial I_d				
Δv_g				
Initial I_d				
Δv_g				
Initial I_d				
Δv_g				

values, or require around 40 millivolts for a 2-to-1 change in device current. This does NOT mean that they are defective. It simply means that the device is probably an IGFET device, and the presence of the capacitance at the gate reduces the effectiveness of the voltage change applied to the gate. Only part of the control voltage gets through the insulation layer and into the channel itself.

Step 4.

Test several of the devices you have measured at these low values of drain current and determine what is the transconductive efficiency of each as the device current is reduced. Also, plot the curve of gate-voltage change for a 2-to-1 current change for each device you have tested. Use a linear scale on the X axis, but let each square side represent a change of 2-to-1. (You can replot using five-cycle semi-log paper, if you wish.) Discuss and explain your results.

EXPERIMENT 5
The Insulated-Gate FET Device as an
Input Amplifier for Scopes

The bipolar transistor makes an excellent unity gain "transformer" in that it can take an input signal and provide an output which is an almost perfect replica of the input, but at an extremely low output impedance. Its main deficiency is that for some applications it draws enough current to load its source excessively. The electron tube, when used as a cathode follower, did not have this difficulty, but it does have a dc bias problem. The FET device and, in particular, the insulated-gate FET device need not have either of these problems, as a depletion-mode IGFET device can operate with both a zero voltage difference between the gate and the source and it can present an input impedance of as much as hundreds of thousands of megohms to the input signal. The result is that the n-channel depletion-mode IGFET is used extensively in the input circuits of instrumentation amplifiers and, also, the horizontal and vertical amplifiers of oscilloscopes. Typically, when an IGFET device is used for this purpose, it has an npn transistor in its source return that is adjusted to lock the device operating current at a value for which the gate-to-source voltage is essentially zero. A typical circuit for this is shown in Fig. 5-17.

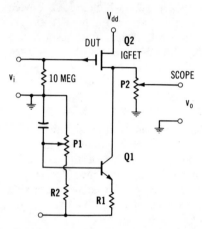

Fig. 5-17. High input-impedance preamplifier circuit.

NOTE: SET P1 SO THAT v_0 DOES NOT CHANGE WITH THE INPUT SHORTED AS P2 IS CHANGED.

Step 1.

Wire up the circuit as shown, using a 2N2222 or similar npn transistor for Q1 and a depletion-mode n-channel IGFET for Q2. Then, using a high-sensitivity differential voltmeter, adjust potentiometer P1 so that the reading from input to output (gate to source) on Q2 is as close to zero as you can get it. Potentiometer P2 is the gain-control potentiometer and is used to control the fine adjustment on the amplifier sensitivity. Once you have P1 adjusted and P2 set for maximum output, you should vary the input voltage on the gate of Q2 over a range of ± 1 volt, and measure the voltage difference from the gate to the source at a series of points. Then, plot your data in the form of a differential voltage (source-to-gate voltage) as a function of the signal voltage. Locate the zero point on the signal-voltage scale midway on your graph paper, and choose the scale on the gate-to-source axis to allow you to plot all of your data. Record your data in Table 5-14.

Table 5-14. Data for Experiment 5, Step 1

v_i				
Δv				
v_i				
Δv				
v_i				
Δv				

Step 2.

How can you determine the "dc" gain of the input circuit from the data you have obtained? Explain. How should these data compare to the data you took on the ac gain of the stage? Should these results be equal? Explain.

The dc gain of the circuit is the quotient of the input voltage deviation less the differential voltage at that deviation divided by the deviation. As is true with all "follower" type of circuits, this quotient must be slightly less than unity. These data are average gains from the quiescent point to the point of measurement, and are somewhat like dc beta in that respect. As a result, the measurement by the two methods cannot be equal.

Step 3.

Repeat your ac measurements on this input configuration, varying the frequency of your test signal from a few hertz up to as high a frequency as you can use and still get a reasonable response. Place your scope first on the input and then on the output, and collect data giving the gain as a function of frequency. Then repeat, using a capacitive load on the circuit which might be typical of what you might expect to encounter from a transistor amplifier stage. A value of 50 picofarads should be suitable and fairly typical.) Record your data in Table 5-15, and then explain what you have learned. Bias the stage at + 1 volt, 0 volt, and − 1 volt.

Table 5-15. Data for Step 3

V_g				
v_i				
f				
v_o				
V_g				
v_i				
f				
v_o				
V_g				
v_i				
f				
v_o				

You will probably find that if you have used a good probe to get your signal to the scope (one with ten times attenuation is probably called for), you can get a surprisingly high-frequency signal through the circuit. But, if you put 30 to 50 picofarads of capacitive load (or more) on the output, you suddenly encountered a severe degradation of the high-frequency response. Plot the data on a four-cycle semi-log chart, with the frequency on the log scale as shown in Fig. 5-18. (Here, the value of K_v is the ratio of v_o to v_i.)

EXPERIMENT 6
The Resistance-Coupled FET Amplifier

You have measured lots of data on FET devices; and you designed some amplifiers in Experiment 3. In this experiment, you will make some further designs of resistance-coupled amplifiers so that you can

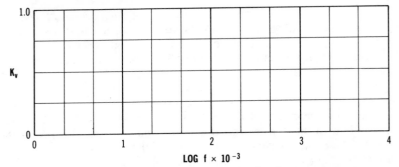

understand some of the further intricacies of the process. Probably the most significant of the considerations is the "compound load line" consideration, in which the signal-frequency load line has a different slope than the static load line. (You have seen one form of this in the transformer-coupled amplifier.)

The design process is relatively straightforward. First, one establishes the static load line on a trial basis just as was done in Experiment 3. Then, one selects the load impedance that the combination load line will show under normal operating conditions. This impedance is usually the parallel impedance for the load for one stage and the input impedance for the following stage. If, for example, the load resistance for the first stage is 1000 ohms, and the input admittance for the following is 0.001 mhos or siemans (the equivalent of 1000 ohms), then the combined impedance is 500 ohms. Under small-signal conditions, this is the impedance that the device will see as a driver.

One plots, as shown in Fig. 5-19, several tentative load lines representing a load of 500 ohms across the 1000-ohm static load line. The device can only swing to zero current (current polarity in the device

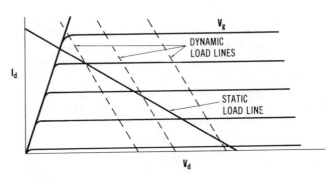

Fig. 5-19. Amplifier combination load lines.

287

cannot reverse), and it cannot go below the saturation line. As in Experiment 3, voltage amplifications may be calculated at a series of points along *each trial combination load line,* and the one that offers the best overall results (as plotted on the output of the second stage) is selected.

Step 1.

Based on the circuit of Fig. 5-20, select a pair of FET devices and also a pair of bipolar devices to use when making trial designs. With the FET circuit, the "load" at the input of the second stage takes the form of a pair of resistors; with the bipolar circuit, it is largely the input admittance of the second bipolar transistor. Using the curves you have prepared on the various devices, make a trial design for each type of device, plotting first a static load line and, then, a series of combination load lines. For the bipolar transistors, you may assume a constant effective input impedance if you wish. It really is variable, of course, and changes rapidly with the collector current.

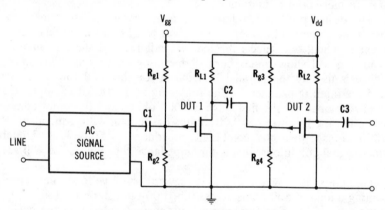

Fig. 5-20. Resistance-coupled amplifier circuit.

Now, tabulate the small-signal voltage amplifications along these dynamic load lines and record the data in Table 5-16. Select an operating load line which will yield a symmetric signal, then, an operating load line which dwells in the high-current region most of the time and, finally, a load line dwelling in the current cut-off region most of the time. Then, set up your circuit on your solderless breadboard. Review what you have done, and run some preliminary tests. Explain what you have learned from this process.

Table 5-16. Data for Experiment 6, Step 1

V_g				
V_d				
I_d				
K_v				
V_g				
V_d				
I_d				
K_v				
V_g				
V_d				
I_d				
K_v				

You should have found that with the use of combination load lines you can extend your capability in signal processing by selectively suppressing parts of signals and amplifying other parts strongly. The only change you have to make is the static value of bias. This is equally true of FET and bipolar devices, although you do not have as good an independent control with the latter. The input impedance and its variation with bipolar transistors can introduce severe problems.

Step 2.

You need to make some detail measurements of your amplifiers. You have a choice of two ways to go. You can read off the effective supply voltage that a combination load line sees and, using the effective impedance, vary the operating point step-by-step along that line. This is the simplest approach *if* there is no nonlinearity in the load. If there is, then you need to pick two frequencies within the operating range of the amplifier—one near the upper limit and the other near the lower limit. You use the lower frequency to describe the actual load contour (you can project it on your scope), and superimpose a small-amplitude high-frequency signal which you can use to measure the point-by-point voltage gain of the combination. You should try both techniques to see how they compare. A circuit configuration you can use for the second approach is shown in Fig. 5-21.

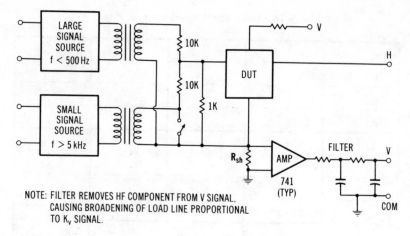

NOTE: FILTER REMOVES HF COMPONENT FROM V SIGNAL.
CAUSING BROADENING OF LOAD LINE PROPORTIONAL
TO K_v SIGNAL.

Fig. 5-21. Load line and gain test circuit.

Test both the FET circuit and the bipolar circuit in this way. Remember that with the latter, the input impedance (or admittance) will vary with the operating conditions, with the result that unless you degenerate its emitter, the variations will alter your response. Record the data in Table 5-17. Draw sketches of the waveforms that are shown on your scope on the graphs given in Figs. 5-22 through

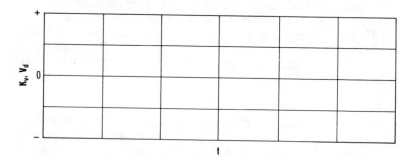

Fig. 5-22. Waveform for Experiment 6, Step 2.

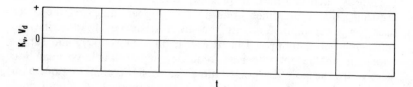

Fig. 5-23. Waveform number 2 for Step 2.

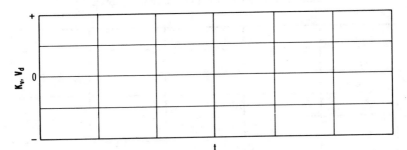

Fig. 5-24. Waveform number 3 for Step 2.

5-24. (Fig. 5-22 is to be a symmetric waveform for n-channel and npn devices while Fig. 5-23 is to show a saturation waveform for the same types of devices. Fig. 5-24 is to display a cutoff waveform for n-channel and npn devices.) Discuss and explain your results based on the data and the waveform sketches.

By varying the static bias, you can "clip" or distort the output waveform on either half of the wave, flattening the sine-wave signal to suit. To control it, you readjust the bias and add emitter or source degeneration to suit your requirements.

Step 3.

It should be interesting to learn a little more about what happens when the load is nonlinear. A combination of bipolar transistors (without degeneration) is an excellent arrangement for examining this. The principal problem is in making an estimate of the kind of nonlinear characteristic to plot on the load diagram. The reason for this problem is the rapid change in conductance with voltage.

You can get an idea of what this does by arranging a load resistance which changes between two values when a diode turns on and turns off. Since such a configuration can be plotted, at least in principle, its use is suggested. To do this, you can work with the FET configuration, and replace the gate load (bias) resistor with a pair of resistors, both of which have a resistance that is equal to the load resistance of the amplifier. However, one resistor is connected in series with a diode before paralleling with the second resistor. This will cause the combined resistance on the amplifier to be one-half of the load resistance with one polarity of signal, and a third of the

Table 5-17. Data for Experiment 6, Step 2

V_g or I_b				
I_c or I_d				
V_c or V_d				
v_i				
v_o				
K_v				
V_g or I_b				
I_c or I_d				
V_c or V_d				
v_i				
v_o				
K_v				
V_g or I_b				
I_c or I_d				
V_c or V_d				
v_i				
v_o				
K_v				

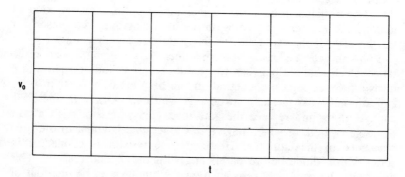

Fig. 5-25. Output-time plot graph for Step 3.

resistance in the other polarity. There will be a break in the slope of the load-line combination; the break will be near but not on the static load line. (The effect of the capacitor prevents the break from being precisely on the waveform.) Apply the combination of two frequencies as the input signal. The high frequency signal must be small enough to make a linear test, and the low frequency signal must be able to swing the signal along the desired operating load line. An input-output plot on your scope can be used to test the combination. The small-signal measurements may have to be made on a conventional output-time plot, however. Trace the curves you get on your scope onto the charts given in Figs. 5-25 and 5-26. It is

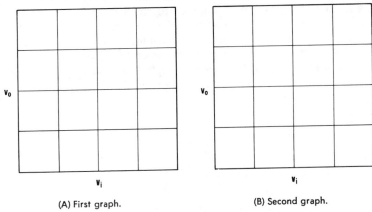

(A) First graph.   (B) Second graph.

Fig. 5-26. Input-output plot graph for Step 3.

possible to get a modified input-output plot which will show both the load contour and the small-signal gain. How would you propose to do that? Try it!

What you have to do is eliminate the high-frequency component of the signal on either the horizontal or the vertical display, so that the small-signal line will not follow the low-frequency load line, but will be either horizontal or vertical. This can be done by filtering out the high-frequency component from one axis. If the load contour is nearly horizontal, you should filter the high-frequency component from that axis, and vice versa. Then the deflection due to the high-frequency signal will be visible by itself. For this to be successful, the ratio of the two frequencies should be at least 100.

Step 4.

Repeat this process with other bipolar and FET amplifier configurations using devices you have already measured. If you need to verify the transconductance value for any device, it can be done by introducing either a source or emitter resistance that is large enough to cut the stage gain in half. Then, take the reciprocal of the resistance value as the transconductance. (It is possible to make an approximate measurement of the base-spreading resistance by making a gain measurement at a frequency that is at least four or five times the frequency at which the loss of gain, as a function of frequency, starts. Then, add a series base resistance that is sufficient to halve the gain at that frequency. The value of the additional resistance is approximately equal to the base-spreading resistance.) You normally will not find a significant spreading-resistance effect in the gate lead of a FET device. Try this method with several device combinations to see how they behave. Remember, though, that you may find it convenient to adjust the bias on one or more of the devices. Discuss and explain your results.

It is helpful to be able to observe both the small-signal and the static characteristics at the same time. Since you usually want voltage gain, you attempt to wipe out the high-frequency current-gain component. In that way, the high-frequency current appears to be zero, and the two signals can be evaluated independently. Measurement of the transconductance by the method just suggested is approximate at best, as you will be able to see a change in both the load line and the small-signal gain at the same time.

EXPERIMENT 7
The Transformer-Coupled FET Amplifier

You are now ready to test a FET device in a transformer-coupled amplifier circuit similar to the one used with the bipolar transistor. You can add one new feature, however, and that is taking into account the fact that a combination load line actually has to be considered. Except for this, we suggest that you use the same basic circuit and techniques as those you used in Chapter 4.

Step 1.

What is the static load impedance that your transistor will face when a transformer is used as the interstage device? Explain, and then measure it with an ohmmeter.

The static load impedance of a transformer is actually the primary resistance and can be measured with an ohmmeter. Even if the secondary is shorted, it has no effect on the primary resistance when it is measured with direct current.

In addition, you should be already aware that the current-input mode that you tested with the bipolar transistor does not exist with FET devices, and you will have to use an input signal voltage at all times. The first thing that is important to determine when using a transformer-coupled amplifier, as with the resistance-coupled amplifier, is the amplitude of the input signal which causes the amplifier to show a significant distortion.

Step 2.

First, based on the value of the resistance that you measured for the primary of your coupling transformer, draw a static load line on a set of curves for the transistor that you are using as the amplifier drive transistor. Next, you must plot the load line that the signal you are amplifying will see. Since the ac impedance of any good transformer is substantially greater than the dc resistance of its primary, the ac load line will cross the static load line at a different position than what you observed in Experiment 6. Draw a series of trial load lines across this static load line at a resistance slope that is defined by the equation:

$$Z_L = \left(\frac{n_p}{n_s}\right)^2 R_x \qquad \text{(Eq. 5-5)}$$

where,

R_x is the resistance placed across the secondary of the transformer, (n_p/n_s) is the ratio of the primary turns to the secondary turns.

Here, Z_L is the reflected impedance at the primary resulting from the secondary resistance load. Quantity Z_L defines the resistance value that is to be used for plotting the load line. These trial load lines may be plotted starting at any zero drain-voltage point (with a current less than the peak encountered on the static load line). Just pick a point near the maximum that you show on your curves and, then, select several other values of current that are uniformly spaced toward the zero current point. Plot trial load lines through these points, and calculate the values of voltage gain that you can expect to obtain at various bias points along each load line. Tabulate your data in Tables 5-18 and 5-19, and select the load line that you wish to use for your tests. Also, describe and explain the differences you observed between this amplifier and the resistance-coupled amplifiers previously considered. (Use a value of 1000 ohms for the secondary load resistance R_x.)

Table 5-18. Data for Experiment 7, Step 2

V_g			
V_d			
I_d			
r_m			
v_1			
v_o			
v_x			
i_o			
K_v			
K_{vx}			
R_x			

In Tables 5-18 and 5-19, the subscript x indicates the secondary side of the transformer, and the quantity r_m is the current-metering resistance used in determining the drain current. (Your transformer has a balanced load rather than the unbalanced load encountered with bipolar transistors.) As long as the load resistance is small enough that magnetizing and leakage effects can be neglected, the circuit should behave almost exactly as you have predicted from your calculations. The only factor which might be significant would be FET nonlinearity. If, however, you remove the load resistance completely, then the circuit will not function at all well. Try it!

Step 3.

Set up your circuit, and fill in Tables 5-18 and 5-19. Determine the gain and the distortion as the input signal is varied in amplitude.

Table 5-19. More Data from Steps 2 and 3

V_g				
V_d				
I_d				
r_m				
v_i				
v_o				
v_x				
i_o				
K_v				
K_{vx}				
R_x				

(You can use a combination signal like you used in Experiment 6, if you wish, so that you can get both the load contour and the small-signal amplification. However, the frequency ratio may have to be smaller.) Also, use the characteristic curves of the device to examine the operating points that you analyzed in Step 1 to see if the data check out. Also, find the amplitude of input signal at which the output distortion becomes noticeable as you make these tests. Record your results and explain what they mean.

Step 4.

The amount of flux required to handle a specified signal voltage in the transformer is a function of frequency, since the voltage induced in an inductance is proportional to the frequency and the peak current. (The rate of change of flux is proportional to the frequency.) As a result, at very low frequencies, the power-handling ability of a transformer is much less than at higher frequencies. Winding and core losses generate a similar limitation at very high frequencies. Distortion is indicative of the limitation on the power-handling ability of the unit. In any case, the amount of input signal required by a FET device for generation of a device-controlled distortion is substantially greater than for a bipolar device under similar conditions, solely because of the relatively lower transconductance-efficiency value for the FET. In fact, the following relation is approximately correct:

$$v_i \text{ (bipolar)} \simeq \kappa \, v_i \text{ (FET)} \qquad \text{(Eq. 5-6)}$$

or, the product of the transconductance efficiency multiplied by the input-signal voltage at the onset of distortion for the FET device is approximately equal to the input-signal voltage at the onset of distortion for the bipolar device. This relation is not exact since, typically, the value of kappa for FET devices decreases as their drain currents increase. This will make it harder to detect the onset of distortion with the FET device. Record the values you measured in Table 5-20, and see how closely they satisfy Equation 5-6. The subscripts b and f in Table 5-20 refer to bipolar and FET devices respectively. How well do they check?

Table 5-20. Data for Step 4

v_i (FET)				
v_i (bipolar)				
v_{ib}/v_{if}				
κ				

Step 5.

In this step, you will explore the ampere-turn effects of your interstage transformer with a FET amplifier. This amplifier is very different in its properties than that of the transformer-coupled bipolar-transistor amplifier, inasmuch as this transformer is required to provide secondary current to its resistance load almost exclusively, and almost none to the FET device. As a result, there is almost no variation of the waveform resulting from a variation of loading on the secondary. You will, however, want to look for the effect of dc flowing in the primary of the transformer. To do this, make measurements, both with the direct current in the primary and, also, with it decoupled from the primary of the transformer. (In the latter case, you will recall that the signal current is truly alternating current, not a varying direct current.)

Wire up your solderless breadboard so that its circuit conforms with the circuit given in Fig. 5-27. Then, using a frequency around 300 Hz, examine the input and output waveforms on the scope as you vary the input-signal amplitude. Note the input amplitude at which degradation of the output waveform first shows on your scope. Do

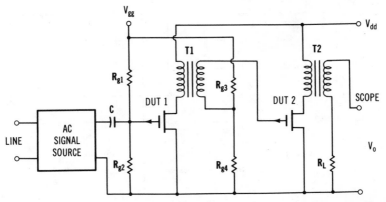

Fig. 5-27. Transformer-coupled FET amplifier circuit.

this both with dc in the primary windings and, then, without it in the primary windings. Then, repeat your tests using frequencies of 150 Hz and 75 Hz. (You can go to as low a frequency as you wish, as long as the amplifier gives you any reasonable output waveform.) This test can be performed with an unloaded secondary and, then, with a series of load resistance values, with each being one-half the value of the previous test. Sketch your waveforms as you go (at the point where degradation of the sinusoidal waveshape first sets in). These tests should be continued at successively lower frequencies and successively higher frequencies (an octave apart) until the output has decreased to about a third of the midrange value. Make sketches of the limit waveforms in Figs. 5-28 and 5-29.

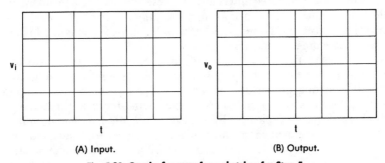

(A) Input.   (B) Output.

Fig. 5-28. Graphs for waveform sketches for Step 5.

When the transformer is operated without a load, you can expect some distortion of the waveform at all levels of signal. When the transformer is operated under proper loading, there may be minor imperfections in the output waveform at very small signal levels. This will be due to low-flux core distortion. The primary cause of

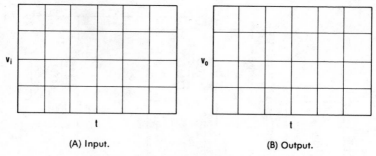

(A) Input.   (B) Output.

Fig. 5-29. Graphs for more waveform sketches.

distortion, however, will be magnetic saturation, which can be encountered nonsymmetrically if there is a direct component of current in either winding. The active device can also contribute to this kind of distortion.

EXPERIMENT 8
FET Devices as RF Amplifiers

It is important for you to know what property of the field-effect transistor has led to their being used extensively in the input rf circuits of high-quality radio receivers. You would offhand expect that this application would be one where the bipolar transistor, with its high transconductance-per-unit-current compared to the FET device, would be outstanding. After all, current noise, which is the principal kind of noise generated internally in these circuits, is dependent on device current. Clearly, a maximum isolation of the signal from the noise, therefore, would be expected with the device having the highest transconductance efficiency.

The reason FETs are used is simply that the level of distortion generated in them is substantially lower and, as a result, significantly less intermixing of the desired signal with both noise and adjacent stronger signals will occur. This is due to the more linear behavior of these devices. Typically, FET devices are used in the rf, mixer, and first if stages in many high-quality receivers for these reasons.

For this experiment, you will use a circuit that can function as a mixer. You will examine how it behaves, both with bipolar and with FET devices. The two basic circuits you will be concerned with are shown in Fig. 5-30. As you will notice, an audio-frequency signal is put into the base or gate of the device under test, and an rf pulse is put into the emitter or the source of the device. You should vary the amplitude of the rf signal and observe the effect on the audio signal that comes through.

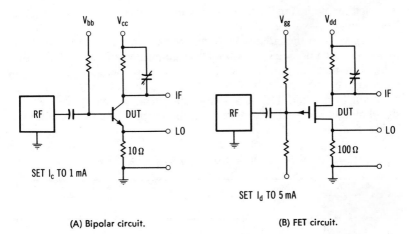

(A) Bipolar circuit.　　　　　　　　(B) FET circuit.

Fig. 5-30. RF mixer circuit.

Step 1.

Wire each circuit on your solderless breadboard so that you can readily shift from one to the other. The resistance in the emitter or the source return, as the case may be, should be small enough that the degeneration it introduces is negligible. Vary the amplitude of the rf signal introduced across the resistor so that you are, in effect, modulating the output by any nonlinearity that is present in the amplifier.

Normally, an LC tuned circuit will be used in the output circuit instead of the RC circuit shown in Fig. 5-30. The reason for not using an LC circuit is that you want to be able to observe any bias shift which may be introduced by the pulses of rf that you are introducing into the emitter circuit; a tuned circuit would filter them out. Only a "low-pass" circuit formed from a parallel RC circuit can make the change visible. You will see a break in the smooth sinusoidal waveform as a consequence of the rf pulse.

It is best to have the rf pulse synchronized to the audio waveform if you can do it without distorting the waveform. By now, you should be able to design a simple circuit to do this, but we will give you some suggestions. If you have one of the waveform generators which can simultaneously generate a triangular wave, a sine wave, and a square wave, you can use the sine wave to excite the amplifier and use the triangular wave to generate a short pulse at the peak of the sine wave. To do the latter, bias an amplifier into "class-C," so that only the top of the triangular wave will turn it on. Amplify that pulse, and use it to turn on an rf signal from a similarly biased "amplifier," which can really be an emitter follower circuit. Whenever the circuit is turned on by the amplified pulse, it can then pass a pulse of rf signal from your rf signal generator.

When you get the circuits working, you can apply both waveforms to the circuit under test and vary the amplitude of each. This will show you that the circuit "works." Then you measure the output amplitude of the circuit as a function of the output current with the rf circuit turned off. This is a small-signal test, so the applied audio signal should not lead to distortion in the output. Record the data in Table 5-21.

Table 5-21. Data for Experiment 8, Step 1

	Bipolar			FET	
V_o					
I_o					
v_1					
v_o					
K_v					
V_o					
I_o					
v_1					
v_o					
K_v					

Step 2.

Once the circuit is functioning as a simple amplifier and you have checked its operating characteristics, set the bias level so that you do not get distortion due to anything but the normal nonlinearity of the device, whether it be a bipolar transistor or a FET device. You will want to make this test on both kinds of devices, as you know. Then, you can introduce varying amounts of rf signal into either the source or the emitter, depending on what you are testing. Increase the amplitude of that signal until the rf signal causes a noticeable effect on the waveform. What happens to the waveform when you do this to the bipolar amplifier? To the FET amplifier?

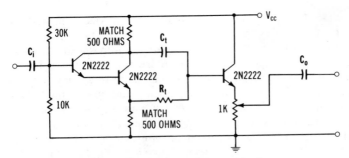

Fig. 5-31. Phase-shifter audio-source circuit.

Look up Experiment 8 in Appendix E. This experiment shows you how to make a phase shifter based on a Darlington compound circuit. Introduce one of these circuits into the audio signal path of the circuits shown in Fig. 5-30, and follow it with an emitter follower configuration, as shown in Fig. 5-31. The potentiometer in the emitter return gives you a way to vary the amplitude of the output voltage from the phase shifter. This combination will give you a way to shift the phase of the pulse with respect to the audio sine wave and, thereby, gives you an easy way to examine the entire waveform for distortion introduced by the rf pulse. Do this for both the FET circuit and the bipolar circuit. Describe what happens.

Both circuits will be affected when the rf pulse is applied, but you can expect that the effect will be detected sooner with the bipolar transistor because of the exponential nonlinearity, that is, with a smaller ratio of rf to audio voltage. The magnitude of the effect will grow more rapidly with the exponential nonlinearity because of the larger number of significant higher-order terms. To show this, sketch on a separate sheet of paper the comparative waveshapes for some examples where the ratio of rf to audio is the same for each device. Note the value of the ratio for each pair.

Step 3.

Next, you can test the bipolar and FET circuits as modulators. If there are any problems in their operation as mixers, this test will reveal them. For this test, introduce a high-frequency continuous wave into the main input (base or gate) circuit where the audio signal was introduced in Step 2. Then, introduce your audio modulation into the emitter or the source resistor where you introduced the pulses before. Examine the output waveform from the circuit, using a very small coupling capacitor to take out the audio-frequency components. Vary the amplitudes of the two signals, and record the waveforms you observe. Make records of your data and sketches of the envelopes of your waveforms. (The envelope of the waveform shows the outer limit of the pattern as observed on your scope.) Did either one of these circuits seem to give you a better modulation envelope than the other? Vary the level of the modulation and the carrier at the inputs, and note particularly what happens at reasonably high modulation levels where the audio almost cuts the carrier off. Explain what you have observed.

You probably found the modulation envelope more sinusoidal with the FET circuit than with the bipolar circuit at the higher levels of modulation. The kind of nonlinearity required for a good modulation waveform is very highly specific!

Step 4.

FET devices can be used with oscillator circuits as well as with amplifiers and mixers. Several possible circuits which may be used are shown in Fig. 5-32. (These circuits are to be used with depletion-mode devices.) In each case, you should set up each basic circuit as an amplifier and arrange it so that at an appropriate time the feedback coupling may be completed by placing a jumper across the gap "x." Prior to closing the loop, you should adjust the voltage gain at the operating frequency to about 1.5. If you use a Lissajous figure for the test, the circuit can be tuned so that it is also zero phase when the voltage gain exceeds unity. Closing the circuit will then initiate oscillation at approximately the desired frequency.

In testing oscillators, it is important to determine the effect of varying the LC ratio of the tuned circuit. Often, as a circuit is initially designed, the effective voltage gain may be much larger than is either required or desirable. As has already been noted, the effective impedance of the circuit can be reduced without changing its operating frequency just by decreasing the inductance and then by increasing

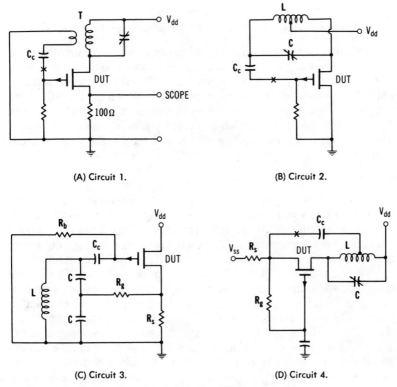

(A) Circuit 1.

(B) Circuit 2.

(C) Circuit 3.

(D) Circuit 4.

Fig. 5-32. Typical FET oscillator circuits.

the capacitance in such a way that the LC product remains constant. Try several LC combinations with at least one of the circuits given in Fig. 5-32, and record your results.

You will have found that the best operation is obtained by selecting a tuned impedance that is just enough to provide the required overall loop voltage gain of between 1.1 and 1.5. A certain amount of juggling is required, as you have several variables at your command with most of these circuits. Factors which will help the frequency stability include high Q, high capacitance, voltage and temperature-stable components, and a minimum variation of loop-voltage amplification that is consistent with a self-starting and stable oscillation condition.

EXPERIMENT 9
Do You Want To Use a FET or a Bipolar Transistor?

There are times when you should select a bipolar transistor for a given application, and there are times when you should select a FET device. There are also times when you will want a combination of both. The purpose of this experiment is to help you gain a little more insight into the ways you can help yourself in making that selection. The principal conditions which can force you to consider the type of device to select are the loading effects and the overall power-output capabilities.

Step 1.

You have just designed the sawtooth-wave generator circuit shown in Fig. 5-33. As you can see, you are using a pnp transistor to act as a constant-current charging source (Q1). This transistor charges the capacitor C1. Transistor Q2 connected across it must be a very low-leakage device, like possibly a 2N1613. You start out with an npn silicon transistor (Q3) used as an emitter follower. This device gives you the output waveform for your generator, and it also provides the switching signal for resetting the sweep. The reset signal is generated by transistors Q4 and Q5. In this part of the circuit, transistor Q5 conducts when the capacitor is charging, meaning that Q1 and Q6 must also be conducting, and Q2 and Q4 must be turned off. When the output from device Q3 (the emitter follower circuit) rises sufficiently to trigger transistor Q4, it will switch on and switch transistor Q5 off. Transistors Q1 and Q6 will be turned off, and Q2 will be turned on. Wire up this circuit on your solderless breadboard

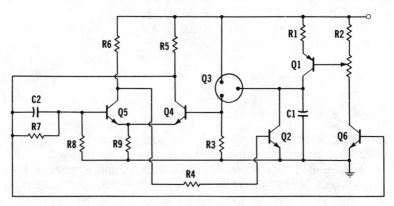

NOTE: Q3 CAN BE EITHER AN NPN BIPOLAR OR
N-CHANNEL FIELD-EFFECT TRANSISTOR.

Fig. 5-33. A sawtooth-wave generator circuit.

using the suggested values, and see if you can get a linear sawtooth waveform out of it. Check the circuit with your scope, and explain what you observe.

Unless you had uncommonly rare good luck, you must have found that the waveform was somewhat curved, or "hump-backed." This definitely is not what you want. It would not give you the linear sweep. What is wrong?

Step 2.

You might first suspect leakage in transistor Q2, since the presence of a leakage resistance across the capacitor could cause such an effect. Actually, if you have selected and used an extremely low-leakage transistor in this position, it is very unlikely to be the cause. The circuit still has an output which is curved. The only other transistor which might be loading the capacitor is Q3 which is operating as an emitter-follower and should have a very high input impedance. It is the only transistor that is conducting when the capacitor is charging. But substituting another npn transistor has no effect. What do you think would happen if you replace it with a good n-channel enhancement-mode field-effect transistor? It, at least, could not leak the charge off as you put it on. An insulated-gate device, with its thousands of megohms of input resistance would be best if this were it. What happens now? Explain.

Lo and behold, the curvature almost completely disappears! The input impedance of the emitter follower just was not high enough. An IGFET source follower is required. The only problem will be in getting a device with a high-enough maximum operating frequency. The results at low frequencies are almost ideal.

Step 3.

Next, you will want to find out what the effect of frequency-response limitations might be, so you increase the operating frequency by increasing the current through transistor Q1 and also reducing the size of capacitor C1. You should examine this circuit with several different IGFET devices in the position of Q3, and experiment with the other devices in the circuit to find which might be critical. The maximum operating frequency is going to be a crucial parameter. Record your data in Table 5-22. Explain what these tests have shown you.

Table 5-22. Data for Experiment 9, Step 3

	Max F	Linearity	Remarks
Bipolar 1			
Bipolar 2			
Bipolar 3			
FET 1			
FET 2			
FET 3			

You probably found that the circuit would generate a sawtooth waveshape at a higher frequency with a bipolar transistor than with a field-effect transistor, but that the nonlinearity problem always plagued you. On the other hand, the field-effect device almost completely eliminated the nonlinearity.

Step 4.

This should intrigue you. Maybe FETs will work better in some of the other positions in the circuit. Try it. Explain what you find.

It was found that the principal place where the substitution of a FET in this circuit would help significantly was at position Q3. At Q2, there were some leakage problems, and the device would not turn ON hard enough to discharge the capacitor fast enough. At Q1, it was found that there was not a wide enough range of control, and the turn-off resistance was not what was wanted. At locations Q4 and Q5, the switching was somewhat degraded because of the low trans-conductance-per-unit-current efficiency of the FET devices used here. A somewhat similar effect appeared at Q6.

EXPERIMENT 10
The FET Device as a Possible Power Amplifier

In this experiment, you should put your devices through essentially the same series of tests that you ran in Experiments 6 and 7 in Chapter 4. Try to get some VMOS-type FETs especially for this test, as they are the best devices presently available. You should plot both trans-conductance and transconductance efficiency as a function of drain current, and find out how you have to adjust your circuit to make it behave in the linear manner that you require. Record your data in Table 5-23 and explain the differences that you find in these devices. You will find it particularly interesting to plot sets of curves on VMOS FETs. Use the graph given in Fig. 5-34 for that. Also, plot curves of

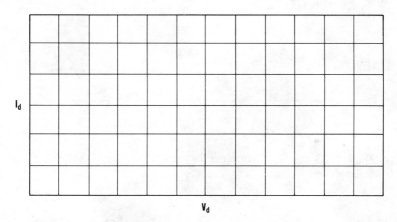

I_d

V_d

Fig. 5-34. Characteristic curves for a VMOS FET.

actual transconductance as a function of drain current (use graph of Fig. 5-35), and the effective transconductance (in-circuit) and the transconductance efficiency as a function of gate bias as well as a function of output current (use graph of Fig. 5-36). If you take the transconductance and transconductance-per-unit-current efficiency with reference to a selected drain current, and take the gate bias at that current as reference zero, how do the curves for the devices compare as a function of the gate-bias change? Tabulate your data

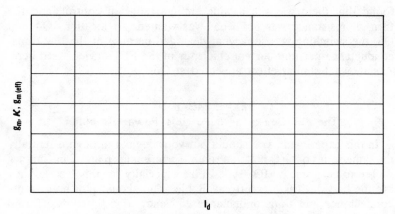

Fig. 5-35. Transconductance vs. drain current.

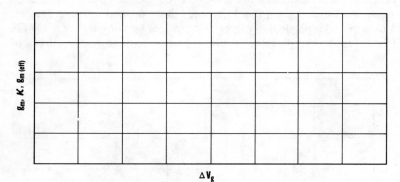

Fig. 5-36. Transconductance vs. gate bias.

in Table 5-23 and plot your device curves in Figs. 5-34, 5-35, and 5-36. Explain what you have learned.

Step 1.

You have measured a set of data for your VMOS transistor, but you may not have used a large enough value of current in your measurements. These devices are power-type devices when used in discrete power mounts like the TO-3 or TO-220 mount. Be sure that

Table 5-23. Data for Experiment 10

V_g				
V_s				
V_d				
I_d				
v_i				
v_o				
R_s				
R_L				
K_v				
V_g				
V_s				
V_d				
I_d				
v_i				
v_o				
R_s				
R_L				
K_v				

you have data in Table 5-23 that goes to at least 100 mA. Suggested currents and voltages include values of 5, 10, 20, 40, 60, 80, and 100 mA, and 2, 5, 10, 15, and 20 volts, if the ratings of the device go that high. Measure the gate voltage and the transconductance as well. Then, plot curves of transconductance against bias deviation from a selected point. Also, plot the value of kappa on the same basis. Using the technique described in Chapter 4, find the proper matching point for a push-pull amplifier, and select the amount of source degeneration required to limit the distortion to a reasonable figure. Plot your data on the graph of Fig. 5-37.

Step 2.

Choose two of the VMOS transistors tested, and balance the drain currents in them to the level that is indicated by your study of the

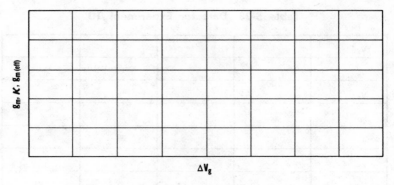

Fig. 5-37. Graph for Experiment 10, Step 1.

transconductance curve, as shown in Fig. 5-37. You will need to find out how to match the effective transconductance plots to minimize the distortion in the overall combination. In the circuit diagram shown in Fig. 5-38, the values of the resistors depend completely on the actual devices used and the amount of forward bias required. You will need to adjust resistances R_A and R_B to the proper value to assure matching of the two devices. (Fortunately, the approximate match point will occur with roughly equal drain currents in the two devices. Taking R_A and R_B as 100,000-ohm potentiometers, the proper approach is to adjust from first one side and then the other to balance the currents at points A and B.) The reference point for gate voltage for each device, as before, should be taken at some convenient current level. With power devices, this value may be between 2 and 10 milliamperes, depending on the maximum current rating

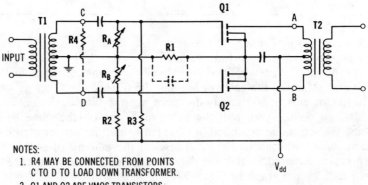

NOTES:
1. R4 MAY BE CONNECTED FROM POINTS C TO D TO LOAD DOWN TRANSFORMER.
2. Q1 AND Q2 ARE VMOS TRANSISTORS.
3. R_A AND R_B ARE VARIABLE.
4. R2 = R3 = 500K TO 5 MEGOHMS.

Fig. 5-38. VMOS transistor power amplifier.

312

for the device. (If you pick too high a current, you will need both positive and negative gate-voltage increments.) Balance can be achieved by adjusting R_A and R_B so that the voltages at points A and B are approximately equal and indicate that the correct level of current is flowing in the transistors. You can use the two-signal source used in Step 2 of Experiment 6 of Chapter 5. With this arrangement, you can again construct both the static and the small-signal load line at the same time. Record your data in Table 5-24, and convert the drain current to gate voltage from your curve set. Next, you again plot the amplification sum at equal increments on

Table 5-24. Data for Experiment 10, Step 2

V_g				
V_d				
I_d				
v_i				
v_o				
ΔV_g				
K_v				
V_g				
V_d				
I_d				
v_i				
v_o				
ΔV_g				
K_v				
V_g				
V_d				
I_d				
v_i				
v_o				
ΔV_g				
K_v				

either side of your chosen quiescent point, searching for the quiescent point which leads to the most constant sum.

The values of resistors R2 and R3 will be equal and will be between 500 kilohms and 5 megohms. They are too device dependent to be more specific about the values. Resistor R1 can be chosen to develop about 1 volt at maximum current through the conducting transistor. Leaving off its bypass capacitor will improve the linearity of the amplifier. The values of R_A and R_B should be as large as possible. A separate transformer output may be used as indicated by resistor R4.

EXPERIMENT 11
Testing Field-Effect Transistors

The technique for testing FET devices that is explained in the preceding experiments is somewhat too involved to set up each time you want to check a device to be sure that it is good. You will want to have a quicker and more convenient method of testing FETs. You need a device where you can plug in your transistor, throw one or two switches, turn a knob or two, and be able to say, "GOOD" or "NO GOOD." A circuit that will make an excellent tester is presented in Appendix B; only the basic circuit will be considered here.

Step 1.

A basic circuit diagram that can be used with all types of FET devices is shown in Fig. 5-39. A 9-volt transistor battery is used as a power supply, and the unit is turned on as needed. The potentiometer provides the bias for the gate (through a current-limiting resistance), and the zener diode provides the required 5 volts or so needed for the FET. The resistor on either side of the zener diode makes the circuit useful for either n-channel or p-channel devices. The source is connected to one side of the zener diode, and the drain

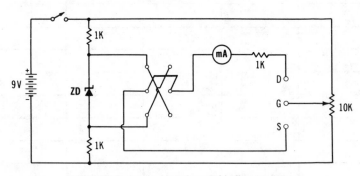

Fig. 5-39. Basic test circuit for a field-effect transistor.

is connected, through a meter, to the other side of the zener diode. The drain current will flow if the gate is sufficiently forward biased; otherwise, it will not flow. Before you attempt to use the circuit, however, you must calibrate it. It is convenient to know where the voltage midpoint is on the potentiometer, and it is also convenient to know the points that correspond to the two ends of the zener diode. You can use either a dvm or a sensitive differential voltmeter to locate these points and mark them on your potentiometer calibration dial. (You use these marks to isolate depletion and enhancement-mode devices of either type.) You will also want to mark a rough current calibration on your meter. A microammeter/milliammeter of normal sensitivity may be used here.

Step 2.

You need to test a series of devices, separating them according to class—n-channel depletion type, n-channel enhancement type, p-channel depletion type, p-channel enhancement type, insulated-gate device, etc. There are plenty of classes for these devices. The goal at the moment is to get so that you can recognize the various types quickly. List the types you have found among the devices you have.

Step 3.

How can you identify an IGFET? The resistance from the gate to the source and to the drain are both infinite for all practical purposes. Use a very-high-sensitivity ohmmeter for this test that will operate on 2 to 3 volts at most. Find the gate, and measure the resistance from it to both the source and the drain. A differential voltmeter can be used, but probably *not* your dvm unless it is specially designed for the purpose. If you use a differential voltmeter, be sure that it has a 1.5-volt battery connected in series with at least a 1 megohm of resistance and the junction. If the device is a junction FET, a slight current of one polarity will flow to both the source and the drain. A fairly substantial current of the reverse polarity will also flow. With an IGFET, you should have difficulty getting any deflection at all, and it will also be independent of polarity. Sort your devices into these two kinds of groupings. Comments?

Step 4.

Test a few bipolar devices using the circuit given in Fig. 5-39. What differences do you note between the two kinds of devices?

The current builds up much more slowly with FETs than with bipolar transistors as the potentiometer is turned.

WHAT HAVE YOU LEARNED?

You have learned what a field-effect transistor is and how it works. You have learned something about many of the members of the FET family that are in common use. You have learned the differences between FET devices and bipolar devices.

You have also learned how to build circuits around field-effect transistors, and you have learned several important distinctions between them and bipolar transistors. You have seen how you can compare these devices in circuits, and something about when you should choose one over the other. You have learned that although the bipolar transistor is stabilized by controlling base current, and the field-effect transistor is controlled by stabilizing the gate voltage, inherently they operate in very similar ways, with the principal difference being the much lower transconductance efficiency that is obtained using a FET unit as compared with a bipolar device. You have learned how to linearize these devices by use of properly applied degeneration, and above all, you have learned how nonlinear many of them really are. Even though there apparently is no theoretical reason why a device capable of amplifying has to be nonlinear, evidently all of our present-day solid-state amplifying devices are extremely nonlinear.

You have seen the results of "degeneration," or "feedback," in improving the linearity of the output of a variety of amplifiers. Possibly, now is a good time to tie the ideas together and explain just what is happening. It is really surprisingly simple, yet at the same time, it is rather unique. In effect, when you introduce degeneration in a circuit, you use the linearity of a device like a fixed resistor to introduce a balancing voltage that is opposing to the applied signal, leaving only a small part of the applied signal across the active device. Since the returned signal is proportional to the output parameter (output current), and the resulting comparison signal is nearly equal to the applied signal, the balancing action forceably linearizes the operation of the device. The new effective input voltage to the device is:

$$v_i - i_fR_f = v_i - K_fv_o = v_g \qquad \text{(Eq. 5-7)}$$

where,

K_f is the feedback voltage gain.

The output voltage is the product of v_g and K_v; giving the result:

$$v_o = K_vv_g = K_v(v_i - K_fv_o) \qquad \text{(Eq. 5-8)}$$

or

$$v_o = \frac{K_vv_i}{1 + K_vK_f} \qquad \text{(Eq. 5-9)}$$

which is the usual feedback equation, but with a $(+)$ sign in the denominator.

Why do you remove a substantial part of the signal this way? By making the part you remove proportional to the output and, at the same time, almost as large as the input (it will only be bigger if the circuit is oscillating), the effects of the nonlinearity are almost completely eliminated, and the effective device transconductance will be approximately the reciprocal of the feedback resistance instead of the much larger value indicated by the current level.

The reason that you want to do this locally in the emitter or in the source lead before applying overall feedback in the circuit is that the overall feedback will be more effective if the amount of nonlinearity that it has to contend with is reduced. If the overall feedback can reduce the distortion by a factor of 10, it is better to have the distortion limited to 1% prior to its application than to have a 25% distortion to correct.

CHAPTER 6

Specialty Devices

In this chapter, some special devices that are commonly used in analog electronics will be considered. These devices are used for a variety of purposes. They have some properties which will probably prove interesting and useful to you.

OBJECTIVES

After studying this chapter and performing the experiments described, you will be acquainted with a variety of special semiconductor devices that are commonly encountered in communications and computer electronics. In your study and measurements, you will have verified the properties and characteristics of the following devices and their associated circuits.

1. Unijunction transistors and how to use them.
2. Programmable unijunction transistors and how to use them.
3. How to use triacs and silicon-controlled rectifiers and switches.
4. How trigger diodes and diacs can be used.
5. How to use stabistors.
6. How to construct constant-voltage reference circuits.
7. How to construct circuits using the above devices.

DEFINITIONS

The semiconductor devices that will be important in this chapter include the following:

unijunction transistor (UJT)—This device is formed from a small segment of semiconductor material (weakly doped) which has two terminals, called bases, on it, and a third, called an emitter, that is of opposite polarity and is diffused onto its surface between the two bases. The emitter is doped more heavily than the base material, and can introduce a surge of carriers into the segment under proper conditions. Usually the segment is doped into an n-type, and the emitter into a p-type.

programmable unijunction transistor (PUT)— A device whose function is similar to that of the UJT, but which utilizes a four-layer device like a silicon-controlled rectifier to perform its switching action.

silicon-controlled rectifier (SCR)—This is either an npnp or a pnpn semiconductor configuration so arranged that it acts like a switching circuit when the applied voltages on its anode and its gate are correct.

trigger diode—A device which behaves somewhat like a zener diode in that it starts conducting at a certain level of voltage. When it starts conducting, it usually experiences a voltage drop and levels off at a reduced voltage. It generates a pulse of current in the process that can turn on an SCR.

diac—A pair of trigger diodes that are connected back-to-back, and that have the same kind of triggering properties with either polarity of voltage.

triac—This device is the equivalent of two separate SCR elements connected in such a way that one is triggered with a positive voltage, and the other with a negative voltage.

silicon-controlled switch—This is a special SCR device having two gate leads, one for an "anode" gate, the other for the usual "cathode" gate. It can be used to operate high-voltage low-current lamps like neon lamps. It also has many other uses.

stabistor—A two-lead, or diode-like, device which stabilizes either a voltage or a current of a specified polarity regardless of the voltage applied to it. It must have a voltage applied that exceeds some predetermined minimum, and it must be operated within its defined dissipation rating.

positive rail—The most positive voltage applied to a circuit is often defined as the "positive rail." Many IC units are described as capable of being operated to within one diode drop either below or above the positive rail.

negative rail—The most negative voltage applied to a circuit is often defined as the negative rail. Many IC units are described as capable of being operated to within one diode drop either above or below the negative rail.

UNIJUNCTION TRANSISTORS

A unijunction transistor is a unique form of switching device; one which, when a certain value of voltage is applied to its control terminal, will initiate a strong conduction to one of its base terminals. The control terminal is called its emitter. The base terminals are at opposite ends of a strip or a ribbon of a semiconductor chip. This chip is usually a silicon material. Fig. 6-1 gives a diagram of a unijunction transistor.

An emitter "diode" is placed adjacent to one of the base terminals, which is called "base 2." The n-type silicon material forming the

Fig. 6-1. Diagram of a unijunction transistor.

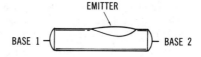

bar or chip itself is relatively lightly doped, making it weakly conducting. However, the emitter is more heavily doped so that it can induce a large number of carriers into the bar when it is properly biased. Normally, base 2 is biased positively with respect to base 1. As long as the voltage applied to the emitter is less than approximately half of the base 2 to base 1 voltage, no conduction can occur from the emitter to base 1. As the bias on the emitter is made more positive, the diode junction becomes forward biased, and the emitter injects large numbers of positive charges into the channel. These charges are then swept to base 1 where they are collected. The only function performed by base 2 is to provide a means to keep the diode nonconducting until it is forward biased by a desired amount.

The unijunction transistor is excellent for generating delayed switching action. It can be used for initiating sweep action, and is commonly used to initiate or trigger silicon-controlled rectifiers. Since the discharge current flows almost exclusively from the emitter to base 1, the output signal is taken from base 1 by placing a small resistor in series with this lead. The duration of this pulse is a function of both the diode resistance to base 1 and the base 1 resistance to ground. Typically, 27 ohms are used for this. Fig. 6-2 shows a typical circuit.

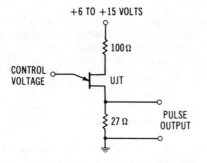

Fig. 6-2. A typical unijunction transistor circuit.

Unijunction transistors, in their simpler circuits involving only one device, are convenient for initiating periodic sweeps for electrical devices such as the windshield wiper on an automobile. The unijunction transistor, in this case, triggers a silicon-controlled rectifier, which is connected directly across the wiper switch. The pulse generated across the resistor in the base 1 circuit is sufficient to trigger the SCR and turn the wiper on. This initiates the wiper, and the internal circuit within it then shorts out the SCR, allowing it to stop conduction. The wiper completes its cycle and awaits the next trigger pulse. When the wiper does turn off, the capacitor, from the emitter to ground on the UJT, is again allowed to recharge, and after the appropriate delay, it triggers the UJT for the next cycle.

The unijunction transistor is particularly convenient for the generation of a series of evenly spaced pulses, from repetition rates that are just less than a few hundred per second to as slow as one pulse every 30 seconds or more. By the use of the circuit given in Fig. 6-3, much longer periods can be generated. Low-leakage devices and components are critical with long time delays. Any low-repetition-rate

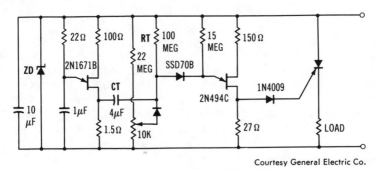

Courtesy General Electric Co.

Fig. 6-3. A trigger circuit for long time delays.

function which must be controlled by a series of pulses can be initiated with the aid of one of these oscillators.

A circuit that is based on the use of the unijunction transistor for controlling the timing of an enlarger is shown in Fig. 6-4. This circuit uses two relays, but perhaps you can work out a way to make it

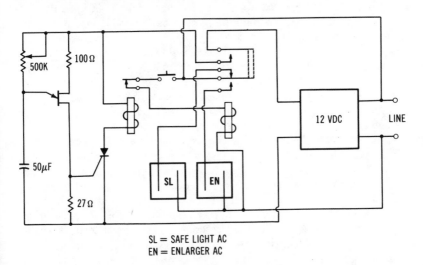

SL = SAFE LIGHT AC
EN = ENLARGER AC

Fig. 6-4. A photo-printer timing circuit.

exclusively solid-state. At least, you know how to determine the correct parts to use for most of the circuit. As you can see, a push-button switch trips the ac relay and applies dc to the unijunction circuit. When the voltage in the capacitor builds up to the trigger point, the UJT fires and causes the SCR (or a power transistor) to trip the power relay back off. The power relay locks itself on until that pulse is generated.

SILICON-CONTROLLED RECTIFIERS

The silicon-controlled rectifier and its near-relative, the triac, are the solid-state cousins of the thyratron. These devices are able to start conduction on command, but they are unable to interrupt it without help. There are various methods of interrupting the flow of current, with the simplest being the application of ac to the device. Most of the methods, in effect, reduce the voltage from the anode to the cathode to zero, in some way, to extinguish the current flow. The windshield-wiper control unit does this by the "return-to-rest" shorting switch feature that is built into the wiper assembly. The relay does it in the enlarger-timer circuit by cutting off the current supply.

The SCR is turned on by placing a forward voltage on its gate. When this voltage is sufficient to cause the initiation of an avalanche in the device, and the load can draw enough current to assure a continuation of the avalanche, the rectifier starts conducting. It is easy to vary the point of initiation when ac is applied by varying the phase of the voltage applied to the gate circuit. Frequently, it is also convenient to use some kind of a "breakdown" device to sharpen the initiation and make it more reliable. For this purpose, a 1/10-watt neon bulb (NE-10) may be used, as between 75 and 125 volts, a typical neon bulb will start conducting and the voltage drop across it will provide the required initiation pulse of 20 or more volts. Trigger diodes are also used for this purpose. The rectified ac power supply that you are using to plot transistor contours will be fine for demonstrating the properties of these devices.

Silicon-controlled rectifiers are often called "four-layer" diodes. They switch much like a flip-flop does, except that when they conduct, they can pass very large currents, depending on their ratings. It is not possible to simulate these devices with transistors, because the flow of carriers within the four-layer diode is altered to create the avalanche and latching that is responsible for conduction. As noted previously, the anode-to-cathode voltage must go to zero and the device current must drop below an initiation level for the device to turn off. These characteristics will be examined in one of the experiments.

SILICON-CONTROLLED SWITCHES

The silicon-controlled switch is similar in structure to the silicon-controlled rectifier, but it has leads on all four of its layers instead of only three of them. As with the SCR, the positive supply is connected to the emitter of an internal pnp transistor array with the emitter being called the anode of the device. The connection to the base of the pnp device, which also connects to the collector of the internal npn device, is called the anode gate. The remaining gate, connected to the base of the npn transistor and to the collector of the pnp transistor, is called the cathode gate. The emitter of the npn transistor is the cathode of the structure.

The load of the silicon-controlled switch is sometimes connected to the anode gate rather than to the anode, especially when it is used to control high-voltage lamps. In other applications, the anode is used as the output. Typical representation diagrams for an SCR and an SCS are shown in Fig. 6-5.

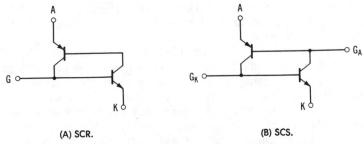

(A) SCR.   (B) SCS.

Fig. 6-5. Schematic representation of an SCR and an SCS device.

TRIACS

The triac is a dual silicon-controlled rectifier, in that it will pass current in both directions in a more-or-less symmetric pattern. Triacs are typically used to control loads which include more reactance than is safe to handle with SCR devices. They are particularly required with circuits involving transformers, motors, or other magnetic devices. SCR devices will saturate the core of a reactor, a motor, or a transformer. This leads to a very high flow of direct current. The triac, like the SCR, turns on as a result of a command from a gate signal and then turns off again when the source voltage passes through zero, but it does this every half cycle, not on alternate half cycles. It behaves like two separate SCRs, one that is triggered "on" when the anode and the gate are both positive, and the other that is triggered "on" when they are both negative.

The shortened conduction cycle that is observed with both SCRs and triacs leads to a reduced dumping of energy into the load—lamp bank, power supply, or whatever. This minimizes the amount of energy which must thereby be dissipated in a resistance. The use of a properly designed phase-control network will lead to a smooth variation of the energy delivered to the load. This control of the input energy results in minimum waste energy, although the switching process tends to generate substantial amounts of rf electrical noise.

The triac has three leads just as has the SCR; its single gate lead gives a fairly balanced control. However, if a balance of the current in the two half-cycles is critical, as it may be with a transformer or motor load, separate SCR units having matched control circuits, or possibly some other special devices, may be desirable. Information on solutions to these kinds of problems may be obtained from device manufacturers.

TRIGGER DIODES

The trigger diode that is typically used with gate circuits of SCRs and triacs to sharpen up the triggering pulse and produce a more reliable operation is somewhat similar to a zener diode, except for the fact that the voltage across the diode drops by several volts when the device is triggered on. The result is a sharp rise in gate voltage on the SCR, and a sharply enhanced ability to trigger it. These diodes are used in the reverse direction as are zener diodes, since in the forward direction they test much like ordinary diodes. It is important for you to obtain a few of these devices and to test them, as their behavior is rather unique and you will want to be able to recognize them by their operating characteristics. An experiment for testing trigger diodes is included later in this chapter.

You will find that trigger diodes behave as if they were low-voltage neon bulbs, in that the diode starts conducting at a voltage somewhat above that which it is able to sustain. They differ in that they only work with one polarity of voltage, however. The drop in voltage on initiation is from 5 to 10 volts, giving a sharp voltage transient.

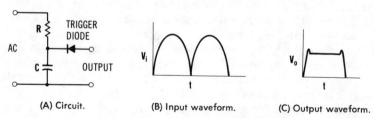

| (A) Circuit. | (B) Input waveform. | (C) Output waveform. |

Fig. 6-6. Trigger diode circuit and waveforms.

A sample circuit for their use is shown in Fig. 6-6. The expected input and output waveforms are also shown.

DIACS

When triggering on both the positive and the negative polarity of the alternating current is required, as in the case of the triac, a bi-directional trigger diode, or "diac" is usually used in place of the conventional trigger diode. This device acts as if it were two trigger diodes connected in series opposing, with the result that it will generate essentially the same waveform regardless of the polarity of the applied voltage. You can distinguish a diac from a trigger diode by testing it in both directions. With the diac, the same kind of triggering action will occur regardless of the polarity of the applied voltage. These devices are always used with triacs, but can be used with SCRs as well, since normally it is unimportant that the gate can swing negative with respect to the cathode of the SCR.

STABISTORS

A stabistor is a device intended to stabilize the operation of a circuit in such a way that will render it insensitive to the effects of some other parameter, such as temperature. Some of these devices are used to stabilize either a base-to-emitter static voltage or a gate-to-source static voltage, thereby stabilizing the level of static current through the device. They are commonly used with power transistor circuits, which must be able to draw variable amounts of base and collector currents, but must maintain a stabilized quiescent point. (They can perform a similar function for FET power amplifiers.) Stabistors can be helpful in cancelling out common-mode signals. Zener diodes are a special kind of stabistor that stabilizes voltage in critical circuits. The feature which makes them useful for this purpose is the exponential relation between voltage and current.

Most of the stabistors that the author has been able to buy for testing (in connection with this book) have proven to be voltage stabistors. The voltage drop across them appears to be about one silicon-diode drop, and the voltage is extremely constant with the variation of current through them. There is an experiment for testing them later. If you operate stabistors from the swept-collector supply, with a series resistance, you will find that the output voltage forms a much more perfect square wave than you can get with a conventional diode.

Current stabistors can be made from either bipolar transistors or FET devices, and they use either emitter or source degeneration to control the current. The controlled current is extracted from the collector or the drain. Most of these circuits work by limiting the voltage across a linear resistor. One of the best current stabistors the author has seen uses a voltage regulator to control the current by placing the metering resistance between the output and the reference point. Since the device is supposed to provide a precisely stabilized voltage between these points, the result is a precisely stabilized current. Several circuits for current stabilization are shown in Fig. 6-7. You will find it interesting to test both voltage and current stabistors, as they can be extremely useful.

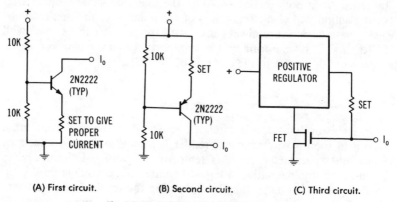

| (A) First circuit. | (B) Second circuit. | (C) Third circuit. |

Fig. 6-7. Typical current stabilizing circuits.

Current stabistors can be used with many circuits to lock the quiescent-point current in the output circuit to a previously selected value. It is possible to use FETs to stabilize the base current of a bipolar transistor so that much lower values of base-supply voltage may be required to assure proper operation of the device over a range of temperatures. The result is that both the base-supply voltage and the collector-supply voltage can be as low as 3 to 5 volts, leading to a reduced power requirement and better overall operation.

Where protection against current surges is important in circuits, it is often convenient to introduce a metal-film resistor (or, in some instances, wirewound resistances) between the collector or the drain return and the voltage source. This resistance should be made large enough to limit the current to a value that will assure that the maximum device-dissipation rating will not be exceeded. Typically, the dissipation in a device, whose current level increases with temperature, can be shown to be maximum when half of the circuit voltage is applied across it. This, of course, assumes that the resistance of the limiting resistor does not change in the process.

PROGRAMMABLE UNIJUNCTION TRANSISTORS

The programmable unijunction transistor is not really a unijunction transistor at all, but is a special form of silicon-controlled rectifier used with appropriate internal and external resistances. With this device, the terminal which would be used as the anode for an SCR is used as a gate, and the adjacent n-type region is used as "base 2." The voltage set on base 2 determines the trigger point, and the pnpn combination assures regenerative switching, which is considerably more effective than that in the ordinary unijunction transistor.

In a sense, one can say that the PUT, or programmable unijunction transistor, is almost an inverse of an SCR. The pnp transistor is controlled instead of the npn, and without the additional resistors, it would turn on with the kind of voltage conditions that are just the reverse to those for a normal SCR. (Probably the second structure in a triac is, in fact, similar to the active element of a PUT.) These units can be used to trigger SCRs in a way similar to that for the UJT. More data on these interesting devices can be obtained from their manufacturers.

VOLTAGE REFERENCES

One occasionally needs a voltage reference which possibly need not be precise, but which should have fairly good regulation. Often one would prefer not to use a zener diode directly, as the current capacity of the diode may not be high enough. In addition, the voltage stability may not be adequate at the required current level, as the stability of these devices can be very current-sensitive. Depending on the current level required and the stability required, there are a number of options available, ranging from a simple zener by itself through high-precision regulated supplies. Several simple but useful circuits will now be described which can satisfy this kind of a requirement.

The first step, in complexity, is the use of a zener diode or a special reference device like the Intersil ICL 8069-series of devices. Where best use of the precision properties of a device like the ICL 8069 is important, these devices are used after a semiprecision reference like a 6-volt or a 10-volt zener diode. This arrangement is fine as long as the circuit is only required to provide a voltage, with a negligible current load. Where a significant range of current is required, the next step is to use the reference with a transistor that is connected as an emitter follower. (Why do we NOT recommend the use of a FET as a source follower here?)

The low transconductance efficiency of the FET when compared to the bipolar transistor is, of course, the reason.

Using the above combination, currents that are from 50 to as much as a 1000 times greater than the possible range that a zener diode can control are available with a surprisingly low source impedance. The penalty which must be paid for this is the base-to-emitter voltage difference that is required to place the regulating transistor in an operating condition (Fig. 6-8). The simple emitter follower may be replaced with a Darlington pair, thus making large currents available from this arrangement, yet with extremely small changes in the output voltage.

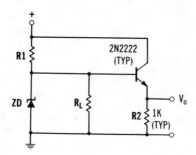

Fig. 6-8. Zener diode augmented cathode-follower circuit.

The change in output voltage for a 2-to-1 change in output current can be as small as 0.018 volt or less with an npn transistor, and may be somewhat more with a pnp transistor. In any case, this represents a very low output impedance for a supply, yet it is one having a wide range of current capabilities.

A more sophisticated regulated power supply would use an amplifier of some kind which could compare the output with the reference and, then, adjust the output voltage to compensate for changes. Depending on the precision which is required, these configurations are used in many applications. Where it is possible, it is better to use circuits which do not require critical control of the supply voltage, as the more complex the circuit, the more likelihood there is of a failure.

Power supplies which are required to provide a greater output voltage precision normally use an amplifier and a comparator, but they also use a chopper stabilization system which turns the difference voltage into an ac signal which is then phase detected. This combination can yield stabilization to parts-per-million if a sufficiently precise reference source is used.

The Intersil ICL 8069 low-voltage reference is typical of the devices which are available for either calibration or control. This particular device, with some 500 to 800 microamperes through it, will deliver a voltage in the range of 1.20 to 1.25 volts with a temperature stability that is between 0.001% per degree (Celsius) and 0.01% per degree, depending on the specific device selected. It has a low inherent noise level of about 5 microvolts over the audio spectrum, and a typical dynamic resistance of 1 ohm. These devices can be used either for low-voltage reference (less than 1.20 volts), or they can be used to control a regulated supply (using a circuit of the sort shown in Fig. 6-9). This circuit, based on one suggested by Intersil, can be used to produce a regulated voltage in excess of 5 volts at a current of 1 ampere or more.

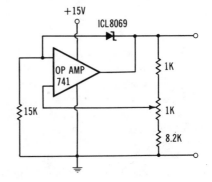

Fig. 6-9. Regulated voltage source using the ICL 8069.

By nature, most solid-state devices tend to drift and have an excess of low-frequency noise. For these reasons, the most precise regulators, as noted previously, usually include some form of chopper stabilization. The development of field-effect transistors has made available switches that have extremely small "off-set" voltages. The result is that they can be used to form solid-state choppers. The main reason chopper stabilization is required is that most other circuits do *not* introduce *exact* zeroing-out of the error voltage, but take a difference which is actually a small difference of two large values. The result is that drift and low-frequency noise enter into the balance process and make the circuit less stable than is desired. If a circuit is set up which will actually *null out* the net error by the use of a small motor-driven potentiometer or some similar device, then some of these problems can at least be substantially reduced, by an order of a magnitude or more. Such a configuration has another advantage in that its memory is not volatile. A momentary interruption of the circuit will not necessarily require that the unit start all over again in achieving balance, and correct operation can be accomplished much faster. This technique can be used with most kinds of regulators, saturable reactors, series-pass devices, chopper supplies, etc.

EXPERIMENT 1
Characteristics of a Unijunction Transistor

Your first experiment in this chapter is to examine and verify the operating characteristics of a unijunction transistor. In doing this, you will observe the currents in both base 1 and base 2 as the UJT operates, so that you can see where the current in base 1 comes from. You will measure the ratio of the voltage on the emitter to the voltage on base 2 at the point of initiation for a series of base 2 voltages. Then, you will set up a simple oscillator, and examine the waveforms on both the bases, and also on the emitter, when it is generating saw-tooth oscillations.

Step 1.

Wire the circuit for a unijunction transistor on your solderless breadboard in accordance with the diagram in Fig. 6-10. Arrange to apply the voltage from your variable-voltage (positive) supply to base 2, and connect a potentiometer across it with the variable arm (the swinger) connected to the emitter of the UJT. Starting at 5 volts on base 2, measure the voltage required at the emitter to initiate your device. (Some devices may not initiate at this low a base 2 voltage, but most of them should.) The required voltage on the emitter should be between 2 and 3 volts.

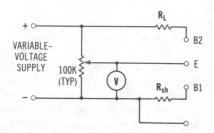

Fig. 6-10. Unijunction transistor test circuit.

Step 2.

After you have gotten your UJT to function, repeat the test for a series of voltages; for example, 7.5 volts, and 10, 12.5, 15 volts, and up to the limit of the maximum rated voltage for your device. Record your data in Table 6-1, and plot a curve of the ratio of the trigger voltage to the supply voltage as a function of the base 2 supply voltage, where V_{bb} is the interbase voltage, V_{eb} is the emitter-to-base 1 voltage, and V_R is the voltage ratio. Plot your curve on the graph given in Fig. 6-11.

Table 6-1. Data for Experiment 1

V_{bb}				
V_{eb}				
V_R				
V_{bb}				
V_{eb}				
V_R				
V_{bb}				
V_{eb}				
V_R				

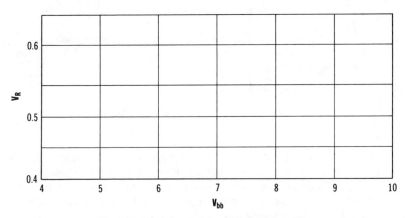

Fig. 6-11. Graph for Experiment 1, Steps 2 and 3.

Step 3.

Repeat Steps 1 and 2 with additional UJT devices. Record your data in Table 6-1 and 6-2. Plot the values of the trigger ratio against voltage for these new devices also on the graph given in Fig. 6-11.

Table 6-2. More Data for Experiment 1

V_{bb}				
V_{eb}				
V_R				
V_{bb}				
V_{eb}				
V_R				
V_{bb}				
V_{eb}				
V_R				
V_{bb}				
V_{eb}				
V_R				

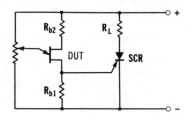

Fig. 6-12. A UJT test circuit.

Step 4.

Next, wire the UJT so that the voltage generated across R_{b1} is coupled into the gate of a silicon-controlled rectifier, as shown in Fig. 6-12. Observe whether you can trigger the SCR, and find the range of voltages over which you can do so. Explain what you have learned from this experiment.

EXPERIMENT 2
The Unijunction Transistor as a Sawtooth Generator

In this experiment you will learn more about the behavior of the unijunction transistor as a sawtooth waveform generator. You had the equivalent of such a generator in Experiment 1, but you did not operate it in that manner. To make it generate a sawtooth waveform, you need a fixed or variable resistance from the emitter to the base 2 voltage source (or perhaps a higher voltage if the linearity is critical), and a capacitor from the emitter to the base 1 return. (The capacitor must discharge through the resistor in series with base 1.) The basic circuit configuration is shown in Fig. 6-13. You will make several variants of this configuration after you have examined the properties of the basic circuit.

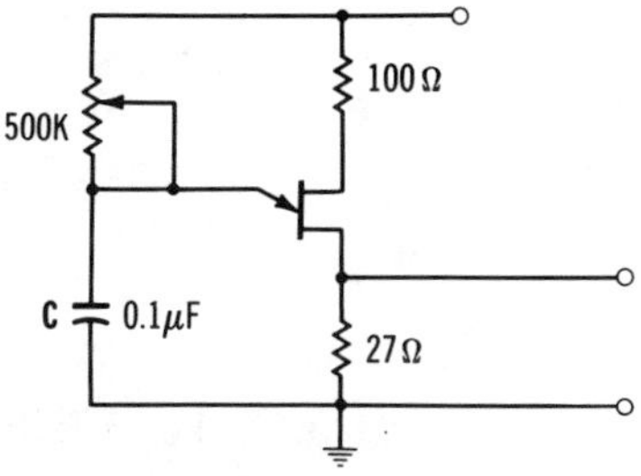

Fig. 6-13. A basic UJT sweep circuit.

Step 1.

As noted earlier, the voltage that must be applied to the emitter of the UJT is approximately one-half the voltage applied to base 2 at the time of initiation. If you use your scope to observe the emitter waveform, what is the maximum resistance you can use between the emitter and the base 2 supply if you want to be sure that the circuit will function properly? Why?

It is necessary that the voltage at the emitter be able to rise above the critical voltage in a finite length of time if triggering is to be reliable. This means that the resistance from the emitter to base 2 must be small enough so that the loading of the scope does not prevent the voltage from rising to the required level. Based on this, one would select the maximum value of the variable resistance from the emitter to base 2 as about one-half the effective input resistance of the scope. Since the leakage in the UJT and the capacitance may also have an effect, the maximum resistance is probably about 0.25 megohm, unless the components used have been selected extremely carefully.

Step 2.

Wire the circuit on your solderless breadboard, and vary the resistance in the circuit to see how the time duration of the sawtooth varies. You are using a 0.1 μF capacitor for the present, so the repetition rate of the sweep is fast enough that you can tell what is happening. If the circuit stops oscillating as you increase the charging resistance, measure the value of the resistance from the emitter to the base 2 supply (disconnect it so the device and the capacitor do not affect the reading), and see how the value compares with the input resistance of the scope. Can leakage from the emitter to base 1 or to base 2 affect the value of the resistance? Try it and find out! Explain what you have learned.

Leakage to base 1 or in the capacitor will affect the operation of the circuit. The presence of leakage to base 2 will increase the effective magnitude of the charging current, and will cause the circuit to trigger faster. Leakage in either the capacitor or to base 1 will retard the charging, and may keep the voltage from rising high enough to trigger the device. Loading from the scope can have the same effect.

Step 3.

Based on the resistance and the capacitance that you are using for your charge/discharge circuit, determine the correct value of parameter A in the following equation:

$$f = A(RC)^{-1} \qquad \text{(Eq. 6-1)}$$

Select different values of capacitance and resistance, and determine whether this equation holds over a wide range of resistances and capacitances. Can you use this method for estimating the shunt leakage resistance that your circuit is loading onto the capacitor? Explain.

This equation should apply if the proper corrections are made for circuit resistances and component tolerances.

Step 4.

So far, you have assumed that the sawtooth waveform you are generating with your UJT is a linear sawtooth. Now, you need to see if it really is, and to determine what you might do to improve its linearity if it is not. Examine your waveform as carefully as you can with different numbers of cycles across the scope face. (You first have to find out about the linearity of the sweep in your scope!) Synchronize your scope with the UJT oscillator with about ten cycles of the UJT waveform visible across the scope display. Then, with a pair of dividers or calipers, measure the width of successive cycles. They should all be equal to the best of your ability to measure them. Are they? If not, see if your scope circuit uses a FET to sense the voltage across the capacitor that is charged and discharged during the sweep cycle. (Either a FET or a vacuum tube can be used as a voltage follower to minimize the loading on the capacitor.) The other principal cause of this nonlinearity is variation in charging current. When you charge a capacitor through a resistor, the voltage across the resistor decreases as the voltage across the capacitor increases, and the charging current of necessity must decrease. To correct this requires that the charging current be made more constant. Either the charging voltage can be increased substantially, or a so-called constant-current circuit may be used as a source. (This is one of the important applications of stabilized current sources.) Using the circuit shown in Fig. 6-14B, a pnp transistor might be used to generate the required constant current. Either resistor R2 or R4 may be varied to generate the variable-current source. Of these, possibly R2 might give the more acceptable operation. If your sweep is not adequately linear, you may want to redesign it. What did you find?

You will probably find that variation of the emitter resistance is a more effective way of controlling the sweep rate than changing the base voltage. The total amount of degeneration remains essentially constant as long as the voltage rise to the emitter is constant. The result is that better control of the charging current will be available (Fig. 6-14). This figure also shows the use of a source-follower based on an IGFET whose function is to eliminate curvature due to loading introduced by the output amplifier circuitry.

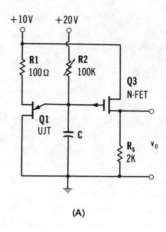

(A)

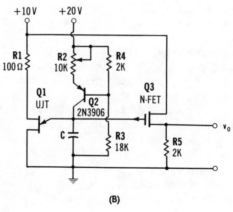

(B)

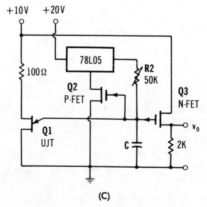

(C)

Fig. 6-14. Linearized unijunction sawtooth generator circuits.

Step 5.

The basic technique for turning a variable resistance circuit into a variable current circuit, in which the current remains constant with time for any setting of the value of resistance, has been described briefly above. A field-effect transistor can be used as well as a bipolar transistor or an electron tube for this function. Any of these devices may be used in the position of transistor Q2 with satisfactory results, but the bipolar transistor will require a smaller overall voltage for a given level of stabilization. It will be useful for you to compare the relative effectiveness of varying R2 and R4 as a means for adjusting the current level used to charge the capacitor. We suggest that you arrange the circuit shown in Fig. 6-14B so that you can vary the charging current from 100 microamperes to 10 milliamperes, with 1 volt across the emitter resistance. Examine the linearity of the waveform by using this signal for your scope horizontal input, and introduce either a sine wave or other waveform from your audio wave generator. This signal should have a frequency approximately ten times the sweep rate so that you can use the cycles as a measure of linearity. Record your data here in Table 6-3, and discuss your

Table 6-3. Data for Experiment 2, Step 5

R2				
L_{Mx}				
L_{Mn}				
R2				
L_{Mx}				
L_{Mn}				
R2				
L_{Mx}				
L_{Mn}				

results afterward. The variation of sweep linearity is the important issue here, and the cycle lengths of the timing waves (the longest and the shortest) should be recorded, with L_{Mx} being the length of the longest cycle of the timing wave, and L_{Mn} being the length of the shortest. The next set of data, Table 6-4, applies to the configuration of Fig. 6-14B with the value of R4 changed. Resistor R2 should have a value of 500 ohms for this test. (V_{e2} is the voltage across the

Table 6-4. More Data for Step 5

R4				
V_{e2}				
L_{Mx}				
L_{Mn}				
R4				
V_{e2}				
L_{Mx}				
L_{Mn}				
R4				
V_{e2}				
L_{Mx}				
L_{Mn}				

resistance R2.) If you wish to measure more than the maximum and minimum cycle lengths, it will enable you to plot a calibration of the linearity of these sweep generator circuits. Discuss what you have learned.

Since charging current through a resistance has an exponential "decay" either on the build-up or the discharge, a good current stabilizer is very helpful. Of the two you have tested, we suspect that you have found that the use of resistor R2 as the variable has proven to be superior to the other.

Step 6.

Now, without increasing the base 2 voltage, replace the resistance in the emitter circuit, and increase the supply voltage to it to twice the voltage on base 2. Examine the effect of that change on the sweep linearity and on the maximum value of series resistance that you can use in the circuit. Compare that change to the value with the emitter supply voltage the same as the base 2 supply voltage. What effect does this have on linearity?

Use of the higher voltage means that the UJT will trigger before the voltage across the emitter-return resistor has decreased enough to allow any significant amount of nonlinearity to develop. You will also find that you can use a higher value of emitter-return resistance because the higher return voltage reduces the effectiveness of any shunt leak in either the device or the capacitor.

EXPERIMENT 3
More on the Unijunction Transistor as a Sawtooth Generator

In this experiment you will find it useful to repeat the tests you made in Experiments 1 and 2 on several other UJT devices to see how they compare to the device you have already tested. Testing of an assortment of devices having different EIA codes is suggested, as you may find significant differences among them. After this has been accomplished, it may be interesting to make some additional tests which will show how you can reliably get long time delays. Perhaps you will then understand better how to test devices for use in long-delay applications as well.

Step 1.

First, repeat your tests in Experiment 1 but with some additional features. This time, as you increase the voltage on the emitter, record the current flow in the emitter both before and after triggering. Use the circuit for Experiment 1, with one addition. Place the current-metering element between the potentiometer and the emitter. Then, take your data again, including the resulting current readings in your data. Record the data in Table 6-5. Plot a graph of V_R against V_{b2} in Fig. 6-15.

Step 2.

Plot the data on all the devices you have tested, and examine the relation between the emitter leakage current and the maximum charging resistance that you can use with a device as an oscillator. Based on the emitter leakage current and the emiter-to-base 1 return voltage, find out how this resistance compares with the maximum charging resistance. What is the maximum voltage to which the capacitor is charged? Plot the charging curve you would expect in

Table 6-5. Data for Experiment 3, Step 1

V_{b2}				
V_e				
I_e				
V_R				
V_{b2}				
V_e				
I_e				
V_R				
V_{b2}				
V_e				
I_e				
V_R				
V_{b2}				
V_e				
I_e				
V_R				

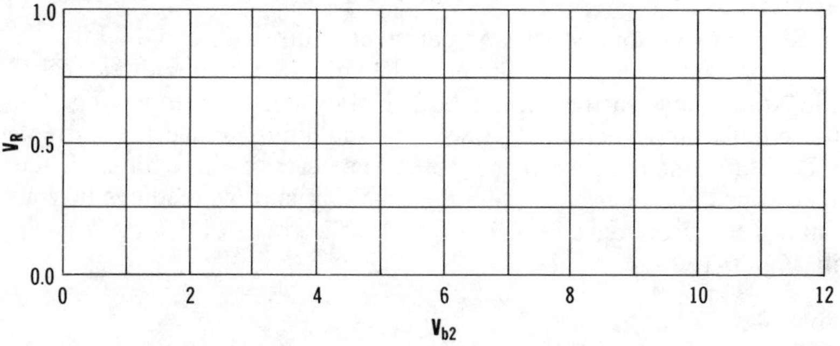

Fig. 6-15. Graph of V_R against V_{b2}.

terms of the ratio of V_e to V_{b2}? Is the limiting voltage greater than the trigger voltage? (Hint: In the absence of leakage, the charging curve will take the form suggested by Equation 6-2, and with leakage, V_{b2} may be replaced by the approximate value suggested by Equation 6-3. This voltage must exceed the value of V_e for triggering to occur. The value of r_e can be approximated from the value of V_e at just below breakdown along with the value of I_e at the same voltage.)

$$V = V_{b2} [1 - \exp (-t/RC)] \qquad \text{(Eq. 6-2)}$$
$$V_{b2} (1 + R_e/r_e)^{-1} \qquad \text{(Eq. 6-3)}$$

Plot the charging curve you would expect in terms of the ratio of the emitter-to-base 1 voltage to the base 2-to-base 1 voltage. Is the limiting voltage greater than the trigger voltage? Include further data as needed in Table 6-5 (Step 1), and see if your data check with experience on the devices.

You probably found that some of your UJT units drew more emitter current than others, and those devices could not tolerate as high a charging resistance as those that were drawing smaller currents. Based on this, which do you think would make superior long-time-delay generators? How could you improve their linearities and the overall time delay if you needed to?

Increasing the emitter supply voltage would accomplish both of these purposes. A higher-value charging resistance can then also be used.

Step 3.

Set up your circuit as it was for Experiment 2, and repeat your tests of these devices as sawtooth generators. Try each device one by one and, then, repeat the tests using a larger storage capacitor (from three to ten times the size previously used). Each time you test a device, be sure you find the maximum value of resistance that will permit an emitter-to-base 2 voltage to cause a generation of sawtooth waveforms. You should either trigger your scope off your base 1 return resistor or use an IGFET follower to give you access to the actual waveform without creating an adverse loading situation. Also, test to see if the point of scope connection affects the maximum value of resistance that can be used by shifting the scope take-off point from one position to another. Record your data in Table 6-6, giving the device identification, the maximum resistance, the capacitance value, the leakage current in the device and in the capacitor, and any other data you may consider essential. Then discuss your results.

Step 4.

Compare the values of emitter currents just before initiation (from Steps 1 and 3) with the maximum value of series resistance that you can use in the charging circuit with any given capacitor. Do you find any connection? Do the characteristics of the capacitor you are using seem to have any effect other than on the sweep-rate change (which would be expected)? For example, does a change of capacitor have any effect on maximum resistance? Explain your observations.

You probably found that the higher the total (emitter plus capacitor) leakage current, the lower the limiting value of the charging resistance. You will also find that this shows up by the fact that the emitter voltage does NOT reach trigger value with high leakage and high resistance.

Step 5.

With a large capacitance, and with the charging resistance near the maximum at which you can get sawtooth generation, measure the period with a stopwatch. Do this six or eight times for each combination of UJT and resistance value. Record your times for the tests with each device, and explain what you have observed. Also, determine the values for parameter A as shown in Equation 6-1 for each case.

You probably found that there is considerable variability in the time required for triggering. This will apply both for an individual device

Table 6-6. Data for Experiment 3, Step 3

Device				
R_{max}				
V_{b2}				
V_e				
I_e				
I_{cap}				
C				
Other data				
Device				
R_{max}				
V_{b2}				
V_e				
I_e				
I_{cap}				
C				
Other data				
Device				
R_{max}				
V_{b2}				
V_e				
I_e				
I_{cap}				
C				
Other data				

and, also, between different devices. If you insert either a normally closed push-button switch in the path of the charging resistance, or a normally open push-button switch in parallel with the capacitor, you can establish a better reference "zero." Also, the time will be somewhat more constant, but it will be somewhat variable nonetheless. You will find that the time interval is longer, as the charging has to start from zero voltage instead of from a partially discharged condition. How much does this actually increase the time interval?

You will get the longest interval by starting from zero voltage on the capacitor. The time delay will also be more stable.

Step 6.

Next, modify your slow-discharge circuit as shown in Fig. 6-3. The additional UJT is used to pulse the base 2 voltage in a negative direction in order to increase momentarily the ratio of the emitter-to-base 2 voltage. Again measure the delay time with a stopwatch. Also, you will want to increase the series resistance to the emitter after you have tested the circuit for the maximum resistance that you could use for a normal free-running operation. Is the triggering more reliable with the pulses applied? Explain what you have found.

The pulsing of base 2, as noted previously, causes the voltage ratio at the emitter to be increased, leading to triggering as soon as the emitter voltage has reached a point where triggering is possible. The initial effect of this is to shorten the timing cycle. It also makes possible an increase of the charging resistance, however, and this makes it possible to get a net increase in measurement time. When the circuit is used with carefully selected components, extremely long time delays can be generated.

Step 7.

Could this configuration be used to make an extremely long-time triggered sweep for an oscilloscope? Discuss what you might do with such a configuration.

A circuit of this kind could indeed be used for a triggered slow sweep. The voltage for the emitter would be run through a constant-current circuit which is clamped by a limiting diode to keep the emitter voltage less than the trigger level. The synchronizing pulse would be applied via the resistor that is in series with base 2 by a transistor which would be pulsed by the trigger signal. This would cause the UJT to fire and start to recharge linearly. A possible circuit for such a device is shown in Fig. 6-16.

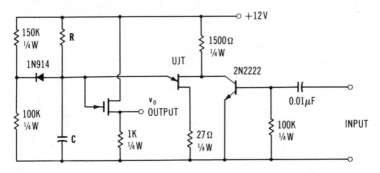

NOTE: R,C SET THE SWEEP RATE

Fig. 6-16. Triggered sweep using a unijunction transistor.

EXPERIMENT 4
The Silicon-Controlled Rectifier

As its name implies, the silicon-controlled rectifier is a form of unilateral, or one-way, conducting device which can be programmed to turn on after a selected set of conditions have occurred. It normally can only be turned off by reducing the voltage from the anode to the cathode to zero by reducing the anode current below a defined level. In this experiment, you will test samples of this device on both direct current and alternating current so that you can see just how they behave and how they can be controlled. In later experiments, you will assemble practical circuit configurations which can be useful either in your car or in your home.

Step 1.

Wire the test circuit shown in Fig. 6-17 on your solderless breadboard using a low-current SCR. This arrangement will enable you to both vary the voltage on the gate of your device, and to observe when a significant current starts to flow through it. Measure the initiation voltage for several values of anode voltage, and record your data in Table 6-7. Also, arrange the circuit so that you can limit the anode current by varying the load resistance in the anode circuit. Start with a resistance which will allow somewhat more than the minimum conduction current to flow and, then, successively

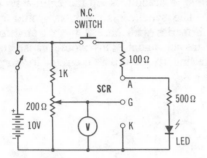

N.C.
SWITCH

100 Ω

1K

SCR A

200 Ω

G

500 Ω

10V

V

K

LED

Fig. 6-17. Test circuit for a silicon-controlled rectifier.

double the resistance until you can no longer keep the SCR conducting when you increase the gate voltage. Then, try halving it from the original value. Vary the anode supply voltage from about 3 volts to 10 volts, or half the rated anode voltage, whichever is greater. Enter your data in the table and describe the effects that varying the voltages has on the triggering characteristics of the device. (Remember to push the closed-circuit push button after each test to restore the SCR to the off condition.)

Table 6-7. Data for Experiment 4, Step 1

V_{aa}				
V_g				
R_L				
I_L				
V_{aa}				
V_g				
R_L				
I_L				
V_{aa}				
V_g				
R_L				
I_L				

The SCR will only conduct if the load current (which can flow) exceeds some minimum value (which depends on the current avalanching in the device). The SCR requires a forward voltage on its gate of a specified value to create the avalanching effect. Internally, it acts as a "switch and latch" which requires special treatment to cause a reversion to cutoff.

Step 2.

Take a device which conducts easily and increase the value of the load resistance to see if you can make it turn off. Did it turn off? Why do you think this happens?

Yes, the unit drops out of saturation, and it cannot maintain a conduction condition. These devices are sometimes said to be the equivalent of an npn and a pnp transistor connected together with the emitter of the npn unit equivalent to its cathode, and the base of the npn unit connected to the collector of the pnp unit. The collector of the npn unit is connected to the base of the pnp unit. The emitter of the pnp unit then becomes the anode of the combination. Can you take two transistors, and connect them in this manner to get the equivalent of an SCR?

It would appear that you should be able to do this, but you cannot. You cannot get the right kind of avalanche conduction unless the devices are integral on a single chip. However, try it for yourself. The author once tried this because he did not have a low-power SCR. It did not work for him either. Current diversion in the chip is apparently the critical requirement that you cannot get.

Step 3.

Now that you have seen the typical behavior of a silicon-controlled rectifier using direct current, it is necessary to try one of them using alternating current. If you have an SCR rated to about 400 volts, you can make this test on your 60-Hz power line using a light bulb. Be sure that the current rating of the SCR is about five or more times the nominal full-load peak current through the bulb, however, as the filament resistance of a light bulb when cold may be only a fraction of the value at full power. (A device with a peak-current rating of twenty times would not hurt if its latching current is less than the normal lamp current!) If none of the available SCR devices have that high a voltage rating, use a low-voltage transformer with them to be sure that you do not destroy your SCR. The peak transformer voltage should not exceed about one-third of the peak rating of the SCR. A transformer that has an output between 12 and 48 volts should be suitable.

Wire up the circuit as shown in Fig. 6-18. If you are using a transformer, replace the neon bulb with a trigger diode connected with its anode toward the gate. Ordinary diodes and zener diodes can be used (but with substantially reduced effectiveness) if you cannot get a trigger diode. The initiation is obtained by a sudden dumping of current into the gate as a result of a substantial voltage jump that is due to the breakdown of the neon bulb or the trigger diode. Ordinary diodes and zener diodes should not be used in practical circuits as the impulse effect of the other devices is necessary for proper operation.

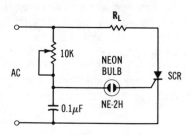

Fig. 6-18. Ac control using a silicon-controlled rectifier.

Vary the resistance in the potentiometer used in the RC network, and observe the SCR waveform output. What happens as you vary the resistance value? Draw sketches showing several examples of the resulting waveform.

Step 4.

Now, reverse the positions of the resistance and the capacitance in the gate-control network. What happens? Does the circuit still work? Can you explain this?

The gate is already positive when the anode turns positive, so you can always expect to get a full half-cycle of conduction this way. For practical purposes, you have no control.

When this circuit is used to control a dimmer function with an electric lamp, either a trigger diode or a diac is usually used for initiation, as at the moment that the device starts conducting, the gate voltage on the SCR receives a strong positive pulse, turning it on hard. If the anode voltage is positive, the SCR turns on instantly. If not, it turns on only when both the gate and the anode next become positive.

Step 5.

Apply a rectified alternating voltage on a trigger diode using the transformer-rectifier combination that you made. This is the device you made to apply rectified sine waves to transistors so that you could plot the current-voltage contours of bipolar and field-effect transistors. Sketch the waveform of the voltage that is across the diode. However, first make sure that there is an adequate series resistance (around 500 ohms) in the supply circuit so that components do not overheat. Try it for both polarities. Repeat the test using a diac. What is the difference between the two devices?

Most of the trigger diodes and diacs that the author tested started conducting in the normally nonconducting direction at between 25 and 30 volts. The trigger diodes started conducting in the forward direction at about 0.6 volt, whereas with the diacs, the forward direction could not be distinguished. With either polarity, conduction started at about the same value. There was invariably a sharp spike at the point of "turn on," as the conduction voltage was possibly as much as 5 or 10 volts less than the breakdown voltage. There was a small transient at the point of "turn off."

Step 6.

The SCR can be used on dc as well as on ac as long as an interruption of the anode current flow can be obtained in some way. This can be done by shorting out the device, or it can be done by reducing the amount of current that can flow below the design minimum. Since windshield wipers (used on cars) have an automatic-return arrangement that can be used to short out an SCR, these devices are excellent for providing delayed-operation circuits. A UJT can be used to turn on the SCR. (The SCR is connected directly across the normal switch contacts on the wiper.) The variable resistance element used with the UJT usually has a built-in switch that will initiate the control unit and will also connect up the SCR. Repeated pulsing of the UJT then operates the wiper for one or two cycles every time the capacitor is discharged. The circuit for one of these units is shown in Fig. 6-19. It uses a push-button switch to represent the return switch; a light can be used to represent the wiper. Wire it on a solderless breadboard and try it out. What range of delays can you get? Do you think you can design a circuit which will cause a flashbulb to flash and which will use an SCR so that a high current will not have to pass through the shutter contacts? Try it.

It is really quite easy. A very small amount of rewiring of the flash-bulb attachment is required. Instead of applying the current from the battery directly through the contacts, the contacts are placed in series with a resistance connected to the plus side of the battery. The minus side of the battery goes through the bulb to the cathode on the SCR. The gate of the SCR is connected to one of the shutter contacts and is also returned through a resistance to the minus terminal. Closing the shutter contacts pulses the gate, and the current flow through the bulb flashes it.

EXPERIMENT 5
An Enlarging Timer

A configuration which performs a function similar to the wind-shield-wiper control and which may be useful for other things around

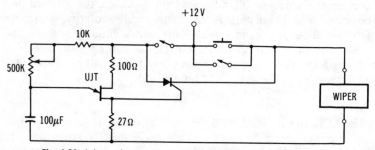

Fig. 6-19. A intermittent sweep-control circuit for windshield wipers.

the house is an enlarging timer for use with photographic timing. This arrangement is designed to go through only one cycle. It starts when a push-button switch is pressed, closing the power relay for the enlarger light or other load, and starting the timer. The power supply of the 15-volt dc source has already been activated. It is connected to the UJT and to the SCR when this relay is closed. The timer times out to the first discharge, and the discharge pulse turns on the SCR. The SCR closes its relay, which opens the power relay. The unit can be calibrated with a stopwatch, and it will give repetitive exposures that are adequately reliable. (See Fig. 6-4.)

Step 1.

Wire up your circuit as in Fig. 6-20. You may wish to try several kinds of devices in the SCR position, so its terminals are labeled K, G, and A. If you wish to try and use an npn power transistor, connect the emitter to K, the base to G, and the collector to A. A Darlington-type power npn device will connect in the same way, except that the internal base-to-emitter connection should be returned to ground through a resistor. This resistor is to keep the unit turned *off* when there is no bias on the G terminal.

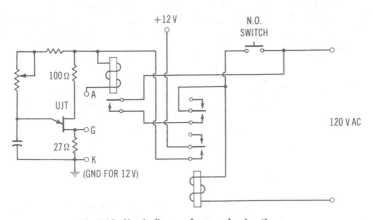

Fig. 6-20. Circuit diagram for an enlarging timer.

Step 2.

Make your first tests using an SCR in the circuit. Find the maximum value of resistance that you can use with the UJT. If it is not at least one-half a megohm, try different UJTs until you find one that will trigger and generate a sawtooth wave with a charging resistance of at least 1 megohm. Then, select a capacitance that will enable you to get time intervals up to nearly 1 minute. A capacitor of approximately 50 microfarads should be close. Wire the circuit and test it. Did you have any problems once you selected the correct parts?

Unless you have a leakage problem in a component, you should not have had any problems. If you have problems, retest each component and recheck your wiring.

Step 3.

Place rough calibration marks on the potentiometer so that you know the positions to get 5-, 10-, 20-, and 40-second time delays. Using this circuit, with it modified to let the UJT free-run, do you get these long time intervals?

When the timer works normally, it charges the capacitor from ground to approximately one-half the supply voltage. When it cycles, it goes from approximately one-quarter of the supply voltage to one-half. The result is that when free-running, the cycle time is approximately one-half of that required for the circuit when it starts with a discharged capacitor.

Step 4.

When the SCR is replaced with either the power transistor or the Darlington power transistor, what happens? Explain.

Our tests indicated that we could not rely on the transistors, because when using them, the circuit did not lock open until the power relay had dropped out. Sometimes the circuit worked, sometimes it went into a free-run mode. The contacts on the low-voltage relay did not stay open long enough to clear the power relay. The SCR, on the other hand, locks on until the power relay has opened and removed the voltage from the SCR, thereby restoring the trigger circuit to its normal condition. At the same time, the opening of the power relay removes power from the UJT and the SCR. The capacitor then discharges through the emitter-to-base 1 junction to the negative return. This gives a maximum charging time.

Step 5.

When assembling the enlarging timer, notice that the duplex convenience outlet plug can be, and should be, separated into two separate plugs. The jumper from section to section has been broken out, and the back contact of the relay connected to the second section. This is so that the one socket can be used for a safe light, and the other for the enlarger. Test it and see if yours works as the author's does. Enter your comments on this gadget, the problems you have met, how you solved them, and what else you might be able to use this circuit for. Do you think you could use electronic relays instead of mechanical ones for this?

EXPERIMENT 6
A Photosensitive Alarm

There are a number of possible uses for a trigger circuit that will sense the arrival of a light beam and will then turn on another light. A simple circuit for this purpose can be devised that is based on the circuit previously described. It needs two additional elements that you can devise for your circuit, based upon what you have learned earlier—a delay circuit that will permit it being turned on and will permit leaving without tripping it and, also, a way to turn the circuit off in daylight. (This can be a power switch, a timer, or an ambient-light detector which will keep the unit inactive until it is turned on.) One possible configuration for the circuit is shown in Fig. 6-21.

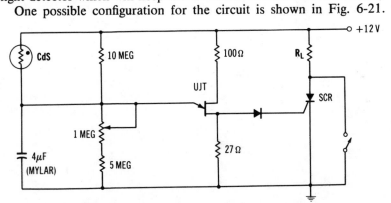

Fig. 6-21. Photoelectric alarm. (Adapted from a GE circuit.)

There are two important differences from the previous circuit. The first is that the charging resistor for the capacitor is now a photosensitive resistance like a CdS photoconductive cell, and the second is that the sensitivity can be controlled by the voltage divider connected across the supply for the emitter. The photocell is in parallel with the base 2 end of the divider and shunts it out when it is illuminated. The capacitor is used to minimize the effects of lightning or noise.

Step 1.

Cover the photocell so that it gets no light, and measure its dark resistance. Depending on the value you measure, you may or may not need a shunt resistor across the capacitor to keep the circuit quiescent in the dark. If possible, pick a unit with a dark resistance near 10 megohms. What is its dark resistance? Also, measure its resistance in a dark room but with a small flashlight illuminating it from a distance of about 5 or 6 feet. (Its resistance should be not more than 100,000 ohms under this condition.) What is its resistance under this condition? Explain how you expect this circuit to operate.

When the light-sensitive element is illuminated, it permits the voltage across the capacitor to increase, to the point where the UJT will be triggered. This triggering fires the SCR, which then activates whatever other things you may wish activated. If the dark resistance is too low, you may compensate by decreasing the resistance from emitter to ground. You want to have the voltage at the emitter sufficiently below the trigger level (in average darkness) that the unit will not "false-alarm," yet, at the same time, sensitive enough that it will trigger from abnormal lighting conditions.

Step 2.

Wire your circuit as shown in Fig. 6-21 on your solderless breadboard. Use the photocell you tested, and adjust the values of the resistances to keep the voltage under dark conditions and below the trigger condition. (The resistance values given are suggestions, as your devices may differ from the ones on which the values have been suggested.) Activate the circuit in total darkness and, if needed, adjust your shunt to provide the required trigger sensitivity at the desired light level. Once you have done this, a flashlight pointed at the photocell should activate the unit after a short delay. The delay is controlled by the capacitor size. The unit will operate and trip the

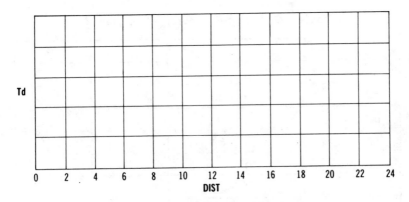

Fig. 6-22. Graph for Experiment 6, Step 2.

relay to perform whatever function you wish. For tests, let it light a light. A push-button switch can be used to release the circuit, and a switch to deactivate it. Wire it and test it. You can get some idea of its sensitivity by using a stopwatch to determine the time needed to trigger the circuit. Record the time as a function of the distance between your flashlight and the sensor. Plot a curve of this on the graph in Fig. 6-22 after you have recorded the data in Table 6-8. (Dist is for distance, and Td is for time delay.)

Table 6-8. Data for Experiment 6, Step 2

Dist				
Td				
Dist				
Td				
Dist				
Td				

Step 3.

Vary the size of the delay capacitor, and find out the range of time delay that can be built in with this arrangement. Again, plot curves of distance against time delay for various capacitor combinations, labeling each contour with the capacitance involved. Record your data in Table 6-9. (Cap means capacitance value.) Explain your results.

Table 6-9. Data for Experiment 6, Step 3

Dist				
Td				
Cap				
Dist				
Td				
Cap				
Dist				
Td				
Cap				
Dist				
Td				
Cap				

Step 4.

You need to set up your unit so that it is unlikely that it will be activated under normal conditions, but so that it will be activated from an unusual source. How might you do this?

You can place a "hood" in front of the sensor in such a way that it will be readily triggered by an unusual light, such as a flashlight held at about waist level, but will not be triggered by more diffuse lights.

Step 5.

Can you suggest a way that the circuit could be made "self-disarming" for indoors use? Modify the circuit for that purpose, and test it. Explain your results.

One way to do this is to have a second sensor located so that it will pick up ordinary room light, but not be affected by a light source like a flashlight. This sensor would be connected from the emitter to base 1 return and would, in effect, disable the main sensor unless it was illuminated much more strongly by diffused light.

Step 6.

Could you use this kind of an arrangement to turn on a garage light at night when you drive up to your garage? Explain how you might do it, listing the precautions you will need to take, and include suggestions on how you might deactivate the circuit in the daytime.

EXPERIMENT 7
The Silicon-Controlled Switch

The silicon-controlled switch (SCS) is possibly the most versatile device in the field of specialty pnpn devices. It has leads brought out for all four layers of the device. It is built essentially like the SCR, having as a result the same kind of avalanche switching characteristics that are observed with SCRs. You will want to operate a sample of these devices in each of its different modes so that you can learn what their properties are and how they respond.

Step 1.

Test an SCS device for operation as an SCR. To operate the SCS as an SCR, you simply ignore the terminal called the "anode gate," and apply your trigger on the cathode gate. The properties of this device will then closely parallel those of a standard SCR. Wire up your sample as shown in the circuit of Fig. 6-23. The switch is used to provide the necessary change to restore initial conditions.

Step 2.

In this step, you will operate the SCS as a programmable unijunction transistor (PUT). A PUT really is an npnp SCR (instead of the usual pnpn SCR), or an SCR with the trigger-control circuit connected to its anode gate. This control circuit provides a voltage divider. When the charge on the capacitor raises the voltage on the pnp emitter above the voltage on the anode gate, it fires just like a UJT or an ordinary SCR. It is called a programmable UJT because of the external control of the exact voltage at which triggering will take place. With the trigger gate returned to the variable arm of a potentiometer as shown in Fig. 6-24, see how much you can vary the

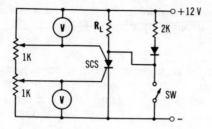

Fig. 6-23. A basic SCS test circuit.

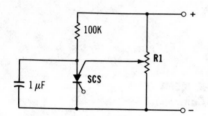

Fig. 6-24. The SCS as a programmable unijunction transistor.

trigger voltage and how the triggering is affected. Describe what you observe, and explain how the circuit operates.

As noted above, the control is established between the anode and the anode gate. When the gate is negative with respect to the anode, the pnp transistor is forward biased and may trigger if the bias is sufficient. The use of a charged capacitor to generate the bias voltage assures that the current available will be sufficient to trigger the device. Then, a surge of current will flow through the device, generating the voltage pulse in the cathode circuit which is used to operate some other device.

Step 3.

Now, test the device as an inverted programmable unijunction transistor. As you can see, this circuit is essentially the same as the circuit in Step 2, except that the positive rail is replaced by the negative rail, and the device is turned over (the anode is in the former cathode location, etc.). The cathode gate is used now. Any standard SCR can be used for this purpose. Explain how the circuit works.

The triggering capability may differ somewhat depending on whether the SCS is used as a pnpn SCR or an npnp SCR. In either case, however, the gate in use forward biases the control transistor and an avalanching of the current results, which locks the device in a conducting condition until either the current drops below the critical level or the device is reset.

Step 4.

You have two control gates with the SCS. It is interesting to see what is the effect of applying trigger signals to both gates at the same time. Try the circuit that is shown in Fig. 6-25. With this circuit, you

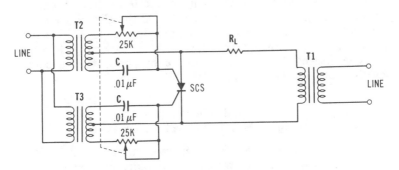

Fig. 6-25. Double control circuit for a silicon-controlled switch.

can vary the phase of the control signal over 180 electrical degrees, and by reversing the polarity of the voltage applied to the SCS, you can get the effect of 360 degrees of shift. What you want to look for is the range of control. First, be sure that the two outputs from the two phase shifters are 180 degrees out of phase, and stay that way. You can use Lissajous figures to determine that (on your scope). Use a dual potentiometer, and put the output of one shifter on the horizontal input, the other on the vertical input. As you adjust the potentiometer, the presentation should show a diagonal line. This diagonal line will lie on a plane that is located clockwise from "10:30 to 4:30." You should know how an SCR will react—with one polarity you can cause a delay in the initiation of conduction, but not with the other. Because of the two control voltages, you may be able to extinguish current flow as well as initiate it. Try this, and vary the amplitude of your control signal as well.

Step 5.

The SCS device is commonly used in a circuit like the one shown in Fig. 6-26. It has the advantage that a relatively high voltage can be applied through a load to its anode gate and still maintain a relatively low operating voltage on the device. For that reason, it is useful with calculators having gas-discharge readout configurations. Set up the circuit, using a relatively high resistance for the anode-gate load or using a neon bulb, and trigger the circuit on and off. Explain how it works.

Placing a forward pulse on the cathode gate turns the SCS on, causing the anode gate to be forward biased and the glow-discharge unit to initiate conduction. This locks the unit into conduction. Current may be extinguished by either dropping the anode voltage to ground, which reverse biases the anode gate, or by applying an adequately negative pulse on the cathode gate.

EXPERIMENT 8
Using a Triac to Control Power Flow

A *triac* is really a pair of SCRs that are connected so that they will control the flow of power with either polarity of applied voltage. One might be simulated by taking a pnpn SCR and paralleling it with an npnp SCR in such a way that the gates can be connected together so that one gate can be triggered with a positive gate voltage, and the other gate by a negative gate voltage. In effect, the anode of one is connected to the cathode of the other, and vice versa. The circuit configuration is shown in Fig. 6-27.

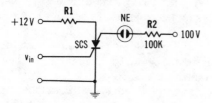

Fig. 6-26. An SCS neon driver circuit.

Courtesy General Electric Co.

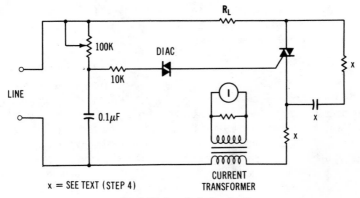

Fig. 6-27. A typical triac circuit.

It is particularly important that you observe the current waveform with this arrangement. The best way to do this is to use a "current transformer" to enable you to observe the current waveform without having to worry about ground coupling. The principal problem is getting a current transformer. Fortunately, you are not worried about precision here, so you can either use a filament transformer backwards, or use a dual-winding filament transformer, or you can make one. The latter is perhaps the easiest and best approach.

Step 1.

The principal thing you need to know about any transformer that is used as a current transformer is that it should be operated with as near to a short-circuited secondary as it possibly can be and still permit observation of the desired signal. A secondary resistance that is between a fraction of an ohm and a few ohms is usually what is needed.

If you do not seem to have a suitable transformer, but do have an old transformer with a defective winding, one can be made fairly easily. (Do not assume that it will be good for precision work as the design and manufacture of a precision current transformer is an art in itself.) Strip off the existing windings and, using well-insulated wire of No. 20 gauge, wrap two windings on the core with each winding having at least 20 turns. A No. 20 wire will carry about 1 ampere; No. 18, 2 amperes; No. 14, 5 amperes; etc. Insulate between the two windings so that you will not have a voltage leak. Then, load one winding with the resistance, and connect the other into the line as shown in the circuit. The scope output is taken across the resistance. You may have to juggle the number of turns to suit your specific situation. If you have some speaker wire that consists of two conductors in a plastic casing, both windings can be wrapped at once.

The overall result will be a better transformer. *Notice that current transformer secondaries are ALWAYS kept shorted when they are not being used for measurements.*

Step 2.

Connect the circuit, including the current transformer and its very-low-resistance load. Connect a scope to the load on the current transformer and observe the waveform output as a current is passing through it. (Any kind of a light-bulb load will be fine for this test— just do not load the transformer winding beyond its current capacity.) After you have verified that the transformer works with the scope, connect it into the test circuit.

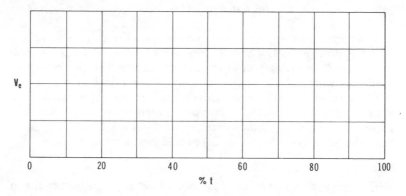

Fig. 6-28. Graph for Experiment 8, Step 3.

Step 3.

Vary the setting of the variable resistance, and make sketches of how the waveform is affected. If a conventional ac ammeter and ac voltmeter are available, you will find it interesting to observe the readings shown on them as you vary the conduction angle. Plot the measured value of output voltage against the percentage of time the triac is conducting. (Remember that you can test a triac the same way you tested an SCR. Fig. 6-23 has the basic circuit. Run the test with one polarity, and then reverse the polarity and repeat.) Record the data readings of your ac voltmeter as a function of the percentage of time of conduction in Table 6-10, and plot the data in Fig. 6-28 where V_e is the effective voltage (or average voltage as indicated by a rectifier voltmeter if nothing else is available), and %t is the ratio of the conduction time to the total period of a cycle. Explain what you have observed.

Table 6-10. Data for Experiment 8, Step 3

V_e				
%t				
V_e				
%t				
V_o				
%t				
V_e				
%t				

If you measure the area under the conduction part of the voltage cycle as you observe it on your scope, it should be proportional to the output voltage as measured by a rectifier-type voltmeter. A true rms meter will give a somewhat different value that will not coincide with your data. (It measures the average of the *square* of the voltage.

Step 4.

Now, add the resistance and capacitance elements that are designated by an "x" into the circuit of Fig. 6-27. You want to observe what effect the presence or absence of these parts will have on a radio and/or television signal reception when the circuit is operating nearby. Both tv "snow" and radio "noise" should be more significant problems without the filtering elements. (Note that where voltage margins are tight, inductive limiting can sometimes be better than RC limiting, since SCRs and triacs are sensitive to dV/dt problems.) Note what you observe.

Properly placed resistances, capacitances, and inductances can limit the voltage spikes that may be generated by the sharp switching action that occurs with either SCR or triac units. Since the prime causes of man-made interference are sharp switching spikes, slowing down the transients can reduce the noise.

Step 5.

Try the circuit shown in Fig. 6-29. This circuit is essentially the same as the previous one, but is made of discrete components. Either two trigger diodes (as shown) or a diac may be used in the gate cir-

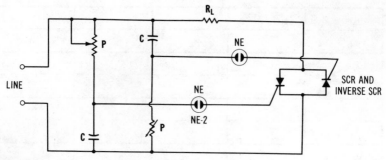

Fig. 6-29. SCR equivalent of a triac circuit.

cuits. Trace out the paths of the current flow, and also the control paths so that you understand how the circuit works. It is much more complicated than is required with a triac, isn't it? Record your comments.

Step 6.

Can you suggest how you might use a current transformer (like the one you made for Step 1) to provide for the cutoff of this circuit if excessive current flow is sensed? (Hint: Can you devise a way to make a current transformer provide a switch pulse that can ground the gate control line or lines?)

To do this, you use the current transformer, but set the turns ratio so that the peak voltage across the secondary load only exceeds one-half a volt when the current approaches the safe limit. Then, you can use a bridge rectifier to produce output pulses every half cycle. These pulses, when present, can be amplified to trigger a flip-flop which will provide the grounding signal. A pair of power transistors, one npn and one pnp, can be used to ground the gate line or lines, based on the signal from the flip-flop. Release can be as desired, either manual or automatic.

EXPERIMENT 9
Making Simple Voltage-Reference Sources

Sometimes, one needs to have a source of an essentially constant voltage which can provide modest amounts of current. For instance, if you are constructing a resistance-measuring circuit for use with a digital voltmeter (dvm or dmm), you may need to use some standardized voltage to control your resistance-measuring function. If your dvm has a 200-millivolt full-scale sensitivity, you may want this circuit to provide 5 or 10 volts. You can then use series resistances to limit the current. You then measure the IR drop across the resistance whose value you are measuring (Fig. 6-30.) You might plan to

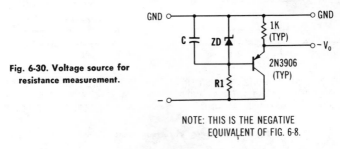

Fig. 6-30. Voltage source for resistance measurement.

NOTE: THIS IS THE NEGATIVE EQUIVALENT OF FIG. 6-8.

draw currents that vary from a few nanoamperes to as much as 5 milliamperes through your measuring circuit. How can you best provide this amount of current with adequate voltage precision?

Since the use of a simple zener regulator would appear to be inadequate, the use of a zener reference diode with a simple emitter follower may be the most direct approach. Two of the configurations which might be used, including an emitter follower, are shown in Fig. 6-31. (These are by no means all of the possible configurations.) In each step, you can try a different alternative and evaluate its properties.

Step 1.

For this approach, you can choose a zener diode alone. It must draw enough current so that when it has a maximum external load, at least 1 milliampere of current will still be flowing through it. (Under no external load, 6 milliamperes would then flow.) Find a zener diode which can carry 6 milliamperes of current with 10 volts across it, and operate it from a power supply that is capable of providing between 12 and 15 volts. Place an external load in parallel with it of

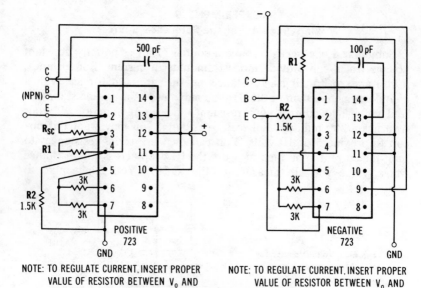

Fig. 6-31. Voltage and current regulator circuits.

about 5 milliamperes, and measure the voltage change with respect to no load with your differential voltmeter. Select other values of current, and repeat the process. Record your data in Table 6-11 and plot the relation of voltage change with current for this combination on the graph in Fig. 6-32. You will add other curves for other combinations in later steps.

Step 2.

Determine the effective source impedance of the zener diode by finding the differential voltage that results from changing the load current from about 2 milliamperes to 4 milliamperes, and from dividing the voltage change by the current difference. The effective internal resistance of the diode is _____ ohms. (The typical range

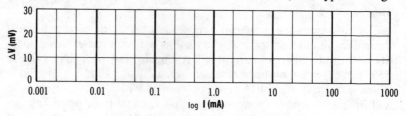

Fig. 6-32. Graph for Experiment 9.

Table 6-11. Data for Experiment 9, Step 1

R_L				
V				
I				
ΔV				

for 500-milliwatt zener diodes is from about 15 to 50 ohms.) What would you do if you wanted a smaller effective internal resistance?

The typical impedance of a zener diode is measured at about a third of the current necessary for full dissipation. Normally, it will increase somewhat as the actual current is reduced. If you assume that an internal impedance of 20 ohms applies at your chosen level of current, the voltage change for 2 milliamperes would be at least 25 millivolts. You can reduce this substantially by use of the emitter-follower configuration of Fig. 6-30. You can in fact easily reduce it to less than 10 ohms, and a voltage change of between 10 and 15 millivolts.

Step 3.

Using the zener/emitter-follower configuration of Fig. 6-8, select a zener diode having a rating between 7 and 10 volts, and choose a value of R1 to give 2 mA through R2 and 2 mA through the zener diode. Then, set up the emitter follower to give a nominal current of 3 mA, but arrange it so that you can vary the load from 2 to 4 mA. How much output voltage change would you expect at each level of current? Measure it and see if it checks. Record your data in Table 6-12.

Table 6-12. Data for Experiment 9, Step 3

I_L				
$ΔV_L$				
R_z				

By arranging for a 2 mA load and 2 mA in the zener diode, you have set up conditions where the variation in current drawn by the base of the npn transistor will not change the voltage at the zener diode. For that reason, all of the change will be at the emitter-follower output. What impedance does this appear to present? The equation to give this is:

$$R_s = \frac{V_o(1 - K_v)}{I_L} \qquad \text{(Eq. 6-4)}$$

$$= \frac{V_o[1 + (q/kT)I_cR_L]^{-1}}{I_c}$$

Using this equation, you will find that the voltage change at the load will be approximately 20 millivolts, whereas, about 40 millivolts would be encountered with the zener diode alone.

Step 4.

Next, introduce an operational amplifier like the LM741 between the zener diode and the transistor as shown in Fig. 6-9. Repeat your series of tests, record your data in Table 6-13, and plot a curve of voltage change against current on the graph in Fig. 6-32 (Step 1). The effective source impedance in this case is _____ ohms.

Table 6-13. Data for Experiment 9, Step 4

I_L				
R_L				
ΔV_L				
R_s				
I_L				
R_L				
ΔV_L				
R_s				

What are your comments on the results of the use of the amplifier?

Step 5.

As you can see, you have quite a variety of options for setting up voltage references or voltage regulators for either experimentation or for instrumentation. Now, we suggest that you set up some IC voltage and current regulators like the LM309, the 340 series, the 320 series, the 7800 series, and the 7900 series. You should test them over a wider range of currents, from perhaps 5% of their rating to 100% of their rating, just to see how well they regulate. A circuit for testing these devices is shown in Fig. 6-33. It is suitable for either

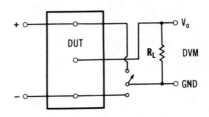

Fig. 6-33. Circuit for testing power-supply regulators.

positive or negative regulators. It is particularly important to remember that the positive and the negative input *always* come in on the same points on these devices, whether they are negative or positive output. The output is taken at the remaining pin to common, whether

Table 6-14. Data for Step 5

I_L				
R_L				
ΔV_L				
R_s				
I_L				
R_L				
ΔV_L				
R_s				
I_L				
R_L				
ΔV_L				
R_s				

positive or negative. Whereas, it is always best to use an input source voltage for these regulators that is as small as will give reliable regulation, you will want to test them to higher voltages to see how they behave at higher dissipation levels. You can test their thermal cutout by providing some outside heat, like from a soldering iron—but, be careful not to overheat them. The data you will want to record for these devices are the same as for the simulated regulators in Step 4. Use Table 6-14.

WHAT HAVE YOU LEARNED?

You have studied about a variety of specialty devices or circuits based on either diodes or transistors. You have built circuits which "exercise" many of these devices, and you have measured the properties of these circuits. You should now be able to apply one or another of these devices and configurations, or use them in combinations, to problems that you may encounter in your own work with computers, control systems, and other kinds of electronic systems.

CHAPTER 7

Some Special Topics

In Chapters 1 through 6, an attempt has been made to provide you with a "road map" to guide you through many of the intricacies in the use of semiconductor devices. As you can tell from looking at typical circuit boards for almost any kind of electronic system or computer, they are extensively populated with both transistors and integrated circuits of several levels of complexity. Also, there is usually a generous population of diodes as well. The fact that many of the ICs have a relatively small output-current capacity (usually called drive capability) means that you must know what to do to be able to provide an adequate interface or coupling between the various ICs and also between them and the output lines or loads. That is why knowledge of the properties of transistors and diodes, the basic building blocks of computers, is so important to you.

This chapter was written with a different goal in mind. That goal is to explain the properties of the instrumentation that is likely to be the most helpful to you. You will use these instruments in establishing the properties of the various devices that are studied in the experiments given in the earlier chapters of this book. Many of these instruments are of equal value in both communications and computer electronics. Since some readers may wish to study this chapter prior to studying Chapters 2 through 6, the material presented herein is designed to require a minimum of prior knowledge of solid-state devices. For example, diodes are treated as if they were lossy switches which close with one polarity of voltage and open with the other polarity. In the closed direction, they have a voltage loss, however. To the extent that transistors are used, they are treated as simple amplifiers with relatively little consideration on how that amplification is achieved. Because use of transformers in electronics circuitry is not explained particularly well elsewhere, it is desirable to clarify some of the uses of transformers as they interrelate with electronic amplifiers. This discussion is important in clarifying some concepts so that the two kinds of devices, transistors and transformers, can be used more effectively in electronic systems.

De Rosa many years ago proved that phase relations were just as vital and important in music as they are today in instrumentation and in such technologies as television. The result is that the properties of Lissajous figures and their manner of use are of extreme importance. For example, in a circuit in which it is impossible to balance out a capacitive component of current, it is necessary to use a Lissajous figure method of measurement. With this method, the circuit balance measurement can be made without being corrupted by the undesired current component.

OBJECTIVES

After studying this chapter and performing the experiments (or constructing the projected pieces of equipment), you will know much more about the different kinds of instrumentation required for the experiments in this book. By better understanding their construction and operation, you will be better able to complete the described experiments. You will have a working acquaintance with the following ideas and methods.

1. The use of either an analog or digital volt-ohm-milliammeter in construction work.
2. Common uses for resistors.
3. Common uses of capacitors.
4. Common uses of inductances.
5. Common uses and special properties of transformers.
6. The nature and properties of alternating current.
7. The Lissajous-figure method of measuring phase relationship and balance measurements.
8. A variety of simple but very useful power supplies.
9. Testing power supplies.
10. How to use a signal generator.
11. How to build high-sensitivity current meters and voltmeters.
12. How to select filter capacitors for power supplies.
13. Some important properties of tuned circuits.
14. Some of the basics of using cathode ray oscilloscopes.
15. Some of the ways to test diodes and rectifiers.

Some descriptive notes are also included that may prove helpful in bridging the gap between discrete electronics and integrated circuits.

There are also some notes on the selection of useful kinds of components. (This subject is discussed in more detail in Appendix C.)

DEFINITIONS

Some of the ideas that will be important to you in both this chapter and in the rest of the book are now defined. Where it is helpful, an "acronym" is given that is commonly used in identifying the word or device.

volt-ohm-milliammeter (*vom*)—This is an indicating instrument designed to measure the three principal electrical quantities in circuits, namely, voltages (either dc or ac), currents (again either ac or dc), and resistance. Usually these devices are multirange, in that they will measure voltages from as small as a tenth of a volt to as much as 1000 volts or more, with currents ranging from fractions of a milliampere to possibly many amperes, and with resistances ranging from less than an ohm to many megohms. The instrument uses multiposition switches and a meter called a D'Arsonval meter (after its inventor) along with appropriate resistances, batteries, rectifiers, and other components.

digital volt-ohm-milliammeter (*dvm* or *dmm*)—This is a volt-ohm-milliammeter that uses complex electronic circuitry to read out in digital form the value of the parameter being measured. These devices are a major improvement over the standard vom units previously described, primarily because of their increased potential sensitivity, increased precision, and increased sturdiness. They are standard items of test equipment that are used with digital systems.

troubleshooting—This is the process of finding and correcting the causes of circuit malfunctions or erratic behavior.

debugging—A vernacular term for troubleshooting.

alternating current (*ac*)—A kind of electrical power in which the magnitude of the voltage and current vary in a periodic and repetitive manner. It has short intervals, called half cycles, in which the voltage will be positive, and also short intervals in which the voltage will be negative. These intervals repeat over and over again.

power supply—A source of electrical energy for operating electronic circuits. It can take the form of a storage battery or a dry-cell battery, or it can consist of a complex arrangement of electrical and electronic components which will convert ac into dc. Most electronic circuits require dc, or direct current, for proper operation. It is usually required that a power supply provide a constant voltage, with the current varying in accord with circuit demands.

transformer—A device for changing the voltage level available in an alternating-current circuit. It usually provides isolation in addition to changing the level of voltage. Often the output winding, or secondary, is tapped at its center. This point is called the center tap.

variable transformer—This is commonly called a variac[R]. It is a specialized transformer whose output voltage can be changed by turning a knob.

oscilloscope—An electronic device which can draw a picture of an electrical signal in terms of a glowing trace on a screen that is similar to a tv picture-tube screen. It has an elaborate complexity of circuits to enable it to perform a variety of useful functions.

signal generator—This device is used to generate test signals which will be required in measuring the important point-by-point characteristics of both active devices and the circuits that use them.

active device—A device that is capable of reproducing an applied signal, usually at an increased power level and also usually at an increased voltage level. The power delivered is drawn from a source that is usually called its power supply. With the circuits considered in this book, the power is drawn from either a battery or a power supply (which converts ac into dc at some set voltage level like five volts).

microammeter—A device capable of measuring with reasonable precision currents between 1 and 500 micromperes.

milliammeter—A device capable of measuring with reasonable precision currents between 1 and 500 milliamperes.

ammeter—A device capable of measuring currents greater than 500 milliamperes with reasonable precision.

voltmeter—A device capable of measuring with reasonable precision voltages in some range greater than a half volt.

millivoltmeter—A device capable of measuring with reasonable precision voltages between 1 millivolt and 500 millivolts.

rectifier test circuit—A special configuration of components designed to test diodes and rectifiers to determine if they function in a satisfactory manner. A circuit for this purpose is required in Chapter 2.

printed circuit—A circuit layout on an insulating board in which the "wires" are copper paths firmly attached to the board. The components are mounted on (usually soldered to) the paths in a way that generates the required operating circuit.

printed-circuit board—An insulating board on which at least one layer of copper is bonded. The copper is etched in such a way as to provide conducting paths for the construction of a circuit (a printed-circuit assembly). The required components are mounted on and soldered to the board.

integrated circuit (IC)—An arrangement that contains within a sealed enclosure a chip that carries a variety of semiconductor devices. These include transistors, diodes, capacitors, resistors, etc., internally connected in a special configuration, and arranged to provide some electrical function at its external terminals.

transistor-transistor logic (TTL)—A form of integrated circuit consisting almost exclusively of bipolar transistors and diodes, and designed to perform switching functions. These circuits usually have a high or one state, and a low or zero state. Many of the most useful TTL elements produce ones from zeros, and zeros from ones. They may be called NAND circuits, NOR circuits, inverters, etc., where the leading "N" is used to indicate that an inversion has taken place.

diode-transistor logic (DTL)—This form of integrated circuit is a forerunner of TTL. It has been largely replaced by TTL since a more reliable operation is provided by the latter.

resistor-transistor logic (RTL)—This is the least efficient form of solid-state logic that has been developed. It is based on use of resistors instead of diodes. It has been almost completely displaced by TTL logic.

PMOS logic—This logic is based on the operation of p-channel insulated-gate field-effect transistors. It consumes much less power than TTL logic, but it is much slower and usually requires a higher voltage.

NMOS logic—This logic is based on the operation of n-channel insulated-gate field-effect transistors. It consumes much less power than TTL logic, but it is significantly slower and usually requires a higher voltage. It has a higher operating speed than PMOS logic.

CMOS logic—This logic is based on combinations of n-channel and p-channel field-effect transistors. It is much more power efficient than either NMOS or PMOS logic, and has about the same speed. It can operate with a wider range of voltages than the other forms of logic, and has a speed proportional to the applied voltage.

Unit under test (UUT)—A device or circuit configuration that has been placed in a special test circuit configuration designed to help determine if it is operating in compliance with certain specifications.

Device under test (DUT)—Another name for *unit under test.*

MEASUREMENT OF CURRENT AND VOLTAGE

In the next few paragraphs, we will discuss the nature of many of the above devices and instruments in greater detail. In some instances, you will need to assemble some of these into unusual configurations to satisfy your task requirements. Some of these units are available commercially, both as assembled units and as kits. Many are extremely useful. They are useful enough that you may want them as permanent test instruments in your shop or laboratory. Others you may wish to have available to build into things you find you need as you continue your activities in the field of computer-related electronics. The initial discussion will be oriented toward voltmeters and current meters, as semiconductor measurement problems are so unique that few available devices can be used without having serious effects on the circuit being tested.

As has been noted repeatedly throughout this book, voltage changes as small as 10 to 20 millivolts can make significant changes in the operating characteristics of solid-state devices like junction diodes and junction transistors. As a result, you will find it helpful to be able to measure voltage changes as small as just a few millivolts. Frequently these measurements will have to be made in a "common-mode" voltage situation; that is, with both terminals of the voltmeter off of the ground circuit. Further, it is sometimes necessary that current measurements be made without introducing any significant voltage changes, with respect to the 10 to 20 millivolts noted above. Yet, few current meters can be operated full scale with less than 100 millivolts across them. The most sensitive 1 milliampere meter the author has seen advertised commercially was a Weston Model 301, having an internal resistance of about 27 ohms. It might not be available any more. However, as this was being written, Keithley announced a digital voltmeter-ammeter that can measure with adequate accuracy and precision, voltages that are less than a millivolt in magnitude. It is intended for use as a low-burden current-measuring device.

Solid-state instrumentation amplifiers can be used either with analog meters like the Weston 301 series or with digital voltmeters to correct this problem. With the supply of relatively inexpensive analog meters available these days, however, you may wish to use a booster amplifier and one of them rather than use a much more precise and expensive digital voltmeter. (However, digital voltmeters have enough extra capability that a person who is able to afford a digital voltmeter will find it well worth his while to have one.) As an alternative to a digital voltmeter for those unable to afford one, the use of surplus microammeters used in conjunction with a simple instrumentation amplifier is suggested. The instrumentation amplifier can be configured from either an LM 741 operational amplifier set or the equivalent LM 4250 operational amplifier. It is best to use at least two of them, so that an offset can be introduced to null the meter.

For the measurement of current, the use of a shunt and amplifier with either an analog or a digital meter having an appropriate sensitivity is suggested. A suitable configuration is shown in Fig. 7-1. If you are using a digital voltmeter having a 2-volt full-scale minimum range with less than three and a half digits, then you will need an amplifier with a gain of at least 100 to get a reasonable sensitivity. For current measurement, you need to be able to read voltages which may be as small as 2 millivolts across your shunts to a precision of 5 to 10%.

You will need a differential amplifier for measurement of voltage changes in addition to the high-sensitivity current metering circuit. Again, an analog meter or a digital meter may be used for the purpose. The circuit for this measurement is somewhat different from the one used for measuring current, inasmuch as you must be able to introduce up to almost 1 volt of offset voltage (to create an imaginary zero point), and then be able to observe voltage changes at that point. The zero-adjust potentiometer is arranged to permit you to introduce such an offset voltage and, at the same time, get the required precision. The amplifier circuit for the voltmeter is shown in Fig. 7-2.

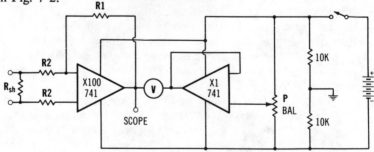

NOTE: THE UNIT LABELED R_{sh} MAY BE
USED IN CONJUNCTION WITH A SCOPE.

Fig. 7-1. Current-metering circuit.

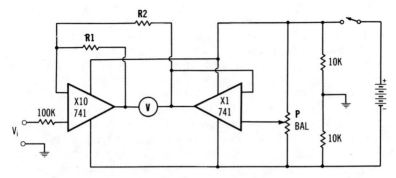

Fig. 7-2. Voltage-metering circuit.

The amplifiers for the two sections of this instrument are required to have different amounts of gain. With the current-metering circuit, the voltage range is about a tenth of that for the voltage-metering circuit. Typical analog meters, as noted above, require about 100 millivolts for full-scale deflection, whereas, the more common digital voltmeters are either ±200 millivolts or ±2 volts. Since you will want to use silicon diodes for clamping rather than germanium diodes, you may want to boost the effective full-scale voltage sensitivity of the analog meter to at least 200 millivolts, and possibly as much as 400 millivolts.

You do not need to know either the exact values of the voltages that the voltmeter reads with respect to reference or the exact values of currents, but you do need to know the current ratios with a reasonable preciseness, and you also need to know the voltage changes with a reasonable preciseness. Basically, the two circuits for the instrumentation amplifiers that are shown are quite similar, in that they use one of the operational amplifiers as an extremely low-burden voltmeter amplifier with a stabilized gain, and they use a second operational amplifier to provide the reference to the meter itself and, in the case of the voltmeter, as the reference voltage for the amplifier. In each case, the reference ground is at the midpoint of a 9-volt transistor battery. With each of these instrument amplifiers when they are used with analog meters, the series resistance is adjusted to provide the correct calibration. With digital voltmeters, it is necessary to adjust the gain to precisely the required value, as the input impedance of the voltmeter is too high to adjust with a series resistor. (A voltage divider can be used, if that is easier to accomplish than the tailoring of the gain.)

The range of adjustment for the reference potentiometers for the two instrument amplifiers is critical. With the current meter configuration, plus or minus 100 millivolts should be adequate, as all you should have to correct for is the zero offset of your main amplifier. (It is assumed that this meter is used in an emitter-return position, and the potentiometer is used to correct for zero offset.) If you wish, you can use the standard zero-balance circuit on the op amp and reduce the voltage range of this control to about 10 millivolts on either side of zero. With the voltmeter circuit, on the other hand, you need a range from about −1 volt to +1 volt so that you can set your effective zero on the voltmeter at any point you wish. With silicon transistors, it will need to be between 0.5 and 0.6 volt, positive or negative, depending on whether you have an npn or a pnp transistor. It is desirable that analog meters used for this function be provided with polarity reversing switches as, then, the scales are spread out more effectively. The use of a zero-center meter will lead to a scale that is too crowded to read very well unless the meter is at least three inches in diameter.

The design of the shunt circuit for the current-metering circuit is rather critical, whether you use a digital voltmeter or an analog meter to read the voltage drop on the shunt. The reason is that you must keep a minimum of resistance in the metering circuit and, at the same time, the circuit must be set up in a way that will give you "four-terminal" metering. This means that there must be no circuit current flowing through the switch contacts that go to the meter. Otherwise, the contact resistance will introduce a variable voltage directly into the meter circuit. A simple circuit which will make this possible is shown in Fig. 7-3. As you can see, the current being measured passes through one switch contact to the shunt resistor, and the opposite switch contact only carries the current required by the meter circuit itself. This does not eliminate the possibility of a contact potential developing, but it greatly reduces the prime cause. The return connection is a soldered connection. A 2-section 6-position switch is ideal for the purpose.

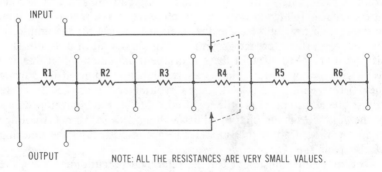

NOTE: ALL THE RESISTANCES ARE VERY SMALL VALUES.

Fig. 7-3. Configuration for a current-metering shunt circuit.

CATHODE-RAY OSCILLOSCOPES

One of the most important instruments that you can have available for use in your experimental and development activities will be a cathode-ray oscilloscope. Whenever you find it important to see just what your circuit is doing, you will need this versatile instrument, as many unusual signal conditions are difficult, if not impossible, to diagnose without one. Ordinarily, scopes are used for the examination of waveforms and for the crude measurement of voltages and frequency. You will find, however, that they can be used as phase-sensitive voltmeters, and they can also be used for estimating the quality of a waveform. In short, an oscilloscope is one of the most useful of all instruments that one can obtain.

In brief, the scope you will need should be able to amplify signals from dc to about 5 MHz with good uniformity, and it should have a sensitivity range from about 20 millivolts to about 100 volts. If you plan to do extensive digital work, your upper frequency limit should be at least 25 MHz, but for much of your work at that frequency, the scope need not be much more sensitive than 100 to 200 millivolts per centimeter deflection. The horizontal sweep, which should be similar to the sweep on a tv set, should be at a uniform speed from left to right with a retrace in, at most, 10% of the total sweep period. You may prefer a driven sweep rather than a free-running one for that kind of an application. (The driven sweep has to be initiated by some component of the signal, whereas, the free-running initiates itself although it can be synchronized to the input signal.)

If measurement of phase relations is important, you will need a scope having identical horizontal and vertical amplifiers. As long as you are operating at a low frequency, however, this may not matter much. Above 1000 Hz and up to 10,000 Hz, it can become important, however. When using the scope to measure currents, be sure that you use the amplifier having the most sensitivity (usually the vertical) for the currents, as you *have* to keep the signal-voltage drop to less than 5 millivolts total. Otherwise, the apparent characteristics of the device you are testing will change as they are presented on your scope.

POWER SUPPLIES

Several kinds of power supplies will be needed for your work. Since you will be working with units that require a positive voltage on their outputs, like npn transistors and n-channel FETs and, also, with units that require a negative voltage on their outputs, like pnp transistors and p-channel FETs, both positive- and negative-regulated voltages will be required. In addition, you will need separate positive or negative voltages, or both, for the input-control electrode on the devices. Some specialized circuits, like the voltage- and current-booster circuits, can be operated from 9-volt transistor batteries. Wherever this is possible, it is highly recommended. (As long as the unit is controlled by a push-button switch, and does not have to run continuously, battery life should be excellent.)

A dual-polarity power supply will find extensive use in many other ways. Whenever you build an experimental circuit based on an operational amplifier, this kind of a supply is ideal. You will find that operational amplifiers are usually listed for operation at ±15 volts. However, unless there are internal high-voltage zener diodes within the chip, they can all be operated at significantly lower voltages. The LM 4250, for example, will operate with two 1.5-volt dry cells, and it can run for years that way.

Many power supply applications will really not require regulated voltages but, because of the availability of inexpensive regulators, it is often more convenient to use them than not. With them, you do not have to worry about power-supply impedance and the ac ripple should be virtually zero. You must know, however, just how much the source voltage must exceed the regulator output voltage for good regulation, and, then, make sure that you exceed that value by at least 1 volt. That is why regulator tests are included in this chapter.

With most power supplies, and particularly those using a regulator, it is essential that sufficient capacitance be placed across the rectifier output so that the input voltage to the regulator cannot dip below the minimum voltage that the regulator requires. Otherwise, there will be dips or drops in the regulated voltage provided to the circuits, and the circuits may malfunction. You may also have to install a small resistance in series with the rectifier input to limit the peak current through it, as the current flows in very short pulses in these power supplies. (You may have a problem getting a high-wattage resistor that has a small enough resistance value to do what is needed.)

As an example, suppose that the peak voltage from a rectifier is 8 volts, and the minimum voltage needed for operation of a regulator (a 5-volt regulator) is 7 volts. If the current required is 1 ampere, the minimum capacitance required to limit the voltage change to 1 volt is determined by the equation:

$$\Delta Q = I\Delta t = C\Delta E \qquad \text{(Eq. 7-1)}$$

If the time increment Δt is taken as a half cycle at 60 Hz, or 0.008 second, the value of ΔQ is 0.008 coulombs, and the required value capacitance is 8000 microfarads, more or less. This gives you an indication of how critical it is that the capacitance ahead of the regulator be adequate.

Relatively inexpensive power supplies can be purchased these days. Some are available in kit form, and some are already assembled. Some circuit boards are available for different types of power supplies, and they can also be built from scratch, either on a chassis form or on a piece of circuit-board material. It will probably be instructive to try building some power supplies using several of these methods so that you can gain some familiarity with how it is done.

WAVEFORM SOURCES

You are going to need a wide-range oscillator for testing transistors, as well as the other circuitry previously discussed in earlier chapters. Several waveform generator kits are available, some of them complete, and at least one that has the circuit board and most of the other components provided for in it. The most readily available unit is probably the Heath IG-5282. It is used with an auxiliary power supply. The Model 2206KB kit made by EXAR Corp is reasonably satisfactory if you are willing to calibrate the unit yourself. Learning how to calibrate such a unit is very instructive, although it does require some patience. Once you know it, you have at your disposal an excellent way to check the calibration of many things. Both of the above units generate sine waves, square waves, and triangular waves over a frequency range that extends from a few cycles to something over 100,000 Hz.

OTHER USEFUL INSTRUMENTS

Most operations in digital electronics are in some way dependent on either a signal source of the kind described above, or on a crystal-controlled signal source. There are many firms advertising crystal-controlled 60-Hz sources based on tv color-reference crystals that operate at approximately 3.58 MHz. Kit prices typically run around $5.00, and they provide both 60 Hz and the basic color frequency as outputs. The high-frequency signal may have to be buffered to permit using it if the unit is being loaded significantly by your selected circuitry. (A 2N2222 emitter-follower circuit with about 100 ohms in the emitter circuit should do. Adjust it for about 5 milliamperes of collector current.) These units operate on voltages that are between about 5 to 15 volts, and the chip apparently is based on MOS technology. After studying Chapters 3 and 4, you should have no trouble making a buffer circuit.

If you prefer, you can use almost any kind of inverter, NAND gate, or NOR gate, using either TTL or IIL (and possibly, some CMOS units as well), to set up a crystal oscillator that is based on a crystal of some decimal frequency instead of the color frequency. A possible circuit based on TTL inverters is shown in Fig. 7-4. The author suggests using a 74LS04 hex inverter IC which contains six inverters, as you can use two of the inverters for the oscillator, use one or two for buffers and, if you wish, you can play with the others to see if you can make a two-phase clock from them. The main reason the author prefers the use of inverters to the more complex circuits for making these oscillators is that you do not extract the signal from the crystal by paralleling at least two inputs, as such loading can possibly cause you trouble. For that reason, you have a wider range of options using the inverters. Even if the circuit is arranged to use only one input, you may be penalizing yourself somewhat when compared to the inverter approach.

However, you do not have to use inverters for an oscillator. Either a quad NAND gate or a quad NOR gate configuration can be used in exactly the same way as the hex inverter. With the NAND units, either both inputs are paralleled or one is tied high. With a NOR unit, both inputs may be paralleled, or one can be tied low or grounded. If you are trying to use the gate near its upper frequency limit, you may have better luck tying off one input rather than paralleling both. You reduce loading that way.

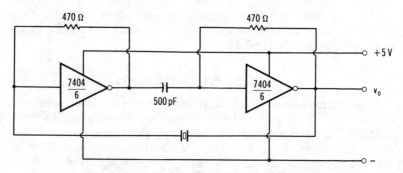

Fig. 7-4. A TTL-based crystal oscillator using inverters.

The resistor that couples the output to the input for the individual element inverters biases the unit in its active, or high-current, region. You will want to choose a value of the resistance for the most effective operation with the crystal. You should try values between 330 ohms and 470 ohms as a starter, but the exact value that is best might vary by as much as 3 to 1 above or below this range of values. A CMOS inverter using FET devices would require much higher resistances. You will probably find it useful to build one of these units. A basic frequency between 100 kHz and 10 MHz should prove optimum.

It would be very helpful to have a series of frequency dividers after your crystal oscillator. If you do this properly, you can obtain a series of frequencies from your crystal, all of them being useful submultiples of the crystal frequency itself. The author prefers using 7490 decade dividers for this purpose, and has several chains that can divide by as much as a million. A crystal oscillator based on a 74LS04 can drive these chips, although a buffer between the oscillator and the divider is preferred. A possible circuit is sketched in Fig. 7-5.

THE NATURE OF ACTIVE DEVICES

An active device can take any number of forms. The first active devices built were triode electron tubes. The first of these was made by Lee deForest in the early part of the twentieth century. Dr. Albert Hull in the early 1920s introduced the screen grid into the tube structure, and the electron tube and electronics were on their way. Dr. Julian Aceves demonstrated how tubes could be operated successfully on alternating current, and shortly thereafter, the cathode-type structure (using a filament inside a sleeve that was coated with a material capable of emitting electrons) was devised, assuring full ac operation of radio receivers and transmitters. Just after the end of World War II, Dr. Shockley and his co-workers invented the point-contact transistor. This was followed by the invention of the bipolar transistor and the field-effect transistor. These inventions made it possible to build the pocket calculator.

All of the active devices that you are likely to use extensively have two ports, or two points of entrance. One of these is known as the input, or control, port, and the other is known as the output port. There is only one kind of *one-port* active device presently known that can amplify. These are the devices that over a modest range of either voltage or current exhibit a decreasing value for one variable as the value of the other variable increases. These devices are known as *negative immittance* devices. (Immittance is a collective term that is used to include both impedance and admittance. It is a standard term in circuit technology.) The tunnel diode and certain types of gas-discharge tubes are possibly the best known examples of these devices, although trigger diodes also have this property. For technical reasons, all of these devices are relatively unimportant in most branches of the electronics that you will encounter.

Theoretically there are at least four kinds of two-port active devices, based on the kind of variable which provides the control of the output, and the kind of variable that is controlled. This condition is a result of the existence of two basic kinds of variables, namely, *across variables,* and *through* (or *flow*) *variables.* Across variables are ones whose values vary along a path, like voltages and pressures. Through variables are the ones that specify a resulting motion, or

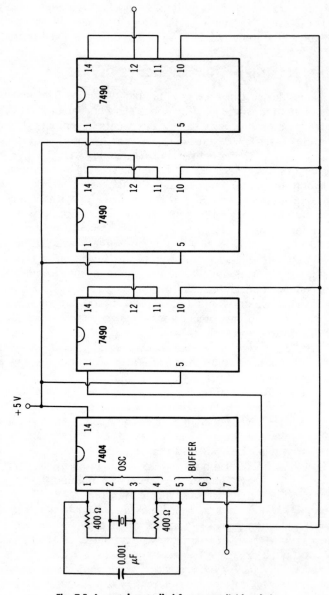

Fig. 7-5. A crystal-controlled frequency-divider chain.

flow. Voltages, forces, pressures, and even height are across variables, and motion or current (flow) are the through variables.

Whereas, theoretically, either kind of variable may be the input, or control, variable and it may control either kind of an output variable, it can be shown that all presently known two-port solid-state active devices used in electronics consist of devices that have the control function performed by an across variable, with this control action affecting a through variable. This is readily recognized from an examination of the basic Ebers-Moll equations that control transistor operation and, also, by the diode equation. In each case, the independent variable(s) in the equations are junction voltages, and the dependent variable is current flow. These basic equations are examined in Appendix A.

SOME PRACTICAL CONSIDERATIONS

One of the problems all users of electronic material must face is the problem of isolating good components from defective ones. In each chapter of this book, therefore, the methods which may be used for testing a variety of typical components have been discussed briefly. Where practical, simple test sets that you, the reader, can build or assemble on a breadboard as desired have been described. Where there is a reasonable probability of a tester being required with any regularity, it may pay you to assemble one based on the described circuits. (Since each user's needs differ, many readers may wish to breadboard the tester and use it for a period prior to final construction. Many of the units the author has described have gone through several iterations.)

When it is recalled that bipolar transistors can be used as diodes, one realizes that a person needs to be able to test diode elements in some way. Also, a person should be able to evaluate the properties of their big brothers, the rectifiers used in power supplies. One needs to know different things about these latter diodes than what is known about the diodes that we usually call signal diodes. For this reason, we have also discussed a technique for testing individual rectifiers and bridge rectifiers that are suitable for power supply use.

Rectifier Testing

As noted above, power rectifiers act as switches; they turn on when the voltage is greater than a specified small value in one direction, but are turned off (up to some maximum voltage) when a reverse polarity is applied. What is important, then, is that these rectifiers pass current with minimal voltage loss in one direction, and block it most effectively to an adequate voltage level in the other direction. Clearly, a tester for these devices should provide a variable voltage, with a current-limiting element, so that the characteristics in the reverse direction can be properly evaluated. (Usually the forward conduction direction does not give much trouble unless the internal resistance of the rectifier is very high.) If the current is properly limited, the voltage will climb until the rectifier approaches its voltage limit. The rectifier will then start to draw current, and any further increases in the source voltage will not lead to increases of the output voltage across the device. It is then said that we have reached the *Peak Reverse Voltage,* or prv. (This is also sometimes called the *Peak Inverse Voltage,* or piv.) If the rectifier is potentially defective, it will often either short-circuit or burn out when this voltage is reached.

This voltage-limit test is possibly easier to apply with the use of a ladder rectifier power supply like that shown in Fig. 7-6. The transformer with the arrow against it is a variable transformer, like the Variac®. The voltage is turned down to zero at the start, and the rectifier is placed across the test probes. The voltmeter across the diode is then watched as the voltage is increased. With a good

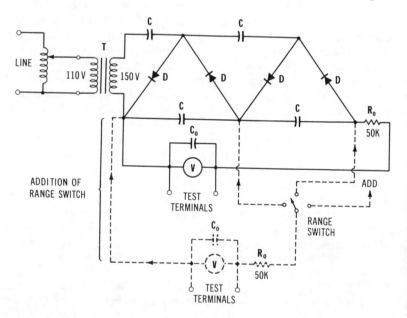

NOTE:
CAPACITORS C = 0.05 μF OR 0.1 μF. 600V.
TRANSFORMER T = 110 V TO 150 V NOMINAL.
CAPACITOR C_0 = 0.01 μF TO 0.05 μF. 3000V.

Fig. 7-6. A ladder-rectifier power-supply test circuit.

rectifier, in one direction the voltage across the device should be less than a volt, but in the other direction, the voltage will increase as the Variac® is slowly turned up to increase the test voltage. If the rectifier shows less than a volt across it in one direction but does withstand 250 volts with the opposite polarity, then one may conclude that the prv for the rectifier is at least 200 volts, one of the relatively standard prv levels. If the voltage goes to the maximum on one range, then the Variac® may be turned back to zero and the range switch changed to a higher-voltage position, and the process repeated. This tester will test rectifiers having prv levels up to approximately 1000 volts. It can be made to go higher if so desired. BEWARE —there is a very real *shock hazard* present when using this power supply!

Zener diodes need to be mentioned at this point, since their use in power supplies is common. They act, to a point, like a rectifier in that they have a forward conducting condition and a reverse non-conducting condition, up to a certain design reference voltage level. At the reference value of reverse voltage, zener diodes switch sharply from nonconducting to conducting, and maintain the reference value of voltage to within rather close tolerances. For applied voltages in excess of the design prv, they try to behave like an almost perfect voltage source. When connected in series with a resistor, they can be used to provide a selected reference voltage. They can be tested with a rectifier tester, and will show the design value of prv for which they were designed. When the applied voltage is reduced below the critical value, they again stop conducting. An ordinary rectifier usually becomes a "leaker" much more gradually than will be observed with a zener diode. It may also either short-circuit or open up instead of maintaining a fixed voltage level.

Testing signal diodes is accomplished in a similar manner, but at lower voltages and currents. Also, a configuration that is intended to detect smaller values of leakage current is used. This process was discussed in detail in Chapters 2 and 3, since it is essentially the same process as used with transistors. You probably will prefer a transistor tester for use with them.

Nature of Printed Circuits

A printed circuit is formed by mounting components, both active and passive, on a special insulating board on which current paths (made of copper strips) have been placed or etched. The board is usually formed by bonding a thin layer of copper to an insulating board. The copper is then etched off everywhere but where the conducting paths are required. Bakelite and glass-fiber laminate are commonly used as the base material, and copper thicknesses of 0.001 to 0.002 inch is bonded to the surface. The laminate itself may have a thickness between 0.02 and 0.06 inch. The desired pattern is printed on the copper either with ink, photographically, of by some other means (such as a silk-screen process). The board is then etched to leave only the required conducting pattern of wires and terminal points to which the physical components may be mounted and soldered. The board with its etched conductors, with or without the components in place, is typically called a pc board. The board may normally have an etched copper pattern on only one side. However, with more complex circuitry like that needed for calculators and microprocessors, there may be a pattern on both sides of the board, (or, possibly, even several layers of conductor patterns). Many holes must be drilled through the board for parts mounting. Some of the holes will be plated through to connect the foil stripes on the two sides. The variety of arrangements is extensive.

You will find rather quickly that working with pc boards is much easier than hand wiring a complex circuit and, furthermore, it is much more reliable as long as the design is correct. The pattern configurations are usually tried out either with soldered wiring or with wirewrap first to make certain that the design circuit is functional before it is finally committed to the "etch pits." Etched mistakes can be rather hard to fix!

Computer and control circuits based on integrated-circuit elements require a combination of discrete components along with the appropriate integrated circuits on a pc board. The integrated circuits themselves are assemblies of diodes, transistors, bipolar or IGFET transistors, resistors, and very-small-value capacitors. They always require at least some external transistors and diodes to assure proper operation. The required resistors are sometimes mounted in IC-type plug-ins, and, sometimes, they are mounted separately. Usually, capacitors are mounted as discrete components.

There are a variety of integrated-circuit configurations that you will want to know about, and which you will be using in other volumes in this series. They are defined in terms of the complexity of the functions they can perform:

1. Small-scale integration or SSI.
2. Medium-scale integration or MSI.
3. Large-scale integration or LSI.
4. Very-large-scale integration or VLSI.

At the time that this is being written, memories with as many as 65,000 or more memory spaces are being made. Others with over 250,000 spaces apparently are in development. Since additional steering circuits are also required, these would be called "VLSI" devices.

It has been found that the fewer contacts and connections that have to be made in the open air, particularly at the low voltages that digital electronics requires, the more trouble-free it can be. Also, the actual construction will be less expensive. Further, the lower the voltage at which the equipment can be operated, the more reliable it will be if the interconnection problems can be solved. Unfortunately, low-voltage contacts in the open air can create serious problems. One finds, for example, that a battery consisting of four AA-type single cells seems to fail much more rapidly than a corresponding package of batteries that are delivered interconnected. The problem is one of maintaining effective contact at the battery terminals. In some applications, the use of small amounts of vaseline on the contacting surfaces will help, but it does not always work. As a result, at these lower voltages, complete encapsulation along with welding or, at least, the soldering of connections would appear to be desirable, whenever possible.

We have been talking about "Unit Under Test" (UUT) or "Device Under Test" (DUT) regularly throughout this book. The reason for this is that we need to know just what our devices are capable of doing if we are going to be able to understand how they function in more complex configurations. Put another way, we need to know enough about our devices and the way they function to be able to recognize, at least reasonably well, when an assembly of them is functioning properly. We must be able to make a fairly reasonable guess where to look for a problem when they are not. Even the best of us gets hung up on problems of these kinds occasionally, but it is important that we be able to minimize these incidents. At least, we need to know when to call for help, and when not to call. It is hoped that in the course of this book we have given you a good enough in-depth understanding of how circuits work so that you can usually make such a judgment with a reasonable accuracy.

COMPONENTS THAT YOU WILL NEED

The definitions probably have given you at least some hints on the devices and instruments you can expect to find helpful as you study this book. Since many of the items that were touched on briefly will prove to be of a continuing use to you, you will probably wish to start collecting at least a small stock in preparation for your future activities. As long as you have a ready access to most of the kinds of material discussed next, you should find yourself in pretty good shape for conducting the experiments found in this book.

One of the most useful adjuncts to any kind of a personal laboratory is the breadboarding socket. With one or two breadboards and the test instruments discussed previously, you will find the construction of experiments to be greatly simplified. You should give some thought as to how you can make these things most useful to you. An offhand suggestion for the socket is to mount it on the lid of a plastic box. Then, make connections to the powering strips through a set of different colored binding posts. You may have the good fortune of picking up a half-dozen colors in addition to the normal red and black. At least five colors of binding posts would appear to be needed, two for positive, one for ground, and two for negative. In addition to that, you may need some posts for signal circuits. Then, as you determine what else you want to install in the breadboarding unit, it will be easier for you to do it.

There is one area in the fixed-resistor domain which has caused no end of trouble. When one is setting up a current-limit circuit or a sensitive current shunt, resistances with values from just a few hundredths of an ohm up to about 1 ohm are needed. For most applications, fractional watt sizes are fine, but occasionally, 1- or 2-watt sizes are better. The author has found it almost impossible to locate resistors in these resistance ranges, either in the ordinary or in the precision types.

Potentiometers are another problem area. Desired values used to be rather easy to find, but not so any more. At least, not the old-fashioned ones that you mount on a panel and can turn with a knob. You have to design around what you can get and find ways to make them do. The principal problem can be "loading" when you have to accept a higher-resistance unit than what you really wanted. Sometimes, a transistor emitter-follower is a lifesaver in this situation. (This technique is used on the multiple-voltage supply that has two variable voltages.) You will want to think carefully about each application, and decide whether current will be drawn from the variable tap, or whether only resistance is really important. Then you can decide whether the use of a different resistance value than what you originally wanted would be a severe problem.

You will also find that you will need lots of capacitors—mostly ones having fixed values. The three principal types of capacitors that you will need include electrolytic large-value types for filtering and storage, ceramic small-value units for coupling and the suppressing of short pulses, and the mica or ceramic very-small-value types for tuning and coupling purposes. The electrolytic types will range in values from 10 microfarads to possibly thousands of microfarads, and the ceramic types will be from about 100 picofarads to possibly 0.5 microfarad. The mica types will typically range from just a few picofarads to about 5000 picofarads.

The sizes of electrolytic capacitors that are required for any application will have to be determined using the formula of Equation 7-1. In power supplies, capacitors with values from 500 microfarads to tens of thousands of microfarads may be required, depending on the required current magnitude. However, your most important ceramic capacitor for use in digital work will have a capacitance of either 0.1 or 0.22 microfarads. Capacitors with these values of capacitance are used extensively with TTL configurations to absorb switching transients. The same equation for determining their values applies.

Transformers are an important item in any experimenter's parts box, and they are equally important to anyone who is attempting to learn more about electronics, either in a laboratory or at home. The different sizes of transformers that you should keep on hand do not really have to be very extensive, but a number of different voltage levels should be kept handy. One very important thing is that a transformer picked for a given application should have just high enough voltage (under adverse conditions) to yield a reliable supply of power. (Adverse conditions, in these instances, are usually the lowest possible line voltages.) As an example, a nominal 6-volt transformer with a large capacitor (following a bridge rectifier) will normally yield enough voltage to provide satisfactory operation for a 5-volt regulator. However, it may not be satisfactory under brown-out conditions. A 7.5-volt transformer would be better. The voltage values the author likes to keep in his parts box include the following. (Each transformer has a center-tap on the secondary.)

$$110/220 \text{ V to } 6.3 \text{ V CT}$$
$$110/220 \text{ V to } 15/18 \text{ V CT}$$
$$110/220 \text{ V to } 12.6 \text{ V CT}$$
$$110/220 \text{ V to } 24 \text{ V CT}$$

In each case, the primary voltage should be selected to match the supply voltage you have available. Since most areas in North America have 110 volts as the basic source, readers in those areas would naturally select the 110-volt type. The use of either 15- or 18-volt centertapped transformers is optimum for high-current ± 5-volt supplies. In all cases, it is best to use the lowest voltage on all parts of your circuits that will assure reliable and proper functioning. (You will sometimes see the statement, "If the circuit puts out more than 'X' watts or 'X' volts, you should readjust it." This statement is an almost certain guarantee that the circuit as it stands is somewhat unstable.) This, of course, means that the lower-voltage transformers are usually the most important ones.

Inductors, or as they used to be called, "coils," are too specialized to be able to say very much about them. The author does not believe that there are very many adequate discussions on their proper use. However, there is some information on inductors in two of his earlier books[1,2] and some in scattered papers and articles. Radio amateur magazines, such as *QST, 73,* and *Ham Radio,* are fairly good sources for useful information of this kind. Perhaps you can find a library that has files dating back to the late 1940s and 1950s. The ARRL Handbook[3] may also have some useful information as well. There has been some useful information in those sources in the past.

OSCILLATOR CALIBRATION

If you choose to build an oscillator based on a partial kit, or decide to build it from a circuit board and components, you have a calibration problem. It used to be extremely difficult to do this in the "never-never land" between about 5 kHz and 500 kHz, but that is no longer the case. Calibration at low frequencies, around 60 Hz, is easy if you have commercial power, as it stays on frequency to within at most a tenth of a cycle. That is closer than you can normally expect a calibration to hold.

Calibration is relatively simple these days, particularly if you have one of the "digital" frequency standards that is based on a 1, 2, 4, 5, or 10-MHz quartz crystal and with TTL chips for the oscillator and frequency dividers. These standards typically use a 7400 or 7404 IC as an oscillator and 7490 chips as frequency dividers. You can use the basic circuits of Figs. 7-4 and 7-5 to build an oscillator and frequency divider chain. You can get frequencies as low as ten cycles or less this way. Then, you use Lissajous figures for the rest.

What are Lissajous figures? In one of the experiments in Appendix E, you will find that you can get all kinds of straight lines, ellipses, circles, and very complex patterns on your scope by varying the setting of a variable resistor. These patterns are forms of a Lissajous figure. When you apply two sine-wave signals on your scope, with one on the horizontal deflection system and the other on the vertical plates, the pattern you get is a Lissajous figure.

When the frequencies are the same, and the phase of one is varied with respect to the other, you can go from a sloping line through various ellipses, through a circle, and on to another straight line. When the frequencies on the two axes differ, you can draw some very intricate patterns. You can also detect frequencies which are complex multiples of each other by using these figures. If you put a frequency on the vertical plate that is twice the frequency on the horizontal

---

1. Pullen, W. A. *Conductance Design of Active Circuits* (New York: John F. Rider Publisher, Inc., 1959).
2. Pullen, K. A., *Handbook of Transistor Circuit Design* (Englewood Cliffs: Prentice-Hall, Inc., 1962).
3. The ARRL Handbook is published by the American Radio Relay League, Newington, CT 06111.

plate, you will find that you have a pattern that seems to rotate like a wheel about the vertical axis as you vary the relative phase of the two signals. The same will happen continuously if one frequency is just slightly more than twice the other frequency. The frequency ratio can be determined from the pattern that is seen by counting the number of loops that appear to be rotating along one of the pattern limits and dividing this value by one more than the number of crossings in the pattern that appear to rotate in the same manner, but which stay fixed with respect to the pattern. If you are calibrating near 60 Hz, you will want to use the cartwheel effect to locate the frequencies 120, 180, 240 Hz, etc. above 60 Hz, and 30, 20, 15, 12, 10 Hz, etc. below 60 Hz. However, it gets very difficult to count the loops above 1000 Hz.

In the higher frequency range, you will also want to use Lissajous figures, but here you should use your crystal standard and frequency divider chain. With the circuit wired as indicated in Fig. 7-5 (with one extra decade counter and two switches), you can get countdowns in the form of 10, 5, and 2 at each decade counter. A single-pole "N" throw switch will serve to select the decade counter, and a double-pole triple-throw switch will select the divide by 2, 5, or 10. (A double-throw switch will be suitable if you only want divide by two and five from this auxiliary counter.) The divide-by-five counter unfortunately is nonsymmetrical. Your counter output will be a square-wave, so your Lissajous figure will be different from that shown with sine-wave outputs, but since you will be calibrating specific frequency points as defined by the counter, this should be no problem. (If you have any doubt, a lowpass RC filter consisting of a series resistance and a parallel capacitance on the scope side of the resistance can be used to cause the square-wave signal to become sinusoidal enough to assure you that you have the right combination of frequencies. The pattern will not be perfect, but it will appear much more like what you have seen before.) This calibration process is really quite a bit of fun as long as you have these kinds of facilities available for performing the job, and you can learn a lot doing it.

With an oscillator kit like the EXAR 2206KB, you will have to obtain an audio-taper potentiometer for your variable resistance. In order to get the frequency scale uniform in octaves, the resistance should follow an equation of the form:

$$R = k \exp (a\theta) \qquad \text{(Eq. 7-2)}$$

where,

θ is the angular position of the potentiometer setting.

The audio-taper potentiometer is a close approach to this scale distribution, but a resistance of about 1% of the maximum value should be used in series to limit the minimum value, or maximum frequency. As viewed from the front panel, the two clockwise terminals should be used.

MISCELLANEOUS

You will need a special soldering iron if you plan to work with field-effect transistors or ICs, as some of them are extremely delicate. If you have only a small amount of soldering to do, a battery-powered iron containing NiCad batteries is very satisfactory. Some of the low-wattage irons like those made by Unger are usually all right but, in many applications, use of either a low-voltage, transformer-operated iron or a battery-powered unit is better. Many IGFET devices are poorly protected, and even a battery-powered unit should be unplugged before use with some of these. *Do not use a high-wattage iron like a soldering gun!*

A few more words should be said about the power supplies that are helpful in building circuits for analyzing transistors, as you should be interested in learning their inherent properties with a minimum of diversion. You will find that you need two sources of voltage and/or current for this function and, while they usually will have the same polarity, it is not always so. For that reason, a supply has been designed and built that will put out the following combinations of voltages, all at the same time and from a single transformer:

Positive 12 volts to common
Positive 5 volts to common
Positive variable to common
Negative 12 volts to common
Negative 5 volts to common
Negative variable to common

You will find that such a supply is a great convenience in much of your work. The 12-volt supplies will normally be used for providing base current through a current-limiting resistor, and the variable supplies will normally be used for providing collector voltage. The 5-volt supplies are needed in generating the variable voltages and are, therefore, brought out for convenience. They are also useful with FET device testing. More information on this supply can be found in Appendix B.

When you operate a transistor in the common-emitter mode, both of your supplies will have the same polarity, whereas, in the common-base mode, the emitter and collector supplies will have opposite polarities. The 12-volt supplies are better for supplying base current, and the 5-volt supplies are better for providing emitter current in the common-base operating configuration. You will find ample use for this supply in your later experimental work.

When testing field-effect transistors, it is helpful to use a potentiometer connected across the +12-volt and the −12-volt terminals to provide the static bias on the gate. It may be helpful to place limiting resistors on either side of the potentiometer in order to increase the sensitivity of the control but, nonetheless, both power supplies will prove helpful with many of these devices.

Sometimes taking data on a point-by-point basis is the optimum approach and, other times, it is helpful to *sweep out* the data. It is common practice to provide *swept* data on both bipolar and field-effect transistors, and it is also convenient to be able to simulate a tester that is capable of doing this with your scope. A "swept" power supply, consisting of a transformer, a bridge rectifier, a polarity-reversing switch, and an adjustable load resistance is ideal. In addition, a means for introducing controlled bias, either voltage or current, is helpful. This bias should only be introduced on command of a push-button switch. The power-supply configuration given in Fig. 7-7 can prove helpful for this, as the controls are consolidated in a single unit.

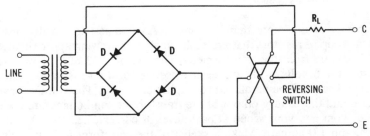

Fig. 7-7. A sweep tester circuit for use with transistors.

EXPERIMENT 1
Testing a Transformer

In this experiment, your goal is to show the variation of output voltage with load, the output waveform with a resistance load and with a rectifier load (half-wave and full-wave), the effect of a capacitor on the output of the rectifier (both with and without a load), and a comparison of these characteristics with half-wave, full-wave, and bridge rectifiers. The various circuits are given in Figs. 7-8 through 7-13.

Step 1.

It is recommended that a transformer capable of supplying either 6 or 12 volts nominal be used for this test. It should have a current

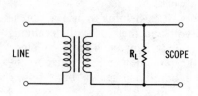

Fig. 7-8. Circuit for a resistance-
loaded transformer.

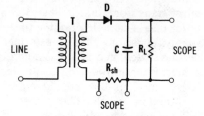

Fig. 7-9. Circuit using a transformer with a
half-wave rectifier and a load.

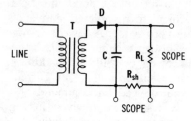

Fig. 7-10. Circuit with a rectifier, a
filter, and a load connected to
the transformer.

Fig. 7-11. Another version of circuit
shown in Fig. 7-10.

capacity of not more than 1 ampere. A good test load will consist
of a group of resistors that are each capable of drawing approximately
0.2 ampere from the transformer. (The use of a 33-ohm, 1-watt resistor for 6 volts, with two each in series for 12 volts is suggested.
Several sets will be paralleled for higher currents.) The voltage waveform and the voltage output of the transformer can be measured with
your vom and your scope at each level of current, i.e., 0.2, 0.4, 0.6,
0.8, and 1.0 ampere. Make a sketch of the waveform for any conditions in which the waveform appears degraded, and include a typical
sine wave for comparison on the graphs shown in Fig. 7-14. Note

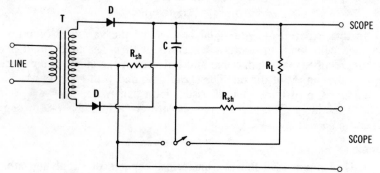

Fig. 7-12. Circuit with transformer connected to a full-wave rectifier.

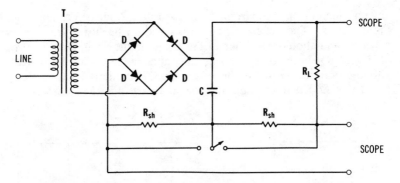

Fig. 7-13. Circuit with a bridge rectifier connected to the transformer.

the conditions under which degraded results were obtained, and explain why.

The waveform is likely to be poorest when the transformer is unloaded. This is because the magnetizing current tends to lead to waveform degradation because of hysteresis effects. This degradation largely disappears under moderate load. However, if the transformer is overloaded, the degradation can be worse because of saturation effects.

Step 2.

After you have completed your tests on the transformer under resistive load conditions, you should repeat the test using a single half-wave rectifier and its load. The test should be made first without using an output capacitor across the load, and then it should be repeated with a capacitor across the load. The two circuit configurations are shown in Figs. 7-10 and 7-11. You will notice that a very small resistance has been added in series with the output to help

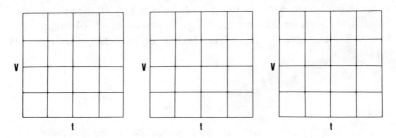

Fig. 7-14. Graphs for Experiment 1, Step 1.

"meter" the current flow from the transformer, both before and after the capacitor. The metering resistances should be just a fraction of an ohm, and should be either in the position shown in Fig. 7-10 or 7-11, but not both. In each case, the capacitance should be at least 330 microfarads, but it may be as large as 2200 microfarads or more. (You will find it interesting and useful to try different sizes in this range.)

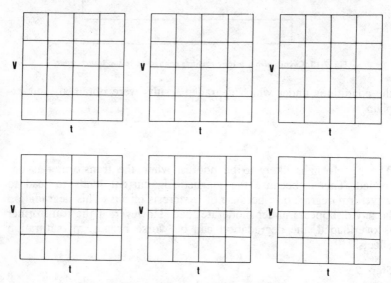

Fig. 7-15. Graphs for Experiment 1, Step 2.

Both the current and the voltage waveforms should be recorded, as you will find it instructive to explain why they appear as they do. The voltage waveforms that are of interest are (1) out of the transformer, and (2) out of the rectifier with and without the capacitor. The current waveforms are taken in the transformer return circuit and, when the capacitor is used, both before and after the capacitor. Draw sketches (or take pictures if you prefer) of all of these waveforms. (Use graphs given in Fig. 7-15.) Then, explain your results.

Step 3.

You should answer the following questions as you are performing Step 2.

1. What is the polarity of the dc output, and what is the polarity marking on the rectifier?

2. Is the transformer waveform degraded under any conditions during this experiment? (If so, label on your graph and explain the reasons.)

3. Explain the characteristics that you observed of the metering current for each of the connections. What are the most notable changes that result from placing the filter capacitor across the output? Explain.

4. You will need to compare these results with the results that you will observe for the full-wave and bridge rectifiers later, so you should take careful notes.

5. Plot the waveforms in a way that you can show the significant characteristics of the rectifier system, and explain what you have learned. (Plots from Experiment 2 will be added on your graph later.)

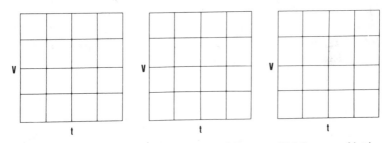

(A) Half-wave and full-wave waveforms, no filter.

(B) Half-wave waveforms, with filter.

(C) Full-wave and bridge waveforms, with filter.

Fig. 7-16. Graphs for Experiment 1, Step 3.

You will notice that the voltage across the capacitor increases when the current flows through the rectifier, and it decays after the rectifier current stops. Without the capacitor, the current follows in step with the voltage. The rectifier stops conducting when the source voltage drops below the voltage on the capacitor. Beyond the capacitor, the flow of current is much smoother than on the input side of the capacitor. When the rectifier does conduct, you should observe a definite degradation of the voltage waveform out of the transformer as a result of the loading effect caused by the current pulse. With this rectifier, the charge cycles are very short, and the discharge cycle is long. Lissajous figures can be used to verify the timing of events. The rectifier only conducts on a very short portion of one-half cycle of the applied voltage.

EXPERIMENT 2
Further Transformer Tests

A full-wave rectifier and then a bridge rectifier can be substituted for the half-wave rectifier and the measurements you have just made can be repeated. You will need a center-tapped transformer for this so, if necessary, make that change. The circuit diagrams for these two configurations are shown in Figs. 7-12 and 7-13. You will notice that two shunt resistances and a switch can be used to simplify the current measurements made in Figs. 7-10 and 7-11.

Once again, the first tests should be made with the filter capacitor removed from the circuit. You should observe the voltage and current waveforms at various points in the circuit, and under varying amounts of load. Plot your waveforms on the corresponding waveform sketched for Experiment 1 so that you can see the effects of the change in rectifier configuration. Identify the rectifier and capacitor configuration so you can recognize the results of the changes.

Step 1.

Repeat the steps of Experiment 1 for the full-wave rectifier, using a varying load and a varying capacitor size. Record your waveforms, as noted above, and discuss the results.

The positive output from the rectifier is taken from the "cathode" end, the end that is usually identified by a band. The positive terminal on a bridge rectifier is usually marked with a plus sign. With the full-wave rectifier, the outer terminals of the transformer secondary are connected to the anodes of the two rectifiers and the negative output is taken from the center-tap point. The positive output, as noted in Fig. 7-12, is taken from the junction of the cathodes. (A negative output with respect to the transformer center-tap can be obtained by reversing the polarities of the two diodes. Using two sets of diodes yields a bridge rectifier.)

With a bridge rectifier instead of a full-wave rectifier (when the bridge is formed from individual diodes), the positive output will appear at the junction of two diode cathodes, and the negative output at the junction of the anodes. Alternating current is introduced at the points where a cathode and an anode are connected. The waveform that you will see when the capacitor is not in the circuit will appear like a series of positive one-half sine waves or negative one-half sine waves. However, with the half-wave rectifier, there is a one-half cycle of zero voltage between each successive one-half sine wave. With the full-wave or bridge rectifier, the waveform is degraded when the current pulse from the capacitor is flowing, but with much less severity as the load on the transformer is lighter and more uniform. Only one-half as much charge must be stored for each conduction cycle.

Step 2.

In the process of making the above tests, you will need to find the answers to the following questions about the operation of a full-wave rectifier circuit and a bridge rectifier circuit.

1. How many diode voltage drops did you encounter with the full-wave rectifier? With the bridge rectifier? With the half-wave rectifier? Explain.

2. How do the charge and discharge times with the capacitor in place compare with the three types of rectifiers, as a function of both capacitor and of load?

3. How do the current pulse durations and amplitudes compare for them?

There is one diode voltage drop with each of the half-wave and full-wave rectifiers, and two diode voltage drops with the bridge rectifier. This makes the source problem somewhat more severe with the latter configuration. If you trace the current paths for each of the configurations, you can easily verify this. The charge and discharge times for the full-wave and bridge rectifiers is shorter than for the half-wave rectifier since the total charge to be stored is only half as much. The peak amplitude is likely to be somewhat greater with the half-wave rectifier because the capacitor will discharge further before conduction is reestablished.

EXPERIMENT 3
The Uses of a Voltage Regulator

Normally with digital systems, a simple filter in a power supply is followed by a voltage regulator. For this reason, you should add a regulator to your transformer-rectifier-filter combination and, then, determine what is important and new about this combination. The regulator may be any positive-output regulator of the three-terminal variety. It should have a current capacity of at least 1 ampere. Any of the following devices should be suitable: LM 309K, LM 340-5K, LM 7805K, or any equivalent unit. The K at the end of these codes means that the regulator is mounted in a TO-3 case with the negative terminal on the case, the input on the "base" lead, and the output on the "emitter" lead.

Step 1.

Connect the regulator to the filter capacitor terminals as shown in Fig. 7-17, remembering that you are looking at the bottom view of the TO-3 case. Connect a small 0.1 μF capacitor, from the output pin to the case. Turn the circuit on and put your finger on the case of the regulator. It should stay cool.

Step 2.

Connect a vom or a dvm across the regulator, first at position A, and then at B. Record the input and the output voltages on the regulator.

Input voltage _____   Output voltage _____

The input voltage should be at least 7 volts and, if you have a 5-volt regulator, the output should be between 4.8 and 5.2 volts. (If the regulator output voltage is more than 5 volts, the input should be 2 to 3 volts more than the expected output.) Tolerances on these regulators are roughly plus or minus two-tenths of a volt at 5 volts. *Continue to check the temperature of the case throughout this experiment.* In a later experiment, you can try to get it hot enough to make it turn off!

Step 3.

Connect a 200-mA load to the regulator, and measure both the current and the output voltage, as well as the change in voltage from

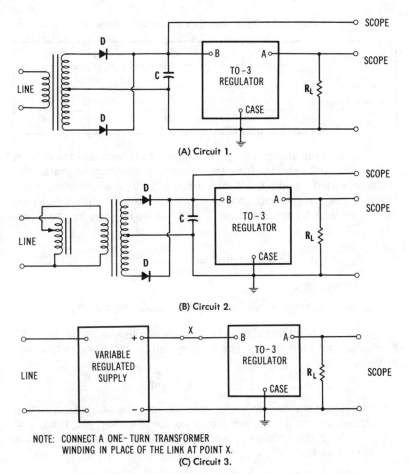

(A) Circuit 1.

(B) Circuit 2.

NOTE: CONNECT A ONE-TURN TRANSFORMER
WINDING IN PLACE OF THE LINK AT POINT X.
(C) Circuit 3.

Fig. 7-17. Circuit configurations using a voltage regulator.

the previous output reading. If you have a digital voltmeter, you will want to use it here, as the voltage change is likely to be rather small. Record your new data.

Input voltage _____ Output voltage _____ Current _____

After you have recorded these data, add another 200 mA of load. *Again, check the case temperature.* Record your additional data in Table 7-1. With these data, you can calculate the effective output impedance of the power supply by dividing the voltage change from one test to the next by the corresponding current change:

$$Z_s = \frac{\Delta V}{\Delta I} \qquad \text{(Eq. 7-3)}$$

Table 7-1. Data for Experiment 3, Step 3

Input voltage				
Output voltage				
Voltage change				
Load current				

As an example, if the voltage change was 0.01 volt, and the current change was 0.2 ampere, then the effective power-supply source impedance would be about 50 milliohms, a very small effective resistance. Voltage regulation, as defined by the equation that is accepted by the Federal Communications Commission, is given by the equation:

$$\text{Regulation} = \frac{V_{nL} - V_{fL}}{V_{fL}} \qquad \text{(Eq. 7-4)}$$

where,

V_{nL} and V_{fL} refer to no load and full load, respectively.

To get the percent regulation, you multiply this quotient by 100. Calculate and record your percent regulation and your output impedance.

Impedance _____ % regulation _____

Now, repeat the impedance calculation and the percent regulation calculation for each increment of load current that you used, and tabulate the corresponding results in Table 7-2. Explain what Table 7-2 shows you.

Usually, even with a regulator, the output voltage decreases slowly as the load is applied to it. This is called a "drooping characteristic." A regulator with either a rising characteristic or one showing no change could prove somewhat difficult to stabilize under a fluctuating load.

EXPERIMENT 4
Further Regulator Tests

With any regulator, there is a minimum voltage which must be allowed to be lost across the device to be sure it will operate. Your project is to measure this loss, and to observe the consequences of

Table 7-2. More Data for Experiment 3, Step 3

I_1				
I_2				
V_1				
V_2				
Z				
% reg				
I_1				
I_2				
V_1				
V_2				
Z				
% reg				

not allowing enough voltage to be applied to the regulator. You can now turn off the power supply and reconnect it back to the power line, but run it off of the variable transformer so that you can change the input rectified voltage.

Step 1.

Before you reapply power to the supply, turn the variable transformer to zero output volts, and carefully check your circuit. Is it functioning properly? Good. Now, turn on both the variable transformer and the supply itself. Connect your voltmeter across the output of the regulator, and increase the line voltage until the regulator output reaches its proper voltage. How much is it? What happens when you increase the input voltage further? Does the regulator output change significantly. At what voltage on the regulator input was the point of proper regulation reached?

The minimum value of the input voltage should be about 7 volts when the output stabilizes. It depends to some extent on the load current, however, and the minimum for a 5-volt regulator could be as high as 8 volts. Above this input voltage, the output voltage should not change significantly.

Step 2.

Leave the voltmeter connected to the output, and again increase the line-voltage setting to a value that gives more than a minimum input voltage to the regulator. Does it change significantly? Decrease the line voltage until the output voltage shows a slight dip in value. Then turn it back up, and measure both the input and output voltages. What have you observed?

You were made to go above the supposedly correct voltage point so that you could be sure that you had in fact passed it. At that point, there is a negligible change in the output voltage. Then, you brought it back down so that you could find the minimum voltage difference that you had to have across the regulator to make it function properly.

Step 3.

Now, connect the scope across the output of the regulator, with the input voltage at least 1 volt or more above the minimum value. How much output ripple can you observe on the scope? Measure the peak-to-peak value of the ripple waveform by changing to ac coupling and increasing the sensitivity until you can read the total swing, which should be, at most, a few tens of millivolts. You may find that it increases significantly when you load the regulator. If you do, look at the input side of the regulator to see what has happened there. You can get an indication of how bad this is by dividing the increase in peak-to-peak voltage by the load-current change. What do you find?

The peak-to-peak ripple voltage will probably be at most 10 millivolts until you load the regulator heavily, and then you may reach a point where it increases significantly. Increasing the source voltage into the regulator will again reduce the ripple magnitude. When properly powered, these regulators are remarkably effective.

Step 4.

Examine the input waveform of the power applied to the regulator. How much peak-to-peak voltage variation do you find there? _____ volt(s). What does this tell you about the ability of the regulator to remove ripple from the power-supply output? Remembering that the ratio of two voltages in dB is stated as twenty times the log of that ratio, how much improvement, in dB, can this regulator give you?

As noted previously, the ripple at the input will probably be significant, and it will increase sharply as the load is increased. It may be as much as 1 volt, but the minimum voltage at the input MUST exceed the minimum required to assure good regulation. If the output ripple is 10 millivolts, then the reduction will be 40 dB. What do you get for yours? _____ dB.

Step 5.
Slowly reduce the voltage into the regulator until the scope begins to show an increase in ripple voltage at the output. This will probably appear as sharp but short-duration dips in an otherwise apparently constant voltage. Measure the minimum and maximum voltage at the input of the regulator, and measure the peak-to-peak ripple at the input. You will begin to notice the dips just before they can be noticed as a reduction of overall voltage output on your meter. You need to know just what the minimum voltage at the regulator input is when it starts to fail to regulate. How can you do this?

There are several ways this can be done. One is to use your variable regulated power supply as a source for the regulator, and to find the minimum input voltage with it. (It should be constant within a few millivolts.) Another way is to calibrate your scope precisely. A third way is to use the regulated supply and a power transformer, and slip ONE winding into the circuit so that you can get a fraction of a volt of ac from it. This voltage may be added to the regulated input voltage, which is then reduced until you can see the "ripple" that you have introduced. Can you suggest other ways?

Step 6.
The most convenient method of checking the minimum input voltage is by the use of a variable regulated supply, either without an ac signal or with it. Connect the variable supply in place of the rectifier and filter capacitor, and connect a single-turn winding in series with its output. Then, put a full load on the regulator under test.

Starting with the output from the regulator at least 1 volt above the minimum, connect a scope and voltmeter to the output of the device under test. Then, reduce the input voltage until you get an evidence of the loss of high-quality regulation. You will need an input voltage that is at least 1 volt more than the minimum value at that point to assure the kind of operation that you require. How much voltage did your regulator require? What else of interest did you observe?

The requirement was probably a shade over 7 volts for a 5-volt regulator. You undoubtedly found that the use of the one-turn winding to inject an ac signal was extremely effective as a means of testing.

Step 7.

With a fixed regulator connected directly to the rectifier (not to the variable regulator), set your variable transformer so that you can just notice on your scope a small dip in the regulator output. Then, place an additional capacitor, at least 1000 μF, in parallel across the regulator input. What happens?

The ripple voltage from the rectifier should have dropped significantly, with the result that the dip in the output of the regulator should have disappeared. The variable transformer must be readjusted downward just a bit more to get the dip back again.

Step 8.

Referring back to the data on the half-wave rectifier and the data on the full-wave and bridge rectifiers, what alternating voltage would have to be applied to each of these rectifier systems (using a 1000 μF capacitor), if the load current is going to be 0.5 ampere? Set up each configuration, and see how closely you were able to estimate the voltage that you needed. Fill in Table 7-3, and discuss the results. Were there any significant discrepancies?

Table 7-3. Data for Experiment 4, Step 8

	Estimated values	Experimental values
Half-wave Rectifier		
Full-wave Rectifier		
Bridge Rectifier		

How large a capacitor would be required with the half-wave rectifier for it to be suitable with the same transformer and the bridge rectifier?

You will observe that you have much more difficulty filtering the ripple voltage with the half-wave rectifier, but that your total rectifier voltage drop will be smaller than for the bridge rectifier. The full-wave rectifier has the same rectifier voltage drop (almost) but with lower peak currents. It requires a transformer having a center tap and twice the total output voltage that is required for the other two configurations. You will probably select the full-wave or the bridge configuration unless your current requirements are very small.

EXPERIMENT 5
Signal Generator Tests

You will need to use a scope extensively in this experiment. You will need to examine waveforms, make approximate frequency measurements, and measure voltages in your circuit. You will, in particular, be concerned with learning the techniques for calibrating the frequency scale on your generator (or checking its calibration).

Step 1.

Set the frequency of the generator at a low frequency—around 100 Hz. How close do you think you can set it using the dial? If you can adjust it to a 60- or a 120-Hz setting, do so. (If your source of power is 50 Hz, use 50 or 100 Hz.) Connect the generator to the vertical input of a scope, and set the waveform so that with 1 volt input, you get 1.4-cm deflection above and below the centerline. Your scope is now measuring about 1 volt per centimeter.

Step 2.

Connect a small transformer delivering about 6 volts to the vertical input of your scope, and see how close the frequency that you set compares to the power-line frequency. How close were you? What does this tell you about the calibration of your oscillator?

The power frequency on an extensive electrical grid is likely to be much closer than the oscillator calibration is likely to be. Most of the utility power in the United States is within 0.02 cycles of being right essentially all the time.

Step 3.

Connect the transformer output to the horizontal input terminals of your scope, and the oscillator to the vertical terminals. Adjust the horizontal and vertical amplitude controls so that the amplitudes on the two axes are approximately equal, and then trim the oscillator frequency until you get a stationary single-line figure like a circle, an ellipse, or an arcuate figure of some kind. When the pattern takes this form, there is an exact integral relation between the two frequencies—1:1, 2:1, 3:1, etc. Draw pictures of some of the patterns that you obtained.

Can you keep the pattern exactly stopped, or does it drift irregularly? (The author obtained figures like those shown in Fig. 7-18.) To what do you attribute any drift you observe? (This is not a steady drift, but a shifting back and forth.) Remember, this pattern is called a "Lissajous figure."

The erratic drift you may have seen is probably mostly in the oscillator, but a little of it could be in the power system. The amount of rotating mass in large generators is so great that frequency changes due to the generator will take place very slowly. The voltages still

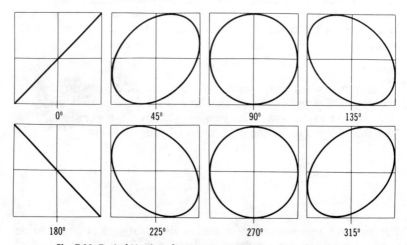

Fig. 7-18. Typical Lissajous figure patterns with the phase angle listed.

may vary enough so that the scope sweep or the oscillator frequency will vary more rapidly than the frequency of the power generator is likely to vary.

Step 4.

If you have one of these little crystal-controlled 60-Hz sources using a color tv crystal that you can buy in kit form, connect it up and compare its output against both the electrical power-system signal and the oscillator signal. First, examine the output of this crystal oscillator on the scope, and observe that its output is nearly a perfect square wave. Then, observe both the Lissajous pattern (which results against the signal generator when it is set at about 60 Hz), and also the power-line signal. Waveforms have been observed which look like those in Fig. 7-19. There will be a very fast near-vertical transfer of the sort that is sketched, and this transfer will move back and forth across the screen at a rate that depends on the frequency difference. You may want to try varying the variable ceramic capacitor on the

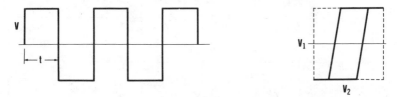

Fig. 7-19. Lissajous pattern for a divider standard.

crystal oscillator to try to stop the motion of the crossover line. If you have a frequency counter available, the crystal output itself should be set at 3.579545 MHz exactly. What do you observe?

The power-system drift is likely to be large compared to the range that you can trim the oscillator. The frequency counter is the better approach. Of course, the frequency counter could be off too so it should be stabilized against WWV. What has this taught you about the relative stability of the frequency from these various instruments?

Clearly, the most stable device is the crystal source. However, there are several things that are more stable than a crystal source. These include lasers and nuclear resonators, which are usually stable to within one part in 10^{12} to 10^{13}. Quartz crystals are usually only good to one part in 10^5 to 10^{12}, depending on the care with which the source has been built. The power system is usually good to about one part in 3000, whereas, ordinary electronic oscillators are less stable. (Long-term stability of the power system is held to one part in 10^{12} or more by special controls.) However, it is possible to make oscillators that are dependent on LC circuits good to about one part in 10,000.

Step 5.

You can check the calibration of your signal generator at a series of frequencies. Since it is easier to use a sine-wave waveform like you obtain from the power line rather than a square-wave waveform like you have from your frequency standard, it is suggested that you connect the transformer secondary to the horizontal input of your scope again. Connect your oscillator to the vertical input, and adjust

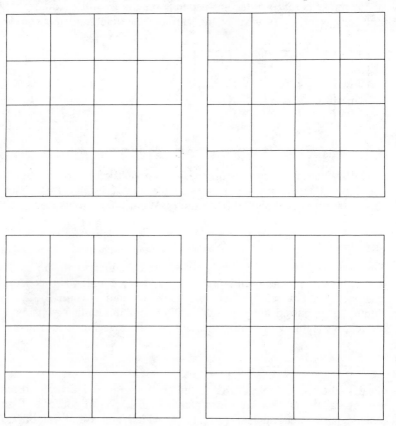

Fig. 7-20. Graphs for Experiment 5, Step 5.

the frequency until the pattern goes slowly through the forms shown in Fig. 7-18. When you stop it exactly, you have a signal on the vertical input that matches the frequency of your power source.

Now, you need to reduce the signal-generator frequency slowly, and watch the "disc" start rolling. As you continue to reduce the frequency, there seem to be crossovers that rotate, sometimes horizontally, sometimes vertically. Decrease the frequency to about 30 cycles (or half the line frequency) quickly, ignoring the crossovers for the moment. Once again, you get a very stable pattern. This time, the circle looks like it has been bent so that you have a parabolic pattern opening either to the left or to the right. (At 120 cycles, it seemed to rotate about a horizontal axis, opening either up or down.) This means that the frequency on the vertical is one-half that on the horizontal. Sketch the various waveforms that you see on the four graphs given in Fig.7-20.

Step 6.

Now that you have seen what the waveforms look like at half power-line frequency and at twice power-line frequency when compared to the line frequency, you should find it interesting to see what the waveforms look like for another series of combinations. (If your power frequency is different from 60 Hz, examine frequencies that have the same ratio to your line frequency as do the following suggested frequencies. The two rows given are for 60 Hz and 50 Hz as the source frequency.) If your frequency differs from these, then take the same ratio to that frequency as these are to 50 and 60 Hz. You should be able to identify the patterns by the number of "humps."

30	40	45	75	90	120	180	240
25	33.3	37.5	62.5	75	100	150	200

Sketch the waveform patterns you observe on the graphs given in

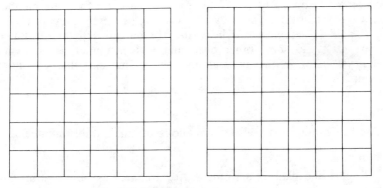

Fig. 7-21. Graphs for Experiment 5, Step 6.

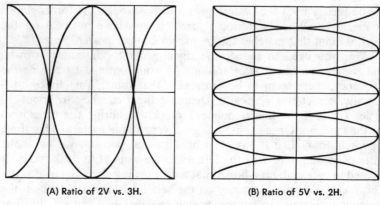

(A) Ratio of 2V vs. 3H.　　　　　(B) Ratio of 5V vs. 2H.

Fig. 7-22. Sample sketches for Step 6.

Fig. 7-21. Some sample sketches are given for ratios of 2V vs. 3H and 5V vs. 2H to help you make your sketches (see Fig. 7-22).

Step 7.

Vary the oscillator frequency from 30 (25) Hz to 60 (50) Hz. Stop at each point where it appears that you may have an exact ratio. Observe carefully as you cross each of these points, and you will find two interesting things. The first is that just either side of the stopping point, the structure seems to rotate. The second is that there will be a series of maxima (both positive and negative), and there will also be a series of crossovers that seem to rotate, with the crossings remaining fixed with respect to the pattern.

For example, examine the 45 Hz (37.5) condition. You will see three positive maxima, three negative maxima, and there seem to be three fixed crossovers that occur at right angles to the direction of rotation. If you take the number of either positive or negative maxima and divide it by a value that is one more than the number of fixed crossovers, you will in fact have the frequency ratio for that pair of waveforms. Check the combinations that you tested previously, and some other ones that you encounter as you vary your oscillator frequency, and note the frequencies that you have found.

The ratios are almost all combinations of small whole numbers!

Step 8.

Repeat the calibration process for frequencies above your line frequency, such as 120 (100), 180 (150), and so on, up to at least

900 (750) Hz. As short a time back as just before World War II, this was about the only means some of us had for calibrating such things as audio oscillators and audio signal generators. You may find it interesting to make a table of frequency ratios against the number of maxima and the number of stationary crossovers that you could observe on the rotating-wheel pattern as you vary the frequency (use Table 7-4).

Table 7-4. Data for Step 8

Positive maxima			
Negative maxima			
Crossovers			
Frequency ratio			
Positive maxima			
Negative maxima			
Crossovers			
Frequency ratio			

Step 9.

How can you extend this technique so that it is useful at still higher frequencies? Explain, and then try it.

The simplest way is to use a reference frequency that is known precisely and is in the frequency range you wish to use. Often, it may be difficult to find such a source. It is common these days to have a frequency standard followed by a frequency-divider chain for use with computation equipment. The problem here is the fact that the output is in the form of a square wave. The simplest thing to do then is to introduce a simple low-pass filter consisting of a series resistance and beyond it a parallel capacitor into the circuit. The value combination of the resistance and the capacitance is determined from $2\pi fCR = 1$, where f is the frequency to be filtered. This will convert the square wave into a roughly triangular wave, and will be an adequate simulation of a sine wave. You can use any output from your frequency-divider chain as a source of a calibration signal simply by putting in the right RC filtering combination.

EXPERIMENT 6
Measurement of Phase Angles by Using Lissajous Figures

This Lissajous-pattern technique is an extremely useful method, as it can be used to measure the relative amplitude of a single component of voltage without having the measurement corrupted by unwanted charging currents or by harmonics. For this reason, you may want to make some sample phase measurements just to see how it goes.

Step 1.

Take a center-tapped low-voltage transformer and wire it up as shown in Fig. 7-23. You will notice that the scope is connected across one half of the winding for your horizontal connection, with the cen-

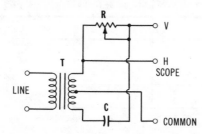

Fig. 7-23. A phase-shifter circuit for making phase measurements with Lissajous figures.

ter tap connected to ground, and the vertical connection is connected to the tap between the capacitor and the potentiometer. This circuit has the interesting property that as you vary the resistance in the potentiometer, you can vary the relative phase between the two voltages by nearly 180° without any significant variation in the magnitude of either voltage. Describe what happens as you turn the knob on the potentiometer.

As you turn the knob, you should see the various waveforms that you observed as two nearly equal frequencies drifted past each other in Experiment 5. You were watching phase drift then, and the turning of the shaft on the potentiometer introduces a similar phase drift now.

Step 2.

Turn the knob on the potentiometer to put a zero resistance in series with the capacitor. Now, the two signals on the horizontal and the vertical amplifiers should be in phase. Adjust the amplitude of one so that the horizontal and the vertical projections of the sloping line are equal in length. Turn the knob on the potentiometer until you get as nearly a perfect circle as possible. The two voltages are now 90° out of phase. If the circle looks somewhat elliptical at all settings, adjust the knob so that the major axis is either horizontal or vertical. Then, a minor trim on one gain control should lead to a nearly perfect circle.

Step 3.

Measure the resistance that is in series with the capacitance with your vom or dvm. (This is the potentiometer resistance that is in the circuit.) The magnitude of the resistance should satisfy the following equation.

$$\omega CR = 2\pi fCR = 1 \qquad \text{(Eq. 7-5)}$$

where,

$\omega = 2\pi f$.

The product of C and R for 60 Hz is approximately 0.00265. What do you find the product to be?

$$C \times R = \underline{\hspace{2cm}}$$

Later you may want to perform this test with an audio or a radio signal generator.

Step 4.

Turn the knob on the potentiometer to insert a maximum value of resistance in the circuit. What happens to the phase based on the behavior of the figure on the scope screen?

Why was the amplitude of the voltage observed between the capacitor and the resistor so constant?

The slope of the line or ellipse was observed to close into a nearly diagonal line with maximum resistance, and it did close completely with zero resistance. The slope of the pattern at maximum resistance was at right angles to the one for zero resistance. The reason for this is that you are examining the voltage on the lower half of the transformer when the resistance is near zero, and on the upper half when it is maximum. This has shifted the phase of the signal you are observing. When the pattern becomes a circle, the phase at the tap between the resistor and the capacitor is 90° ahead of the phase of the voltage in the bottom half of the transformer and 90° behind the top half.

Since the voltage across the capacitor is 90° out of phase with the voltage across the resistor that is in series with it, the combination is like a right triangle inscribed in a circle. The distance, or voltage, from the midpoint of the hypotenuse to the intersection of the capacitor and resistor is constant, and the output of the combination will be a constant voltage with a variable phase.

EXPERIMENT 7
Corner Frequency Measurement

The experiment you just completed showed you that you can get phase shift with a capacitor, but that the amplitude of the signal stayed almost constant because of the way the circuit was used. The reason it stayed constant was because the voltage was measured with respect to the midpoint of a balanced voltage configuration. You cannot count on the voltage being constant in general, however. Now, a typical RC circuit can be examined to see how it behaves in a more normal situation.

You may be interested in tracking both the amplitude and phase in a typical circuit, and the best way to do this is with the same Lissajous figure that you have been finding useful so far. Usually you will have the resistor in series to the scope and the capacitor in parallel with the scope input. Instead of varying the resistance, however, you now vary the frequency, observing the comparative effect at the input and the output of the RC network. (Use the Lissajous figure to make the comparison.)

Step 1.

Take the values of resistance and capacitance that you found in Step 3 of Experiment 6. Vary the frequency of your audio signal generator from about 500 Hz down to about 30 Hz. As you do this, plot the amplitude of the output against the frequency, and sketch the waveforms on the graphs in Fig. 7-24. Select frequencies of about 500, 250, 120, 60, and 30 Hz for your test. (You can use Lissajous figures with respect to your power line if you wish.) Describe how the voltage varies on the vertical for a constant amplitude on the horizontal (input), and notice in particular the shape of the ellipse when the vertical amplitude is 70% of the initial value. Sample patterns are shown in Fig. 7-25. Use Table 7-5 to record your data. (The value of X is the "x" coordinate of the maximum value of the output signal on the Lissajous figure.)

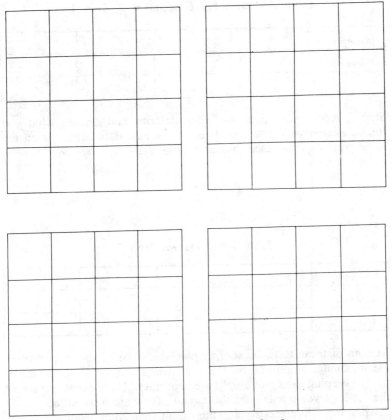

Fig. 7-24. Graphs for Experiment 7, Step 1.

Step 2.

Now, interchange the resistor and the capacitor so that the capacitor is in series and the resistor is across the scope terminals. Repeat

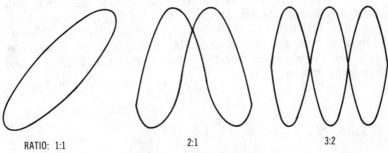

RATIO: 1:1   2:1   3:2

Fig. 7-25. Assorted Lissajous patterns.

Table 7-5. Data for Experiment 7, Step 1

Frequency				
Amplitude				
X				

Step 1, record your data, and then describe and explain what you found. Sketch your new Lissajous figures in a different color on the same graphs as you used in Step 1 (Fig. 7-24).

Table 7-6. Data for Step 2

Frequency				
Amplitude				
X				

In both of these tests, be sure to note the slope of the ellipse when the maximum amplitude on the output is about 70% of the initial value (keeping the input amplitude constant). These points represent the "half-power points" for the circuit. They are also called the "3 dB points." They are the accepted limit-frequency definition points for RC circuits. The voltage gain is said to be "down 3 dB" at that point. You can find 3 dB points for all kinds of circuits, including transformer-coupled circuits, so they are very important to you.

Step 3.

Repeat this experiment with another capacitance value, but with the same resistance. Select the capacitance so that the expected RC products is first 0.00132 approximately, and then 0.0053. Tabulate the data for both tests. Do these combinations have 3 dB frequencies where you would expect them? Tabulate the data in Table 7-7 and discuss the results.

Table 7-7. Data for Step 3

Frequency				
Amplitude				
X				
Frequency				
Amplitude				
X				

If the capacitance values were what you thought they were, your 3 dB frequencies were first double and then half the previous value. You can find out what the correct value of the capacitors were, in terms of the value of the first capacitor, by finding out just how far off the results seem to be.

Step 4.

Select a frequency and, also, a value for both R and C that will give a product for Equation 7-5 that is equal to unity at that frequency. Then, test the effect of scaling by doubling the resistance and halving the capacitance, and then repeat the test by halving the resistance and doubling the capacitance. Find the 3 dB frequency for each case. Discuss the results and list the actual frequencies that you obtained.

Any variations you may have noted would have to be due to variations in the actual values of the components compared to the correct values. If you have access to accurate capacitance and resistance measuring equipment, measure the correct values and see if the results check with the measurements.

EXPERIMENT 8
Properties of an Inductor

In this experiment, you will demonstrate how an inductor can be used with a capacitor to form a tuned circuit, and what you can do with a tuned circuit to vary its properties. But, first, you should note that you will be using the inductor in the *driven* mode of operation, and that the exact frequency of operation is dependent to some extent on just how you use the inductor. As compared to the driven resonance, the natural resonance of a circuit is that frequency which will result if energy is stored (in either the inductor or the capacitor) and then the circuit is permitted to decay (dissipating the stored energy).

Since the operating frequency of a tuned circuit in the driven mode depends on the manner in which it is excited by the driving energy, this energy must be introduced in a manner that will minimize the disturbance of the circuit. The use of an extremely small value of resistance (that has a negligible inductance) in series with the inductor of the circuit is one of the better ways for doing this. If possible, use no more than 0.1 ohm for this purpose. If you do not have a resistance this small, parallel one or more of the smallest that you have. A small ammeter shunt might work quite well, as a 1-ampere shunt has about 0.1 ohm resistance.

Presumably you know what the approximate inductance value is of your sample, so select a frequency that you can easily use with your oscillator and your scope, and select your capacitance by the use of the equation:

$$(2\pi f)^2 LC = 1 \qquad\qquad \text{(Eq. 7-6)}$$

or

$$\omega^2 LC = 1$$

The resulting value of capacitance should enable you to obtain the kind of operation you desire, based on the circuit of Fig. 7-26. You do face one problem here. The operating frequency that you choose MUST be high enough that the inductor is willing to behave as an inductor. How can you assure that?

Since you cannot tell that an inductor will behave properly just by looking at it, you have to do it experimentally. Essentially, what

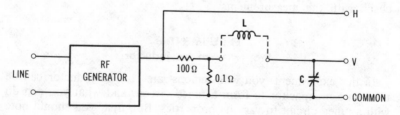

Fig. 7-26. A basic circuit for measuring circuit Q.

you have to do is to set up the equivalent of a "Q" meter, a circuit that will show the buildup of voltage resulting from the exchange of energy that was discussed earlier. Such a circuit is shown in Fig. 7-27. For this test, you need to compare the input and the output voltages of the tuned circuit, and to observe how they behave as you tune across the resonant frequency. You will need to proceed as discussed in the following steps.

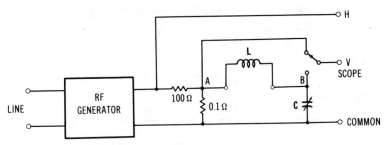

Fig. 7-27. A practical circuit for testing circuit Q.

Step 1.

Connect the circuit as shown in Fig. 7-27, and sweep the chosen range of frequencies with an oscillator. Observe with a scope in normal sweep if there is any significant variation in the voltage across the 0.1-ohm injection resistance (point A). If there is, you will have to readjust the drive voltage from the oscillator to keep it relatively constant as you make your test. Then, transfer the vertical input lead to point B, and couple point A to the horizontal input. In the neighborhood of the expected resonance, the voltage should peak sharply and should reach a value of at least ten times the voltage present at point A. (If you do not find this peak, revert to normal horizontal sweep and search the full frequency range of the generator and the scope.) Now, you can evaluate both the amplitude and the phase behavior of the inductor, and can continue your test. If the Q is apparently too low, try a higher frequency using either a smaller capacitor or another inductor.

Step 2.

Measure the approximate voltage multiplication factor observed at resonance with your inductor. As noted previously, this factor is the approximate Q-factor for the tuned circuit, which includes the Q of the capacitor in addition to the effect of the injection resistance and any loading due to the scope. Make notes on what you have learned, including the various Q values that you have measured. Test the coil at different frequencies by varying the capacitance, and record how the Q behaves. Record the data in Table 7-8 (f is the frequency).

Table 7-8. Data for Experiment 8, Step 2

f				
Q				
f				
Q				
f				
Q				

Step 3.

Because the voltage maximum of a tuned circuit generates an approximately circular Lissajous figure when the voltage gains are properly adjusted, the 3-dB point will again correspond to an ellipse that has intercepts on the horizontal and the vertical axes that are approximately 70% of the maximum deflection on each axis. The frequency difference between the upper and the lower points again gives a way of measuring the Q of the circuit. If you can, measure these two frequencies and determine the Q from the equation:

$$Q = \frac{f_o}{\Delta f} \qquad \text{(Eq. 7-7)}$$

where,

f_o is the center frequency,

Δf is the frequency difference between the upper and lower 3-dB points.

Record your corresponding measurements in Table 7-9.

Step 4.

Set up the equations. Show that the correct intercepts for the 3-dB points on a Lissajous ellipse are at the values of $\sqrt{0.5}$ times the maximum vertical signal, and that the peak phasor value (length of the semimajor axis of the ellipse) is 0.924, based on unit peak values.

Table 7-9. Data for Step 3

f_o				
Δf				
Q				
f_o				
Δf				
Q				

Step 5.

You now know the Q for the coils you have tested. It turns out that the effective impedance of the tuned circuit is given by the equation:

$$Z = Q \times \omega L \qquad \text{(Eq. 7-8)}$$
$$= \frac{Q}{\omega C}$$
$$= \frac{L}{CR}$$

Find a carbon resistor having approximately this resistance for one of the tuned circuits that you tested, and connect it in parallel with the tuning capacitor. Then, retune the circuit to resonance, or trim the frequency to resonance. Repeat the measurements for the circuit Q, and explain what you have found. (Record the data in Table 7-10.)

Table 7-10. Data for Step 5

f				
Q				
f				
Q				

You will find that the Q has been reduced to approximately one-half its previous value, and that the frequency of resonance may also have been shifted somewhat. The shift is due to two things: first, the capacitance inherent in the resistance and, second, due to detuning as a consequence of the shift due to a change in the point of resistance loading of a driven resonance.

Step 6.

Replace your capacitor with a capacitor pair connected in series. The capacitor connected to the circuit return (ground) is to be designated C2 and it will have at least ten times the value of C1, the other one of the capacitor pair. Retune the circuit to resonance, as the effective value of the capacitance is decreased, and the resonant frequency will be increased by a few percent. Find the new center frequency and the new 3-dB points. Compute the new Q, which should differ but little from the previous value.

The new frequency is _____; The new Q is _____

Step 7.

You will want to try varying values of load resistance across capacitor C2, and observe the behavior at the inductor-capacitor junction as done previously. Plot the variation of the peak voltage (retrim the frequency as needed and note changes), and measure the circuit Q as you reduce the value of the loading resistance connected across capacitor C2. The load you apply should appear at the monitor point as if it had the value:

$$R = R_L[(C1 + C2)/C1]^2 \qquad \text{(Eq. 7-9)}$$

where,

R_L is the load across capacitor C2,
R is the apparent value at the point of scope measurement.

This equation applies *as long as* the current drawn by the load resistance is *small* compared to the circulating current resulting from the energy exchange between the inductance and the capacitance chain. Otherwise, the tuned circuit no longer responds as a resonant circuit. It is interesting to note that this problem does *not* occur when a tapped coil having a reasonable amount of magnetic coupling between turns is loaded in a similar manner, as transformer action provides the required increased current.

Test your tuned circuit using varying values of R_L, measuring the circuit Q as you go to get a value of R. Then plot a curve of the product of RC1 as a function of the product of R_LC2. At what ratio of R to R_L does this begin to depart from a slope of 45°? Find the ratio of the loaded Q (at this point) to the capacitance ratio. At what ratio of these two numbers does the break from 45° occur? (Record the data in Table 7-11.)

Table 7-11. Data for Step 7

R_L				
Q				
R				
RC1				
R_LC2				
R_L				
Q				
R				
RC1				
R_LC2				

You can expect to find the break when the ratio of RC1 to R_LC2 begins to depart from the ratio of C2 to C1.

EXPERIMENT 9
Tests of an Interstage Transformer

The purpose of this experiment is to demonstrate some properties of interstage transformers that should be self-evident, but are typically misunderstood even by many experts in the field. These properties result from the fact that transformers are not sources or sinks for energy in the usual sense, but serve to convert the form of energy so that it can be used more effectively. To show this, you need to make some measurements on the transformer as a device and, then, see how it behaves as you modify its manner of utilization.

It might be noted at this point that when the power grid used by electrical public utilities were first built across state lines, an attempt was made to keep these lines out of interstate commerce by installation of transformers on the state line. The courts rejected this ploy as an artifice to avoid regulation. This is not a comment either for or against the legality of federal regulation, but just a statement of the decision reached by the courts that the transformer did not sufficiently change the nature of the product being processed to justify avoidance of the concept of interstate commerce, in this instance! This demonstration can be made most graphically by causing the unit to operate in the output circuit of a transistor amplifier. The reason why this is so was demonstrated in Chapters 3 and 4, but for now it can be stated that the basic reason is that the output circuit for a transistor is a current source when it is operating as an amplifier. You can operate a transformer under varying load configurations, and examine both the waveform available from it and the overall amplification of the amplifier as a function of frequency.

Step 1.

Determine the effective input impedance of the transformer with its secondary open circuited and, also, the corresponding value for a short-circuited secondary. The circuit required can be set up on a solderless breadboard as shown in Fig. 7-28. You need this informa-

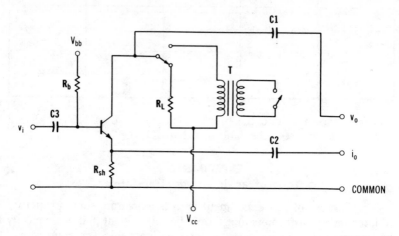

Fig. 7-28. Circuit for measuring transformer reactance.

tion so that you can learn something about the imperfections in the transformer. A perfect transformer would show infinite primary impedance with its secondary open circuited, and zero impedance with its secondary short circuited. An estimate of the coupling factor between the windings can be obtained from the equation:

$$\frac{L_{ps}}{L_{po}} = (1 - \kappa^2) \qquad \text{(Eq. 7-10)}$$

where,

κ is the magnetic coupling factor.

Ideally, κ would be 1.000000. In this equation, the inductance can be determined by measuring the voltage (signal) across the winding and dividing it by the signal current and, then, dividing that quotient by $2\pi f$. (A more complex procedure may be set up to separate out the resistance component using Lissajous figures if you wish.) Remember that the voltage across the inductance will be 90° out-of-phase with respect to the voltage across a resistance carrying the same current, as you may wish to check to see if the resistive component of voltage across the transformer winding is important. It could well be when the secondary is shorted. Also, be sure that the test signal into the base of the transistor is not more than 5 millivolts unless you have an emitter resistance in use to linearize the operation of the transistor. A good starting frequency is 500 Hz. Record your data in Table 7-12.

Table 7-12. Data for Experiment 9, Step 1

V_s				
f				
R_s				
R_L				
V_{Lq}				
V_{max}				
I_L				
V_{Lq}/I_L				
L				
V_s				
f				
R_s				
R_L				
V_{Lq}				
V_{max}				
I_L				
V_{Lq}/I_L				
L				

The voltage V_s is the voltage across the shunt, and this scale must be calibrated. This can be done by applying a known amplitude of signal voltage and setting the gain control on the horizontal axis to give a known pattern width. The ratio of the voltage to the known test resistance gives the signal current magnitude. Application of the signal at the collector gives a means of getting the phase angle of the voltage on the transformer directly, and throwing the switch to the resistance side and adjusting the resistance R_L to give the same vertical deflection and measuring the corresponding resistance gives the impedance magnitude for the transformer winding.

Getting the phase angle of the transformer-winding impedance out of the Lissajous figure involves getting the sine of the phase angle from two intersections on the figure. Symmetry is first established when the resistance load is in use, by adjusting the centering so that both the horizontal and vertical deflections are symmetric with respect to the center of the scope tube screen. The ratio of the vertical deviation of the ellipse (from the center of the screen) to the maximum deviation in the vertical direction (from the high-point on the ellipse) gives the sine of the phase angle of the voltage across the transformer winding. The reactance of the transformer winding, then, is the product of the impedance magnitude (obtained by resistance substitution) multiplied by the sine of the phase angle.

You may wish to verify the linearity of both the horizontal and the vertical deflection systems on the scope before you make the transformer test. The same basic procedure can be used for calibration of both the horizontal and the vertical axes by interchanging the waveforms. Two sinusoidal or triangular waveforms may be used for this calibration, one on the horizontal and the other on the vertical axis. The frequencies should be so adjusted that the calibrating axis has a frequency approximately an odd number of halves times that on the axis being calibrated. For example, if the horizontal axis is being calibrated, a 50- or 60-Hz sine wave or triangular wave might be used on it, and a frequency that is 3.5 or 4.5 times it might be used on the vertical axis. Then, the frequencies are brought into almost perfect synchronism, so that the pattern drifts very, very slowly. The sizes of the loops between the horizontal crossovers can then be tracked as they cross the screen, and any variation in the length of spacing between successive crossovers is an indication of amplifier distortion. A simple interchange of horizontal and vertical will permit a similar calibration of the vertical axis.

Step 2.

Next, place a short circuit on the transformer secondary and repeat the above tests. The apparent reactance of the primary should now appear to be very small, and the Lissajous technique must be used here to separate the resistive and inductive components. The data to be recorded are the same as in Step 1, and may be recorded in Table 7-12. Use Equation 7-10 to calculate the value of the coupling coefficient κ. ($\kappa = $ _____.)

Step 3.

Load the transformer secondary with a resistor that has a value small enough to cut the measured signal voltage at the transformer input in half. You will need to be able to make this test with the resistor both connected and disconnected, so that you can observe the waveform at the secondary both with and without the resistor in place. You will also need to be able to vary the input frequency so that you can observe the effect on the frequency response that is caused by switching the load onto and off of the transformer. As you vary the input frequency, find where the output decreases to 70% of the midband value at the low- and high-frequency end. While you are doing this, observe the waveform so that you can detect any degradation of your applied sine wave. (You will want to observe the amplitude and waveform on the primary side of the transformer as well as on the secondary, as the mode of operation can affect the input as well as the output.) Make your tests both with Lissajous figures and with the conventional sweep. Tabulate your upper and lower frequency points as a function of load resistance (use Table 7-13), and explain what you have observed.

Table 7-13. Data for Experiment 9, Step 3

R_{Ls}				
f_{lo}				
f_{hi}				
R_{Ls}				
f_{lo}				
f_{hi}				
R_{Ls}				
f_{lo}				
f_{hi}				

Was the waveform better with R_{Ls} in place or with it disconnected?

You should have had a better secondary waveform with the resistor in place, as the degrading effects of hysteresis are minimized then.

As you lower the resistance value further, you should find at least some further improvement in the waveform. Sometimes the Lissajous figure presentation will tell you more about the quality of circuit operation, and other times, the conventional sweep. That is why it is valuable to be well-versed in the use of each.

Step 4.

Repeat Step 3 with a series of values of secondary load resistance, each between 50% and 70% of the previous value. Again find the upper and lower 3-dB points, and record the data in Table 7-13. In each test, compare the waveform with and without load, and make any comments which may be pertinent in the following answer space. You may also find it interesting and useful to vary the amplitude of the input signal near the upper and lower frequency limits to find out how the distortion level is affected by proximity to the frequency limits.

You may well find that distortion sets in at the frequency limits with smaller input amplitudes. This will depend on the quality of your transformer. In any case, without a secondary load, the output voltage from the transformer will vary significantly as a function of the frequency, but when the transformer has a resistance load, the effect is much smaller, and the overall operating frequency range for satisfactory operation will be significantly extended. Incidentally, when you get finished with this series of tests, do not be surprised if you find that you know more about transformer behavior than do most of your friends.

Step 5.

There is one additional significant question that must be answered —can a resistance load be placed on the primary side as well as on the secondary side? Repeat Steps 3 and 4, but "loading" the primary side. Observe both the flatness of the frequency response and, particularly, the quality of the waveform on the secondary side. Enter your data in the table, and your comments. As long as the load is on the secondary side of the transformer, a definite improvement of the waveform and bandwidth is noted as the load-resistance value is reduced. This is easily explained with transformer theory. When the load was placed on the primary side, a modest improvement was noted, but the waveform was not improved as much as on the secondary, nor is the frequency response.

Table 7-14. Data for Step 5

R_{Lp}				
f_{1o}				
f_{h1}				
R_{Lp}				
f_{1o}				
f_{h1}				
R_{Lp}				
f_{1o}				
f_{h1}				

These characteristics can only occur if the transformer is an energy-transfer device. You can turn the transformer around, and use the secondary as a primary, and although the values of the required load will change, the behavior will be similar. (Try it.) You will find that the lower frequency limit at which the transformer will function satisfactorily is approximately that frequency at which the open circuit input reactance (ωL_{po}) becomes roughly equal to the transformed load impedance from the secondary. If you wish the transformer to operate at lower frequencies, you will need to lower the value of the secondary load resistance. You *must* have the load-current component large compared to the magnetizing-current component for proper operation of the device. The overall phase characteristic will also be improved by the lowering of the secondary load-resistance value, and the tendency for high-frequency resonance is damped in the same way.

EXPERIMENT 10
The Ladder Power Supply

This power supply, which is described further in Appendix B, will be used to make a few measurements in order that you will know approximately what kind of equipment you have and what it can do when you need to use it. The circuit diagram is shown in Fig. 7-29. BEWARE—THIS SUPPLY CAN GENERATE DANGEROUS VOLTAGES! As you can see, there is a series of taps taken off at the junction points of a series of capacitors and an array of diodes that are cross-connected in a rather unusual fashion. With a transformer putting out approximately 150 volts rms ac, you can expect a circuit like this to be able to generate 300 to 400 volts per stage.

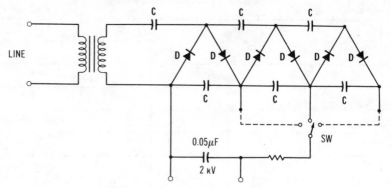

Fig. 7-29. A ladder power-supply circuit configuration.

Therefore, with the four stages shown, the output can be as high as 1200 to 1500 volts. You will also notice that there is a series resistor on each side between the final capacitor, which stores the total voltage, and the test terminals. The final capacitor should have a voltage rating of at least 2000 volts, and the other capacitors should have a rating of 400 to 600 volts, with the latter value preferred. The steady-state current available is relatively small, but that does not mean it is not dangerous. A possible circuit that can be used for testing is shown in Fig. 7-30.

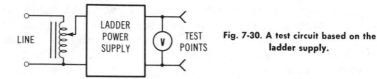

Fig. 7-30. A test circuit based on the ladder supply.

You should operate this supply off a variable transformer since you are going to want to have smooth control of the voltage that you are using in your testing. The series resistors provide an additional function besides reducing the shock hazard somewhat, in that they provide a means of detecting when a device under test is not blocking current flow as might be desired. (Such devices are sometimes called leakers.) They also prevent damage due to excessive current flow so that unless a device is actually marginal, it will survive the test. Many devices have been observed which were not damaged by such a test, and devices were also observed that either shorted or open circuited when tested in this way. In short, if a device seems to draw current and then stops, it should be retested in both directions.

It is instructive to buy a lot of untested diodes (some people call them "junk" diodes) to put through this kind of a test. If one were testing thousands of diodes, this configuration could be automated, but such is not practical here. In each case, one starts with the lowest voltage setting, and finds the reverse-voltage direction first. Warning —do NOT test germanium diodes on this unit, and test germanium rectifiers with extreme care using ONLY the lowest voltage range. Separating germanium diodes and rectifiers from silicon is discussed in Chapter 2. When you know the diode or rectifier is inserted in the reverse direction, be sure that the supply is set for the lowest voltage, and the variable transformer is set to minimum voltage. Turn on the supply and slowly increase the applied voltage with the help of the variable transformer. Stop your test when the voltage no longer increases on your monitoring voltmeter as you increase the applied voltage using the variable transformer. The author usually rates a diode or a rectifier at one-half to three-quarters of this voltage. The device should be able to withstand the limit voltage for a short period of time, but do not let it overheat.

SUMMARY

In this chapter, you learned how to use many of the accessories that are vital in performing either routine or experimental work in electronics. The ideas apply strictly to linear or analog electronics, but most of the principles are important in digital electronic systems as well. This chapter has been placed last because many of the readers will be familiar with many, if not most, of the ideas presented. Since application techniques in many cases are at least somewhat unique, however, even those experienced in the art may find some useful thoughts.

The preceding chapters have presented a rather thorough grounding in the basics of solid-state electronics devices, including some experiments designed to demonstrate the validity of the ideas. Since some readers may need further guidance in preparing material for experiments, however, the Appendices have been designed to fill these needs and will help in preparing for the laboratory or shop work that it is hoped will follow the study of this book. As is always the case in any compilation of information and ideas such as this book, it has been necessary to omit many topics considered of interest. It is hoped that this book has stimulated your interest and whetted your curiosity so that you will proceed on your own.

The Ebers-Moll Model for an Active Device

The basic equations for the Ebers-Moll model for the bipolar transistor takes the general form of Equation A-1, when converted to the common-emitter configuration. (This includes the back-to-back diode model, which can be applied to field-effect transistors as well.)

$$i_b = I_{b0} + I_{b1} \exp{(\Lambda V_b)} + [I_{b2} + I_{b1} F_b(V_b)] \exp{(-\Lambda V_c)} \quad \text{(Eq. A-1)}$$
$$i_c = I_{c0} + I_{c1} \exp{(\Lambda V_b)} + [I_{c2} + I_{c1} F_c(V_b)] \exp{(-\Lambda V_c)}$$

where,
$\Lambda = (q/kT)$,
q is the charge on an electron,
k is Boltzman's constant,
T is the absolute temperature.

In these equations, the values of several of these current parameters are what one might call "unstable," that is, they vary from device to device and they vary extensively for individual devices as well. The most stable parameter is I_{c1}, and the least stable is I_{b1}. The instability of I_{b1} really is not surprising, inasmuch as it is dependent on the small difference between the collector- and the emitter-current components that depend on the base-to-emitter voltage (or the emitter-to-base voltage, if you prefer). The unstable parameters are sensitive functions of minority-carrier lifetime, which represents a small difference in the number of carriers that are entering the base region from the emitter and those that are leaving the base into the collector region.

The existence of the high order of stability for I_{c1} leads one to examine the value of the derivative of the collector current with respect to the base-to-emitter voltage. That is the forward conductance or the transconductance, as you prefer. This derivative takes the form:

$$(\partial i_c/\partial V_b) = \Lambda I_{c1} \exp{(\Lambda V_b)} + I_{c1} \exp{(-\Lambda V_c)}[\partial F_c(V_b)/\partial V_b]$$
$$\text{(Eq. A-2)}$$

The second term on the right can usually be neglected because of the exp

$(-\Lambda V_e)$ term, with the result that the equation can be simplified by neglecting it. When this is done, the equation can be changed to read:

$$g_m = y_f = \Lambda I_{c1} \exp (\Lambda V_b) = \kappa \Lambda i_c \qquad \text{(Eq. A-3)}$$

approximately. Frequently, the kappa may be taken to be unity; when more precision is required, however, its value may be used as a correction factor. Under high-injection conditions with an npn transistor, it may have a value as high as 1.5, and with an pnp transistor, it may be as small as 0.6.

In all of these equations, the fact that the capital lambda, Λ, may have a coefficient, usually somewhat greater than unity, has been neglected. The actual value of Λ then will take the form:

$$\Lambda = (q/nkT) \qquad \text{(Eq. A-4)}$$

where the parameter "n" usually has a value less than two, depending on the material and the operating conditions. Strictly speaking, the "n" parameter should be carried in the equation for transconductance but, fortunately, under normal operating conditions for a bipolar transistor, its value is essentially unity.

The Ebers-Moll equations CANNOT be solved for i_c in terms of i_b in a convenient closed form, with the result that in order to get an equation for beta, it is necessary to determine $(\partial i_b/\partial V_b)$ as well as the partial derivative giving g_m, and taking the ratio of the two. But this places a function of the I_{b1} in the denominator, and clearly shows why beta is such an unreliable parameter. Examination of any standard data sheet for a transistor verifies that this is true in practice, as it is common to find beta ranges that vary from 20 to 60, or 40 to 160, etc., on these sheets.

Needless to say, the dc beta, which is defined directly in terms of the ratio of collector to base current, is no better than the small-signal beta as far as stability is concerned. It is indeed fortunate that sound design principles only dictate that the minimum value of beta be high enough to minimize loading.

When a derivation for the value of kappa (the transconductance-per-unit-current efficiency) is made in terms of these equations, it works out that the resulting equation is:

$$\kappa = 1 - \frac{I_{c0} + [I_{c2} + I_{c1}F_c(V_b)] \exp (-\Lambda V_c)}{i_c} \qquad \text{(Eq. A-5)}$$

Strictly, one can derive an equation of this form for the base-current equation, so there are really two kappa factors. Principally, you will only be concerned with this one, however.

When the values of i_c and I_{c0} are nearly equal, then rather small values of kappa result, and devices capable of handling substantial amounts of power become available. In fact, the power-handling ability is roughly inversely proportional to the value of kappa, as the supply voltage and the required input-signal voltage are both roughly proportional to $(0.026/\kappa)$.

Interestingly, the Ebers-Moll equations are slight modifications of a set of equations known as the equations for an "admittance two-port," and they are related to a set of equations sometimes called the "indefinite admittance matrix." These equations are for that reason equally applicable to electron tubes and to field-effect transistors. Electron tubes are still used in applications where devices having low values of kappa are essential—for example, in high-power transmitters. A device having a kappa of roughly 0.0001 can be used with supply voltages as high as 5,000 to 20,000 volts. The input-signal voltages required in those cases will approach 50 to 1000 volts, as these equations indicate. These equations are an important way to show why the future for the VMOS transistor is so bright!

In terms of these equations, the voltage gain for an amplifier takes the form:

$$K_v = -g_m Z_L \qquad \text{(Eq. A-6)}$$
$$= -\kappa \Lambda I_c Z_L$$

where,

Z_L is the load impedance in the circuit.

Since $I_c Z_L$ clearly is a voltage, this equation can be converted to the form:

$$|K_v| = \kappa \Lambda |i_c Z_L| \qquad \text{(Eq. A-7)}$$

Since the magnitude of the voltage $|i_c Z_L|$ must be related to the output supply voltage, this fact can be introduced in the form:

$$|\eta V_{cc}| = |i_c| Z_L \qquad \text{(Eq. A-8)}$$

Solving for the collector supply voltage gives:

$$|V_{cc}| = \left| \frac{K_v}{\eta \kappa \Lambda} \right| \qquad \text{(Eq. A-9)}$$

where,

η is a small numeric having a value between 0.2 and 0.5.

Its value is dependent on whether the coupling from collector to collector supply is via a resistor or an inductor and is the mode of anticipated operation for the circuit. Since at a given output current and allowed voltage gain, the optimum value of $|V_{cc}|$ is inversely proportional to kappa, the optimum power input and power output are also, typically, inversely proportional to kappa.

The fact that these equations are essentially derived from the equations for the admittance two-port means that they apply to all solid-state active devices (and include electron tubes) that are dependent in any way on the quantity (q/kT) for their operational characteristics.

As is evident from the preceding, the voltage-gain equation, as derived, neglects all of the spreading resistances as well as possible degeneration resistances. The more complete form for Equation A-6, including these factors, is:

$$K_v = -\frac{(g_{f'} - \Delta g R_e) Z_L}{1 + g_{i'} r_{b'} + g_o Z_L + \sigma g R_e + \Delta g [R_e Z_L + r_{b'}(R_e + Z_L)]}$$
$$\text{(Eq. A-10)}$$

where,

$\Delta g = g_{i'} g_o - g_{f'} g_r$,
$\sigma g = g_{i'} + g_{f'} + g_r + g_o$

In these definitions, the presence of the prime means that the parameter is measured inside of spreading resistances, and the subscripts i, f, r, and o indicate the input, forward-transfer, reverse-transfer, and output parameters in the Ebers-Moll model. The terms, $g_{f'}$ and g_m, are equivalent commonly used definitions for the transadmittance.

Substitutions to convert to common-base or common-collector configuration, as is sometimes required, can be based on the indefinite admittance matrix, which has a row and a column for each terminal of the device. In using it, one crosses out the row and column representing the common or grounded terminal of the device. The matrix takes the form:

	E	B	C
E	σg	$-(g_{i'} + g_r)$	$-(g_r + g_o)$
B	$-(g_{i'} + g_{f'})$	$g_{i'}$	g_r
C	$-(g_{f'} + g_o)$	$g_{f'}$	g_o

The structure of this matrix should be self-evident. As before, the definition of σg is the sum of the four parameters.

Some Useful Circuits

In this book, you have been introduced to a series of circuits, some of which you will find extremely useful and others that you will probably use only once. Some of the circuits can be put together to form a test instrument that you can use on a routine basis. The general characteristics of most of these circuits have been described and discussed, either in connection with their application or in a brief application-oriented discussion. In this appendix, you will find some constructional details that will help you assemble the circuits that, because of their usefulness to you, you may wish to keep in some kind of permanent form. You will also find some observations which, in some instances, should seem obvious but which often manage to escape recognition unless they are specifically noted.

The circuits that are described in the following paragraphs are arranged largely in their probable order of importance and usefulness. A discussion about laying out circuits on prepunched circuit-board material is also included. (It is an ideal way to work the "bugs" out of a circuit.) Normally, this technique is used with extremely simple circuits where the design of special circuit boards is not practical or worthwhile. It is also used in assembling more complicated circuits that do not fit on any kind of pre-etched board.

THE "RECTI-REGER"

This is a power supply that is designed so that it can be used with either a center-tapped transformer (Figs. B-1 and B-2) or with one that is not center-tapped (Figs. B-3 and B-4), depending on the desired application. It can be used to provide one regulated voltage in accordance with the following:

1. Regulated positive voltage based on a full transformer winding.
2. Regulated negative voltage based on a full transformer winding.
3. Regulated doubled positive voltage with respect to a half winding.
4. Regulated doubled negative voltage with respect to a half winding.

The latter two might appear to be superfluous but they really are not since they can be used to provide both a positive and a negative voltage with respect to a transformer center-tap. Two of these circuits can be used with the "RECTI-

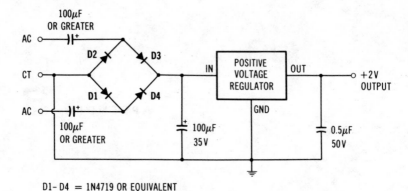

D1–D4 = 1N4719 OR EQUIVALENT

Fig. B-1. Positive voltage supply using a center-tapped transformer.

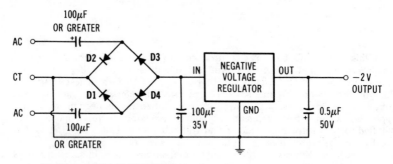

D1–D4 = 1N4719 OR EQUIVALENT

Fig. B-2. Negative voltage supply using a center-tapped transformer.

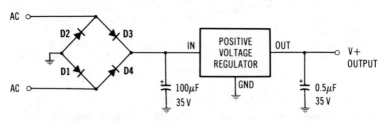

D1–D4 = 1N4719 OR 50V 3A EQUIVALENT

Fig. B-3. Positive voltage supply. Transformer is not center tapped.

REG 2" to give a set of voltages such as +2V, +V, −V, and −2V, with respect to a transformer center-tap, where V is the average voltage across half of the winding. It should be noted that the exact voltages depend on the regulators and transformers that are selected for the circuits.

The "RECTI-REG 2" rectifier-regulator configuration is a dual output-voltage arrangement suitable for use with operational amplifiers. It is designed to generate a regulated positive and a regulated negative voltage from a center-tapped

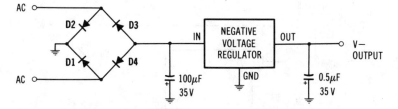

D1 - D4 = 1N4719 OR EQUIVALENT

Fig. B-4. Negative voltage supply. Transformer is not center tapped.

transformer. Its circuit diagram is shown in Fig. B-5. Both the positive and negative output of the rectifier are filtered with a large filter capacitor and passed through the appropriate positive or negative regulator. One of these circuits is used in the "TRANS-SUPPLY" that is described next. It is used to provide the required basic regulated voltages (+ and −12 to 15 volts) for the development of the required fixed and variable output voltages.

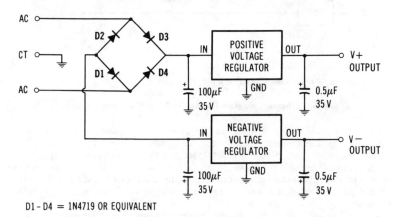

D1 - D4 = 1N4719 OR EQUIVALENT

Fig. B-5. Schematic diagram for the RECTI-REG 2 power supply.

THE TRANS-SUPPLY POWER SUPPLY

The TRANS-SUPPLY power supply is intended for testing both bipolar and field-effect transistors, and will prove to be a versatile unit for general construction work as well. It is powered by a RECTI-REG 2 circuit, or its equivalent, with either positive and negative 12-volt or 15-volt regulators. Its circuit diagram is shown in Fig. B-6. This circuit appears to be rather complicated but that is only because it has at least four different functions it must perform. It takes the input positive and negative voltage, both regulated and unregulated, and passes each polarity through two independent regulator systems. The first set of regulators is used to generate a positive and negative 5-volt supply system for use with TTL as well as transistor testing. Two potentiometers are installed, one from −5 to +12, and the other from +5 to −12 volts regulated, to provide variable voltages for control of an additional pair of regulators. The variable arm on the potentiometer, sometimes called the

443

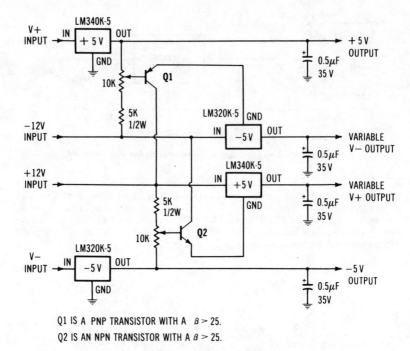

Q1 IS A PNP TRANSISTOR WITH A $\beta > 25$.
Q2 IS AN NPN TRANSISTOR WITH A $\beta > 25$.

Fig. B-6. Schematic diagram for the TRANS-SUPPLY power supply.

wiper, is in each case connected to the base of a transistor whose emitter is connected to the common return for an appropriate 5-volt regulator. This causes the additional 5-volt regulator to generate a variable voltage, either from about 1 volt to 12 volts or from -1 volt to -12 volts, depending on the setting of the appropriate potentiometer. A pnp transistor is used for the return on the positive regulator, and an npn transistor is used for the return on the negative.

In application, either the 5-volt regulated supply of appropriate polarity or the 12- or 15-volt supply may be used for supplying base current (or gate voltage) for the transistor under test, and the variable-voltage supply for providing the collector (or drain) voltage. This supply is ideal for making the small-signal tests on transistors described in Chapters 3 through 5.

THE TEEDEETESTER

A simple and convenient tester is needed for testing both diodes and bipolar transistors. The circuit for this unit should enable one to determine the polarity of the devices tested and also to identify the kind of solid-state material used. It should also make possible the location of common defects such as low beta and high spreading resistances. It should also be able to detect the fact that a device is really a Darlington compound. The circuit diagram of the unit developed for this function is shown in simplified form in Fig. B-7. (Use the parts list given in Table B-1.) It is helpful to make the unit into a multirange sensitivity unit, one capable of testing small-signal devices and at least moderate-power units as well. This can be done by providing a means of changing the

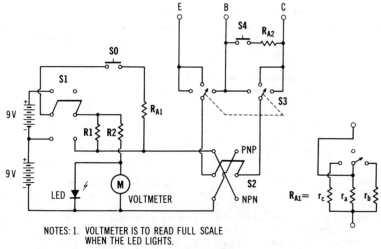

NOTES: 1. VOLTMETER IS TO READ FULL SCALE
WHEN THE LED LIGHTS.

2. ADDITIONAL SWITCHES TO INTERCHANGE
FIRST E & C AND, THEN, B & C MAY BE USEFUL.

3. $r_b = r_a/10$; $r_c = r_a/100$.

Fig. B-7. Basic circuit diagram for the TeeDeeTester.

values for resistance R_{A1} and R_{A2}. In principle, it should at least test some FET devices as well as bipolar devices, but it is easier to use a tester that is designed especially for the FET application. The voltage-control requirement for FETs is incompatible with the base-current control required with bipolar devices.

This tester will test, in addition to diodes, ordinary bipolar transistors, classifying them as npn, pnp, silicon or germanium types. It will identify and test Darlington compounds as well. It can be used to test signal diodes and zener diodes and also some specialty type diodes where the critical voltages are less than the available supply voltage. (This was 18 volts with the author's unit.) However, other techniques are better for testing rectifiers, higher-voltage zeners, and most other specialty devices.

When the polarity switch is in the correct position (S2 in Fig. B-7), and the device terminals are correctly aligned, the LED diode, in parallel with the meter M and its calibrating resistor R2, will only light when the switch S3 is in its center position. The meter then reads full-scale. (This requires that switch S1 is in the 9-volt position, and that switch S0 is pressed. The remaining switches control polarity and device connection, with S2 providing the choice of npn or pnp polarity. Push-button switch S4 forward biases the transistor to test for conduction, indicating beta value and possible high spreading-resistance conditions.

Switch S3, in its three positions, connects power across the emitter-to-base diode, from emitter to collector, and the collector-to-base diode respectively. When the polarity switch is correctly set, the first and third positions lead to diode conduction, and the second operates the transistor when the push-button test switch is pressed. The LED diode will be dark and the meter deflection will decrease to less than one-half scale with a good transistor.

High values of leakage current in a transistor will cause the LED diode not to light and the deflection on the meter to be less than full scale. An abnormally

445

Table B-1. Component requirements for the TeeDeeTester

Components	Quantity
meter, dc, 100 μAmps	1
Switch, dpdt	4
Switch, spdt (with center position off)	2
Switch, push-button	2
Switch, rotary, 2P3T	1
Battery, 9-volt transistor	2
Diode, LED	1
Binding posts, assorted colors	3
Resistors, assorted, as needed	—

low beta or a high series resistance in the emitter or collector region will cause the loss in voltage when the transistor is forward biased to be less than normal, indicating that the transistor should be tested further or discarded. If in doubt, sweep out some of the curves for the device with your test circuits and your scope. You will learn quickly what the problem is.

Provision has been made for applying both 9 volts and 18 volts to a device under test. The higher-voltage position must be used with care on very-high-frequency diodes and transistors, as some of them cannot withstand that much voltage. In particular, many high-frequency transistors cannot withstand more than 3 or 4 volts from the base to the emitter. The 18-volt setting is particularly useful with ordinary diodes, low-voltage zener diodes, and for ordinary transistors. If a transistor tends to avalanche, symptoms often will show up at the 18-volt test level.

It is unfortunate, but manufacturers do not arrange transistor leads so that they are all oriented the same. Some transistors have lead orientations that read EBC from left to right, and some read ECB. Since it is also possible to have the emitter and collector terminals interchanged, reversing switches must be provided to permit a quick rearrangement of the lead positions.

THE FETESTER

This unit is intended primarily for testing n-channel and p-channel field-effect transistors. It, as does the TeeDee Tester, operates from a 9-volt transistor battery. Since some of these transistors require reverse bias and some do not, it is necessary that the range of bias be from zero to 9 volts. Two reference points about 2.5 volts from either end are also required. These points are obtained by the use of a zener diode and the resistors R1 and R2, as shown in Fig. B-8. Depending on the polarity of the FET device, the source return and the drain return go to opposite ends of the zener diode.

The meter, with its sensitivity resistors and its clamping diodes, is used to measure the current in the FET. This combination will severely limit the rate-of-change of the device current if it is used with a bipolar transistor, but it normally will not introduce serious changes of sensitivity with FET devices. It is wise to use germanium diodes rather than silicon diodes to minimize the problem, however. When this is done, R7 can be shorted out. The meter should have about 100 microamperes full-scale sensitivity, and the shunts R5 and R6 should change the sensitivity to 1 milliampere and 10 milliamperes full-scale.

Switch S1 reverses the polarity of the applied voltage from the source to the drain to take care of either p-channel or n-channel devices. Switch S2 is used to introduce a load resistance into either the source or the drain side of the device. The potentiometer provides the gate bias through a 10,000-ohm series resistor; it is used in calibration too. One test point is attached to the three-position switch which gives a test voltage at either end of the zener diode or at a point midway between. A voltmeter is inserted between the two test points to help in calibrating the potentiometer for the three values of voltage defined by the switch S4.3

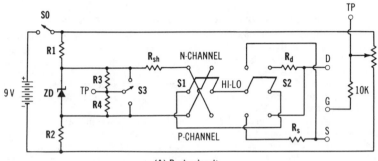

(A) Basic circuit.

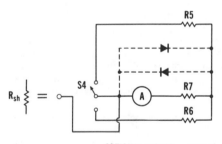

OPTIONAL DIODES = 1N914

(B) Shunt equivalent circuit.

Fig. B-8. Circuit diagram for a FET tester.

This unit is specifically intended for use with field-effect transistors, but it can also be used with bipolar transistors. Field-effect transistors are easiest tested with the use of a variable gate voltage. The first step is determining whether the device is an n-channel or a p-channel device. With the former, the drain is biased positive with respect to the source, and the potentiometer will initiate a current flow as the gate voltage is made more positive. Depletion-mode devices require reverse bias with respect to the source to cut off the current flow, and will show themselves by the potentiometer setting that is required to choke off the current flow. Devices requiring forward bias can be either enhancement-mode FET devices or bipolar transistors, but the latter will show a much more rapid rate of rise of the output current, with respect to control voltage, than the former. The forward initiation point for the start of conduction need not be the same for a FET that it is for a bipolar device, whose forward bias is established by the semiconductor material used and its doping properties.

Under normal conditions, the transconductance of a FET device is small enough per unit current that the voltage lost across a milliammeter or a micro-ammeter will not significantly affect the apparent transconductance of the device in its circuit. The device draws little if any gate current (this is especially true of the insulated-gate variety of the devices) with the result that the measuring circuit has little effect on the measurement made. As you already know, that is not true with bipolar transistors.

THE ULTRASENSITIVE VOLTMETER-MILLIAMMETER-MICROAMMETER

This unit has been designed around the 741-type operational amplifier, but it can equally well use any improved op amp, such as the LM4250 units. These latter devices are largely interchangeable with the 741, but they require one additional resistance (from the negative supply to pin 8). They can be operated at much lower voltages, and should have less drift. A resistance of 1 megohm connected from pin 8 to the negative supply is adequate for the proper operation of the LM4250 unit.

The basic circuit diagrams for these meters are shown in Fig. B-9. With the voltmeter section, one op amp is used as a fixed-gain unit, with either a voltage gain of 10 or 100. The second op amp is used to provide the reference voltage for both the meter and for the negative return for the main amplifier. (Usually a gain of 10 is best for voltage measurement.) With the high-sensitivity micro-ammeter section, the second input is returned to ground (across a shunt), and the amplifier usually has a voltage gain of at least 100. Again, the second op amp is used to establish the zero reference, and is set with the meter input shorted. It cancels bias errors.

This unit, or its equivalent in a dvm, is critically necessary because of the heavy burden that conventional meters place on some points in a bipolar transistor circuit, and because the voltages to be measured (particularly the changes in the base-to-emitter voltage) are so small. The magnitudes of the voltage changes are between 10 millivolts and 50 millivolts, as these magnitudes

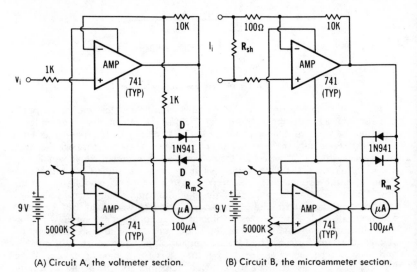

(A) Circuit A, the voltmeter section.   (B) Circuit B, the microammeter section.

Fig. B-9. Circuit diagrams for the ultrasensitive meters.

448

of voltages correspond to output changes of as much as 2 to 1 and more. The full-scale setting for the millivoltmeter at approximately 50 millivolts may be established by using a suitable voltage divider connected to a voltage reference like the Intersil ICL 8069. The ICL 8069 low-voltage reference has a nominal voltage of 1.23 volts and typical limits of 1.20 to 1.25 volts. Its effective source impedance (over a range from 50 microamperes to 5 milliamperes) is typically about 10 ohms. It should be operated at approximately 0.5 milliampere current. A voltage divider drawing about 100 microamperes to 1 milliampere may be placed across this reference. The resistors that are selected are chosen to give the desired calibration voltages. These values are adjusted to the correct values with a digital voltmeter. Some potentially useful values for calibration voltages are 5, 20, 100, and 200 millivolts, and 1 volt.

You may prefer to calibrate your high-sensitivity meter against a digital voltmeter like the Keithley 170 series or the Intersil 7106 series DVM units. Or you may wish to use one or the other of them for these measurements. The principal disadvantage in doing this is the small percentage of the full-scale range that you may be using. The current-meter element of this combination may prove to be indispensable even if you do have a DVM. This unit should be adjusted to require not over 5 millivolts full-scale, as noted previously. A shunt system should then be set up on a double-pole multiple-throw rotary switch. (It should be a shorting switch.) The most difficult part of this problem is making up the required shunts with adequate precision.

COLLECTOR SWEEP-VOLTAGE TESTER

One of the important things you need is a way of examining a range of operating conditions at the same time, so that you can get a more realistic view of how a device functions. Transistor testers like the Tektronix Model 575 and 576 curve tracers are very helpful for this purpose, but they are extremely expensive. Fortunately, it is possible to get some of the advantages of these devices fairly simply and inexpensively. All that is required is to build a special swept-voltage power supply and arrange things so that an input control bias can be applied for a short time using a push-button switch. The peak magnitude of the voltage for the swept-voltage supply can be controlled in steps by switching, or it can be smoothly varied using a variable transformer. Two possible circuits for this are shown in Fig. B-10. A circuit for providing base or gate bias, along with metering circuits for both the base and the collector, are included.

You will find this unit handy with both bipolar and field-effect transistors. It also can be used with trigger diodes and diacs, and probably with SCRs and similar devices. This tester can also give you an indication of the existence of serious spreading-resistance problems and this makes it an extremely helpful device.

The circuit for the unit is extremely simple. Several taps are required on an ac power transformer to give the variety of ranges as desired. (A variable transformer can be used ahead of the power transformer to vary the voltage.) The lowest voltage is taken off a potentiometer, which draws sufficient current so that it can be used with tunnel diodes. After passing the transformer and the tap switch, the ac is rectified with the bridge rectifier. The correct polarity, metering, and load are provided prior to the test terminals. You probably will find this circuit useful for many things.

ZENER/TRIGGER DIODE RECTIFIER TEST SET

The TeeDeeTester can be used for testing some of these devices, but normally it will just indicate that a device can or cannot rectify. The available test

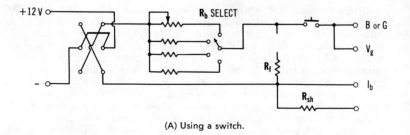

(A) Using a switch.

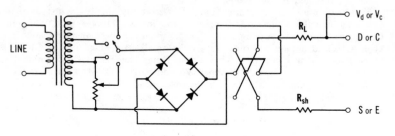

(B) Using a variable transformer.

Fig. B-10. Circuits for the collector sweep-voltage tester.

voltage is too low, and the current level is also too low. The collector sweep-voltage tester can be used for testing many of the devices whose behavior can be observed when the voltages are in the 10 to 50 volts range. With higher-voltage devices, however, a different kind of tester is essential. A ladder power supply that operates from a variable-voltage transformer is used by the author for this function. A suggested circuit design is shown in Fig. B-11. The two rows of capacitors are criss-crossed by a chain of rectifiers in a fashion that leads to a voltage totem-pole across the capacitors. With a configuration like this, you can get voltages from 100 or so volts up to as much as 1200 to 1500 volts. The unit is best used with a variable transformer. WARNING—THESE VOLTAGES CAN BE VERY DANGEROUS! A rotary switch will let you select segments of the ladder and, as a result, you can use any selected part of

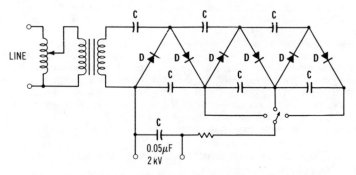

Fig. B-11. A voltage-multiplier ladder power supply.

the voltage developed by the supply. The transformer is one that is intended to operate neon bulbs. It puts out about 150 volts.

The series resistor to the output capacitor and the output terminals is there to limit the steady-state current. The output capacitor can pack a substantial wallop, so be careful. The ladder capacitors should have values of $0.1\mu F$ 600 volts. The output capacitor should be at least a 2000-volt unit.

This supply is particularly useful in testing and sorting higher-voltage rectifiers to help classify them by voltage rating. You can connect some kind of a terminal set to the unit that will enable you to lay diodes across the terminals, and permit connecting up a voltmeter too. Place a diode across the terminals and turn up the variable transformer slowly. If the voltage stops rising as you increase voltage, or if it increases sharply, you probably have a defective diode. Remember, one way the diode conducts! A shorted diode will not build up voltage in either direction, whereas, an open-circuited diode will build up a voltage in both directions.

DIGITAL VOLTMETERS

Some readers may prefer to buy one of the Keithley, Fluke, or Sinclair digital multimeters, or possibly a kit like the Intersil 7106 or 7107 units. The Keithley 178 DVM is an extremely useful unit, and seems to be very precise in transfer from range to range. Their new high-sensitivity Model 177 unit looks very attractive too, although it is very expensive. It apparently has a full-scale sensitivity of 20 mV. The Model 178 unit, with a 4250 preamp, plus an adequate nulling circuit, will do just as well, however, for the present application.

You want to be careful in using these instruments as ohmmeters, however. In most of their other modes of operation, they are rather well protected (you can burn out a shunt, of course) but, in some units, the test resistance is placed in parallel with an operational amplifier as the feedback resistor. The problems that can result from an accidental application of high-voltage there do not need further explanation.

There are at least two basic techniques for measuring resistance with these meters. One method is the use of feedback resistance, as noted previously. The author prefers the use of a current source that is independent of the internal power source in the DVM, one that is based on a zener diode and an emitter follower to produce the regulated voltage and that uses a series resistance to limit the current through the test resistance. (He uses a 9-volt transistor battery as his source of power.) It would be better to use a controlled-current source that was adjustable.

Used with appropriate voltage dividers and protective devices, digital multimeters can be used even with high-voltage tv probes to measure voltages up to 20 kV. It should be noted, however, that a voltage divider is required that divides the DVM sensitivity to the sensitivity that is appropriate for the probe used. This voltage divider should have a 10-megohm output impedance, and should divide the approximately 20-kV range (that the probe is intended to be used with) into the full-scale sensitivity of the DVM. Clamp circuits should also be used at the meter input.

MAKING DEVELOPMENTAL CIRCUITS

It is unwise for a variety of reasons to start building either an analog or a digital circuit of any complexity directly onto a pc etched board, as there are usually either minor mistakes or serious oversights in the circuit that require correction. For this reason, it is usually wise to build up a trial circuit breadboarding the circuit on perforated board, using press pins, sockets, etc. Laying out such a board is a problem. If high frequencies are involved, leads need to be short; at the same time, enough space is required to fit in all the parts and install the wiring in such a way that you can make any changes that might be needed.

There are perhaps a half-dozen different models of punched board that you can use, and there are at least three different hole sizes, along with the appropriate component mounting pins. You should standardize your circuits on two different sizes of perfboard, with one having 1 mm holes spaced about 2.5 mm apart, and the other having 1.5 mm holes on about 4.5 mm spacings. The former is best for use with IC materials like the DIP (Dual-Inline-Pin) integrated circuits. Regular DIP sockets can be mounted directly onto this board, and the Vector T42-1 component-mounting pins can be pushed directly into the holes for mounting discrete components. There is also a T-46-1 wirewrap pin that can be used in a similar way. The punched board with the larger holes is convenient for use with power circuits like power supplies. The T28-1 size terminal is designed for use with this board material. Your requirements for this style of board will be much less than for the material with the 1 mm holes.

At least one, and possibly two, rows of holes MUST be left between the ends of integrated sockets. Insert some typical IC units into sockets and try them on some boards to determine your preferred spacings before you start mounting anything permanently. Check side-to-side spacing the same way. Remember, you DO want leads short, but not so short that you *cannot* work on the thing. Remember that a power-supply bypass (0.1 μF) capacitor is required for every half-dozen IC units that you install, particularly if you are working with TTL.

The pins for the nominal 1-mm holes in the punched board have a nominal dimension of 0.042 inches, and for the 1.5 mm board, 0.063 inches. The author has used Molex pins for mounting IC units, although they are better used on etched pc boards, where the connection end can be soldered down more effectively. They can be mounted adjacent to T42-1 terminals in such a way that they can be soldered firmly to the terminal, and can be used on punched board in that way. Because they are flexible, however, they are perhaps better used with IC units.

Perhaps the best advice that you can be given for this kind of work is to make haste slowly! The first model of any test configuration probably is best assembled a small section at a time on a solderless breadboard. As you get each section working the way you want, transfer it onto a piece of perfboard, interfacing it with previously designed and debugged sections. After you have had all of the sections working, piece by piece on the breadboard and then have transferred them to perfboard and have them working there, you can then consider whether you want to dress your circuit assembly up by making an etched board for it.

When you start thinking about an etched board, you will want to consider whether you will want one copy, or will want to be able to make repeated copies. With a single copy, you put etch-resist directly on the board itself, whereas, with multicopy plans, you will want to consider whether you want to make the copies photographically or by use of silk-screen techniques. Each method is an art and may give you some headaches before you get a method developed that meets your satisfaction.

You may wish to make modifications in one or more of the units that have been described, but you should observe that the circuits used are all conventional. Circuit diagrams that are provided should help you understand how the unit is supposed to operate. If you have not done much electronic assembly work before, putting these units, and any other kits you may get, together should put you to a position where you can handle the simple tools of electronics with a modest amount of facility. Assembly of a Heathkit or some other kind of oscilloscope, oscillator, etc., is strongly recommended, as the assembly instructions and hints given in the kits and the kinks you will discover will prove very helpful to the inexperienced.

APPENDIX **C**

Useful Standard Instruments and Components

In addition to the various circuit configurations described in Appendix B, you will find it helpful to have a number of other instruments that can be bought either complete or in kit form. Of necessity, the author will limit his discussion to the kits he has seen or to the ones whose specifications he has had an opportunity to examine. For that reason, the author will try to give enough specific information on the specifications so that you can successfully evaluate any comparable unit. The following information is, therefore, for your guidance as there are many choices for many of these instruments.

THE OSCILLOSCOPE

One of the most important instruments in a lab or shop will be a scope. You should select an oscilloscope either assembled or in kit form very carefully, unless you want to gain experience on a simple unit and trade up to a more sophisticated unit at a later date. If you wish to buy a really good scope in a ready-to-use condition, the author recommends either a Hewlett-Packard, a Tektronix, or an equivalent unit, as there is little doubt that they set the standards of excellence for this type of equipment. As a result, however, they are also very expensive.

If you, like many of us, cannot afford to purchase a fully assembled, high-quality scope, you may wish to turn to a Heath or EICO scope. (Do not doubt them—many of these units are excellent buys for your money! The author has a Heath transistorized scope, and he has also seen some of their older tube-type units. He considers them to be completely adequate if carefully chosen.) Just be sure that they meet as many of the specifications on the following list as possible.

1. Vertical amplifier

   Maximum sensitivity: 1 mV per centimeter is desirable.

   Minimum sensitivity: At least 10 volts per centimeter (may be obtained using a 10× probe).

   Minimum frequency: To dc.

   Maximum frequency: To at least 3 MHz, not off more than 3 dB; 10 MHz is desirable for computer work. (You may have to add a 10× gain with an op amp, with the resulting loss of top frequency for some measurements.)

Calibration: Either variable and stepped or stepped sensitivity calibrations. Vertical calibrating circuit is very desirable.

2. Horizontal amplifier
   a. For best phase-shift and Lissajous measurements, it should be identical with the vertical amplifier.
   b. Compromise horizontal amplifier.
   Maximum sensitiviy: At least 100 mV per centimeter.
   Minimum sensitivity: At least 10 volts per centimeter.
   Minimum frequency: To dc.
   Maximum frequency: To at least 1 MHz, not more than 3 dB down.
   Sensitivity Control: Variable and calibrated is desirable.

3. Sweep Generator
   Sweep period: From 0.1 sec/sweep to 2 μsec/sweep in decade ranges, with either continuous or stepped increments.
   Linearity: It is desirable that the width of each cycle of a signal, consisting of ten cycles across the screen, be uniform to better than 10%.

4. General Characteristics
   a. Both amplifiers should be provided with stable balance controls that will assure that spot crawl (with a shorted input as a function of time and as the amplifier gain is varied) can be set and maintained at less than a millimeter after a warmup period of ten minutes.
   b. Sweep reset blanking should be provided.
   c. Sweep enhancement is desirable.
   d. Dual-input amplifiers for the vertical sweep are desirable.
   e. Focusing, intensity, and astigmatism controls should function properly.
   f. Sweep lock should be effective with weak input signals.

Whichever scope you choose, whether kit or assembled, you want to review the specifications carefully and, then, if you have to compromise for financial or other reasons, know what you are doing and why.

VOLT-OHM-MILLIAMMETER

An analog meter of this kind is helpful even if you can afford a digital multimeter like the Keithley 177 or 178, as it is usually more versatile and more compact. It is also more independent of the state of charge on the batteries. The minimum voltage scale for a vom may be as low as 0.3 to 0.6 volt; its normal maximum is around 1 kV. Digital multimeter units can be obtained that have full-scale sensitivities from about 200 millivolts to 2 kV, but most of the lower-priced units (other than kits) have maximum sensitivities of about 2 volts full-scale. The kit units usually have to be calibrated by the user.

AUDIO OSCILLATORS

There are a number of choices with these instruments. You can buy one of the Heath IG 5282 kits with a 5280-1 power-supply kit, or you can buy one of their more expensive kits. You also can buy an EXAR model 2206KB oscillator kit, which includes the pc board and many components as well, but which does not include a calibrated panel, the required switches and frequency selective elements, or a power supply. Calibration is either performed or checked with a frequency counter, but the techniques given that describe using Lissajous figures work very well. Comparison with a frequency from a crystal-standard divider chain can be very effective, particularly when a division by 2 and 5 is available, in addition to a division by 10. You set the comparison frequency on the divider chain and tune the oscillator to match. (Frequencies from 10 Hz to 1 MHz usually can be calibrated in this way.) The experience acquired in constructing and calibrating these kinds of devices is really very valuable.

SIGNAL GENERATORS AND FREQUENCY COUNTERS

Calibration in the higher frequency ranges, as well as testing in those ranges, is significantly helped with an rf signal generator, and it can be helped even more with the use of a frequency counter. (The author built such a counter from scratch.) Generators have outputs from about 100 kHz to 100 MHz and, with scalars based on the Fairchild 11C90, counters can be operated to at least 500 MHz. Intersil, Inc. has just announced a counter kit can be used for both frequency and time measurements.

Heath Company has several signal generator kits, including one in their 5280 series. However, they also have more elaborate ones, so you will want to study their catalog. Signal generators usually operate over the frequency range indicated and, normally, will provide signals from microvolts to a tenth of a volt.

Very careful tailoring of the high-frequency coils and their associated circuits is vital if the calibration of these units is to match the scales provided with the instrument. Both the inductances and the capacitances must be trimmed very closely and the placement of the lead wires must be precise, if the scale distribution is to be correct. It is probably best to adjust the inductance at about 20% less than maximum capacitance, and the capacitance at about 5% above the minimum value. This will yield the best all-around precision on calibration. Since precise knowledge of frequency is so important these days, very great care is required for this.

The use of a counter for frequency and time measurements is of growing importance. You will probably need one in your work. There are several counter kits available, and a few modest-priced counters are now available on the market. The author cannot recommend a counter, however, as the only units he has knowledge of are very expensive ones.

FURTHER NOTES ON POWER SUPPLIES

It may seem that a large amount of discussion and reference to power supplies might be out of order but, unfortunately, that is not so. Power supplies are a continuing requirement in building both test instruments and all kinds of computer peripherals. Some instruments are provided with built-in power supplies, and some are not. The author has actually built several dozen power supplies for various purposes over the last two or three years, and the designs he has given are the results of that experience. The RECTI-REGER and the RECTI-REG 2 designs are particularly useful.

With many power supplies, it is sometimes convenient to connect diodes across the regulator output, and the input-to-output terminals on the regulator as well, to protect against reverse voltage. Needless to say, in normal operation of the circut, the diode has no effect. It only conducts if the applied voltage is reversed due to a failure on the input side. If the output capacitor is larger than 1 μF, these diodes should be used, as they can prevent back-biasing of the regulator.

DISCRETE COMPONENTS

It is convenient to have at least a moderate stock of discrete components on hand to help you with your activities. You can buy these items a few at a time, or as you need them. You should write for catalogs from some of the mail-order sources as you can save a lot of money on a modest supply of parts that way. The following list is based on that kind of an approach. It should be noted that such things as DIP sockets, transistor sockets, or other items that may be required in making "permanent" circuits are not included. (The author usually does install ICs in sockets because they do fail sometimes. Thus, they are easier to replace.)

You will need at least one or two solderless breadboards, and you will want to have at least one mounted either on some kind of an instrumentation "platform" or mounted on a chassis in such a way that will enable you to wire up circuits rapidly. With this in mind, the author suggests that the mounted one be wired to a set of binding posts that can be used for powering the power points on either side of the test area. If you only expect to use two voltages and ground, you can wire the inner circuit on either side to ground. If, however, you may need to use more than two voltages and ground, the inner power points should be used for the lower-magnitude voltage, the outer for the higher voltage, putting the positive polarity on one side and the negative polarity on the other. In that case, an independent ground is required. A double-pole double-throw switch can be used to switch one or more lines from "voltage" use to "ground" use. Properly wired and used, these solderless breadboards can be invaluable. The author has one wired into an ROM tester that has proven very convenient.

Since it is not possible to predict all of the components you are likely to need in constructing the circuits given in this book, the following list is provided only for guidance. Some items are not included because the author, himself, does not know where to get them at a reasonable price. In particular, the fractional-ohm resistances between zero ohm and 0.75 ohm are not given. You will need them for shunts in some of your current-measuring circuitry.

1. Resistor, ¼ watt, (at least five to ten of each is desirable).

1 ohm	2.2 ohm	4.7 ohm
10 ohm	22 ohm	47 ohm
100 ohm	220 ohm	470 ohm
1K	2.2K	4.7K
10K	22K	47K
100K	220K	470K
1 megohm	2.2 megohm	4.7 megohm
10 megohm	22 megohm	

2. Capacitor, ceramic, 50 volt (at least five to ten of each is desirable).

100 pF	220 pF	470 pF
1000 pF	2200 pF	4700 pF
0.01 μF	0.022 μF	0.047 μF
0.1 μF	0.22 μF	0.47 μF

3. Capacitor, electrolytic, 35 volt (at least five of each).

1 μF	2.2 μF	4.7 μF
10 μF	22 μF	47 μF
100 μF	220 μF	470 μF

4. Capacitor, electrolytic, 25 volt (five of each). You may require higher-value capacitors, but you should buy them piece by piece, as needed.

1000 μF	2200 μF

5. Potentiometer (five of each type is desirable). You may need 500, 1000, 10,000 ohm, and other values, but probably should buy them as needed.

5000-ohm linear	50,000-ohm linear
5000-ohm audio taper	50,000-ohm audio taper

6. Circuit board (buy two to five pieces of different sizes).

Vector 64P44-062EP or equivalent

7. Component mounting pin
   Vector T42-1 or equivalent (100 pieces minimum)
   Vector T-46 or equivalent
8. Solderless breadboard (two different size boards are desirable).
9. Binding post, five-way, colored black, red, green, blue, white, and yellow (ten to twenty-five of each color).
10. Relay (obtain an assortment with 6V, 12V, 115V ac coils).
11. Switch (a minimum of five each is suggested. If you use them on some special test instruments, you could require dozens!).

spst, 3 amp (subminiature)	spdt, on-off-on, 3 amp
spdt, on-on, 3 amp	dpst, 3 amp
dpdt, on-off-on, 3 amp	dpdt, on-on, 3 amp

12. Rotary switch (minimum of three each is suggested).
    One-pole, 12-throw, nonshorting
    One-pole, 12-throw, shorting
    Two-pole, 6-throw, nonshorting
    Two-pole, 6-throw, shorting
    Three-pole, 3-throw, nonshorting
13. Knobs (choose an assortment that is suitable for use with potentiometers and switches).
14. Meter, dc (two to five pieces each).
    200 microamperes, full scale
    100-0-100 microampere, either side of center, full scale
    50 microampere, full scale
    1 ampere, full scale
15. Instrument boxes, various sizes.
16. Instrument cases, various sizes.
17. Assorted hardware such as machine screws, nuts, lockwashers, heat sinks, sheet-metal screws, grommets, power cords, tie points, etc.
18. The following specific semiconductor devices are suggested *only* because they should be easy to obtain. (RS and 276 are identifiers for Radio Shack. MJE identifies a part manufactured by Motorola.)
    a. Bipolar transistor, npn silicon
       2N2222, 10 to 25 pcs
       2N3055, 5 to 10 pcs
    b. Bipolar transistor, npn germanium
       RS2001 or equal, 10 pcs
    c. Bipolar transistor, pnp silicon
       2N2907, 10 to 25 pcs
       MJE2955 or equal, 5 to 10 pcs
    d. Bipolar transistor, pnp germanium
       RS2003 or equal, 10 pcs
    e. Field-effect transistor, n-channel
       RS2028 or equal, 5 pcs
       RS2036 or equal, 5 pcs
    f. Field-effect transistor, p-channel
       RS2037 or equal, 5 pcs
    g. Diode, silicon, signal
       1N914, at least 25 pieces
    h. Rectifier
       110 V, bridge, 276-1152 or equal, 5 each
       200 V, 276-1102 or equal, 10 each
       1000 V, 276-1114 or equal, 5 each

Suggested Supply Sources for Parts

You probably will find that you have a wide range of part sources in your work. Most readers will find regular radio-electronics wholesale stores that sell parts to the trade within reasonable distances of their homes or shops. These shops specialize in television, amateur, and CB parts. They will probably have most of the things you will want. Perhaps the main exception will prove to be the very-low-resistance resistors, those under 2 ohms. Resistances in the range of 1 to 2 ohms can be obtained from Digi-Key Corporation. The author has not found another source for them elsewhere.

The following list of other sources is by no means all-inclusive, but represents some of the firms the author has dealt with in the past. Since different people find one or another of the concerns to their liking, I cannot really recommend any of them, but am listing them for informational purposes only. Many of you will find that you have local branches of Radio Shack, Heath Company, or Lafayette Radio in your neighborhood, and you can examine the merchandise they are selling before you purchase it. You may wish to make one of them or some other radio-electronics wholesale outlet your first choice.

It is suggested that you write to the following concerns for their mail-order catalogs as they are convenient to have for price reference and, sometimes, for fast service. Concerns that the author has found that DO make an effort to provide fast service have an * before their names in this list. (Jameco Electronics stock the EXAR 2206 KB kits for sine, triangle, and square-wave generator units.)

1. Olson Electronics
    260 Forge St.
    Akron, OH 44327
2. Digi-Key Corporation
    P. O. Box 677
    Thief River Falls, MN 56701
3. *Jameco Electronics
    1021 Howard Ave.
    San Carlos, CA 94070

4. *Quest Electronics
   P. O. Box 4430
   Santa Clara, CA 95054
5. Poly-Paks
   P. O. Box 677
   Lynnfield, MA 01940
6. *Solid State Sales
   P. O. Box 74
   Somerville, MA 02143
7. *Radio Hut
   P. O. Box 38323
   Dallas, TX 75238

You will find many other sources in hobby electronics magazines, hobby computer magazines, and amateur radio magazines. A subscription to one or more of the magazines will be very worth your while.

Additional Useful Experiments

The experiments that were presented in the main part of the text are ones that are important in explaining how and why solid-state devices work. There are a variety of other experiments besides those directly associated with device operation that can show you some important considerations when applying these devices. Some of the principles that need to be demonstrated are described in detail in the following paragraphs. Afterwards, the detailed experiment is described.

INDUCTIVE-KICK EFFECTS

This effect is important in that it shows the magnitude of voltage that may be developed in the interruption of a current flowing in an inductance. This is of major concern to users of solid-state devices in that the voltages generated by this kick can cause failures in semiconductor devices.

When a circuit that includes an inductor or a transformer winding is connected to a source of dc so that a current starts to flow, the current builds up slowly. This is because the increase of the current is opposed by the magnetic field that is also increasing in the core of the inductor device. The total voltage is limited to that which is applied to the circuit. When the circuit is interrupted sharply, however, the situation is quite different. In this instance, there is no longer a limitation on the voltage across the inductor, and the magnetic field resulting from the current flow attempts to prevent interruption of that current. The result is the generation of a very high voltage.

The experiment that you will perform will demonstrate this high voltage impulse through the use of a small neon bulb that will light when the inductor circuit opens (Fig. E-1). You will find that nothing happens when the push-button switch is closed, but that every time you release the button with the inductor in the circuit, the bulb will glow momentarily. With a resistor in place of the inductor nothing happens. The battery you will use to light the bulb is a single 1.5-volt D-size flashlight battery.

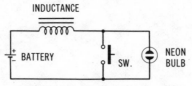

Fig. E-1. Inductive-kick test circuit.

TESTING TUNNEL OR ESAKI DIODES

The general characteristics of the tunnel diode have already been discussed in Chapter 2. What you must do now is find the most effective way to show the properties of these unique devices. Just showing that they do exhibit characteristics which can be called "negative immittance" (in this instance, negative admittance) is actually not enough. It is important to show not only this property, but how it varies with operating conditions in order that the behavior is really confirmed. To do this, it is important to vary the effective source impedance of the circuit that supplies the tunnel diode so that the switching characteristic can be detected, and the maximum negative conductance can be measured.

If one wishes to examine the characteristic curve of the tunnel diode in its totality, it is essential that the test voltage applied to the diode have an extremely small effective impedance. Otherwise, switching will occur. The typical shape of the curve for a tunnel diode is shown in Fig. E-2, along with a circuit that can be used for testing the device.

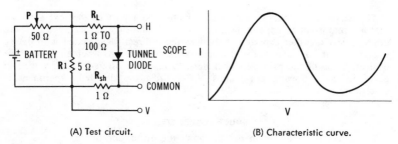

(A) Test circuit.   (B) Characteristic curve.

Fig. E-2. The tunnel diode.

In this circuit, a rectified ac signal is applied to the device through a voltage divider and a variable resistance. The voltage divider is of the sort shown in Fig. E-2A. The function of the variable resistance is to control the operating environment of the diode itself. You will find that when you use very small values of resistance for R_L, your scope will trace out the full characteristic curve of the device. As this resistance is increased, however, a point is reached where the slope of the load-resistance line is just tangent to the characteristic curve of the diode at its most negative point. *This is the point of initiation of oscillation.* As the resistance is increased further, an oscillation of increasing amplitude will result or, in the absence of a tuned circuit, switching will occur. When this condition (in the absence of a tuned circuit) is reached, it will be noted that the center of the negative section of the characteristic curve will disappear. As the resistance is further increased, more and more of this section will disappear. When the series-resistance value has increased sufficiently, *all* of the negative-slope section of the characteristic curve will have disappeared, and switching will then be taking place as indicated in Fig. E-3. Here, the current rises with increasing voltage until the slope of the characteristic curve matches the slope

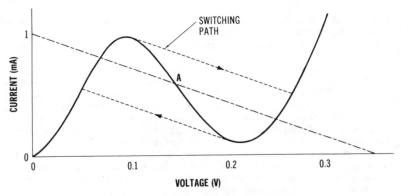

Fig. E-3. Tunnel diode switching characteristics.

of the load line; at that point, transition to the upper segment of the curve occurs essentially instantaneously. From there, the voltage and current will both rise along the upper positive segment of the characteristic curve.

When the voltage starts to decrease, the voltage at the diode will follow along the upper positive segment until the voltage is low enough so that the load line is again tangent with the characteristic curve of the device, but at the low-current high-voltage end of the negative-slope segment of the characteristic curve. At that point, operation will "jump" to the intersection with the lower positive-slope segment, and will continue. The cycle continues and repeats over and over again.

When a tuned circuit is connected correctly into a circuit with a properly biased tunnel diode as in Fig. E-4, the combination will oscillate at the appropriate driven resonant frequency. Since the tunnel diode approaches its negative-conductance region by going through a zero slope (or zero conductance) region that is represented by a zero change of current for a finite change of voltage, it is a negative conductance (or a negative admittance) device. This means that it is capable of cancelling out a certain amount of positive conductance. A tuned circuit used with it must have less than that amount of positive conductance if it is to be made to oscillate by the tunnel diode.

If the circuit is to operate at only one frequency, it should have *more* than the minimum value of conductance for all frequencies except the one which will be selected by the tuned circuit. Only a parallel tuned circuit has that

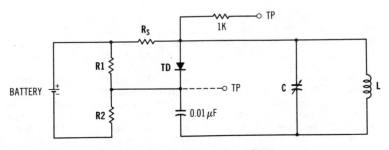

* SELECT VALUES DESIRED FOR L AND C.

Fig. E-4. Tunnel diode oscillator.

property. It is this circuit that must be used, therefore, with the tunnel diode in the formation of an oscillator, and the circuit must be so selected that its minimum positive conductance is less than the negative value available from the diode.

Unfortunately, it is not possible to vary the magnitude of the maximum negative admittance for a given diode, with the result that the amount of control available is rather limited for a given diode. (One can select a different diode, however.) Some variation can be obtained by shifting the static bias point to one that has a lower net negative conductance magnitude, but this tends to increase the distortion content, particularly the second harmonic. A parallel resistance can be used to increase the positive conductance of the tuned circuit, thereby reducing the oscillation amplitude, and giving a way of estimating the negative conductance of the diode or the positive conductance for the tuned circuit. The technique is somewhat difficult to use and not overly reliable, with the result that it is not in common use.

One of the interesting properties of the tunnel diode is the fact that, in the reverse direction, the current increases extremely rapidly with voltage, significantly more so than in either the forward direction of the diode or in the forward direction for a corresponding conventional diode. This property is important in two ways. First, it means these diodes can be used in detectors and mixers with significantly reduced levels of input signal and, second, the small-signal conductivity in the reverse direction is high at a given current level and it also changes rapidly with an applied signal and minimal bias. This phenomenon is used in devices known as *backward diodes,* in that they have the properties of a tunnel diode in the forward direction but draw very little current, and have the rapid increase in current in the reverse direction that is typical of tunnel diodes. This apparent reversal of normal behavior is the reason they have been given the name of "backward diode."

Possibly the most important property required in diode detectors and diode mixers is a maximum rate of change of conductance per unit of current in the device for a minimum of static voltage and current-bias levels. When this condition exists, then the maximum signal-to-noise ratio conditions potentially, at least, exist for a device. (Junction capacitance is also significant in that it affects the figure of merit for the device.) The backward diode does have a small peak and valley in current in the forward direction but, typically, the current at peak will prove to be less than 50 microamperes, far less than that for the normal tunnel diode.

Biasing the conventional diode to get it into a region displaying a maximum rate-of-change of conductance has proven to be difficult at best. Stabilizing the voltage for a mixer admittedly is simpler than for a square-law detector, as the current tends to average to a constant value and shifts the operating point when a signal is applied without a local-oscillator injection condition also existing. This shift in the operating point can lead to nonlinear detection. The greatly reduced bias which is required with the backwards diode can lead to better detection characteristics, including less central clipping when the signal amplitude is small,

As has been noted in Chapter 2, there are also negative resistance devices. With these devices, the increase in current leads to a region of increase with no voltage change, followed by a region of voltage decrease with a further increase in current. In other words, these devices go through a region of "zero resistance" in getting from positive resistance to negative resistance. These devices can, therefore, cancel out positive resistance and should be used with tuned circuits that show a minimum of resistance when tuned to resonance. This resonance frequency is the frequency at which the resistance and impedance are both minimum. The series-tuned circuit is required for this negative resistance device in the development of an oscillator. You will find it interesting to make a switching diode respond in this way.

THE EMITTER-COUPLED AMPLIFIER

A circuit which has been carried over from the days of the electron tube and which has proven very useful is the emitter-coupled amplifier. Technically, there are two forms of this circuit, both of which are commonly used with discrete circuitry. The first of these circuits uses one transistor as an emitter follower to operate another as a conventional common-emitter amplifier. (This is basically the Darlington pair.) The second uses the two devices with their emitters coupled together as a means of transferring the signal from the first stage to the second. Our principal interest here now is with the second configuration, as the Darlington circuit will be discussed in a later paragraph. The circuit configurations for both of these circuits are shown in Fig. E-5.

The emitter-coupled amplifier uses the first transistor (Q1) as an emitter follower. It is directly connected to a common-base amplifier (Q2) by way of

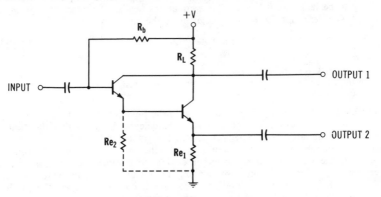

(A) Darlington-pair configuration.

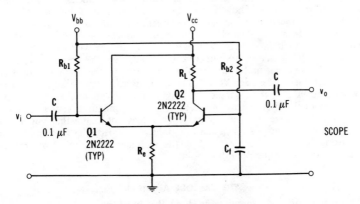

(B) Emitter-connected configuration.

Fig. E-5. Circuits using emitter coupling.

a common-emitter connection. This configuration has high inherent stability and a wide bandwidth, although its overall gain will tend to be lower than that of a Darlington compound. The amplification from base to emitter for transistor Q1 can range from near zero to near unity, but will have a maximum value of between 0.3 and 0.5 where the collector currents in the two devices are approximately equal. The approximate effective transconductance from the input of the first stage to the output of the second may be defined adequately by the equation (based on Fig. E-5B):

$$g_{me} = \frac{g_{m2}\sigma g_1 R_e}{1 + (\sigma g_1 + \sigma g_2) R_e} \qquad \text{(Eq. E-1)}$$

$$g_{me} = \frac{\Lambda^2 I_{e1} I_{c2} R_e}{1 + \Lambda (I_{e1} + I_{e2}) R_e} \qquad \text{(Eq. E-2)}$$

where,
    subscripts 1 and 2 refer to devices 1 and 2,
    subscripts e and c refer to emitter and collector respectively.

Where I_{c2} is small compared to I_{c1}, this value approaches g_{m2}; otherwise, it is likely to be less than $0.5\ g_{m2}$. (These equations neglect the effect of high injection.) The definition for sigma g is given by the equation:

$$\sigma g_i = g_{ik} + g_{mk} + g_{rk} + g_{ok} \simeq g_{mk} \qquad \text{(Eq. E-3)}$$

where,
    subscript k has the value of 1 or 2,
    initials i, m, r, and o refer to the input, forward, reverse, and output conductances.

The simplified form applies in particular to the high-beta transistor.

If, in this equation, the value of R_e is sufficiently large so that the "unity" in the denominator can be neglected, and the values of I_{e1} and I_{c2} are roughly equal, then the overall transconductance to the second stage will be roughly one-half the nominal value for either device. In addition, as one current increases, the other will decrease, leading to a nearly constant value for the denominator. Under these conditions, for small current changes, the product in the numerator will also be relatively constant and a nearly linear circuit will result, one which has a rather constant transconductance. This linear region is of limited extent, however, as the correction term of the form $(1 - \Delta^2)$ does appear in the numerator as a consequence of the $I_{e1} \times I_{c2}$ product, and this does lead to a gradual reduction of transconductance as the currents vary from the point of balance. This variation is ideal for use with oscillators.

Where a high degree of nonlinearity is desired, then the value of R_e should be selected to be a small value so that the term in the denominator has a value which does not exceed "two." The currents can also be adjusted so that they are unbalanced, for example, by making the current in the second device larger than in the first. This combination gives a configuration that will work rather well as a mixer or a modulator. (This does not assure that you will get linear modulation, however.) When this unbalance exists, the transconductance is a strong function of Δ, where Δ is the incremental variation of transconductance from the static operating point.

THE CASCODE AMPLIFIER

This is a version of a circuit (Fig. E-6) that has been used extensively as an input amplifier for radio and radar receivers. It has, in principle, a somewhat lower noise figure than other commonly used circuits. (Both the common-base

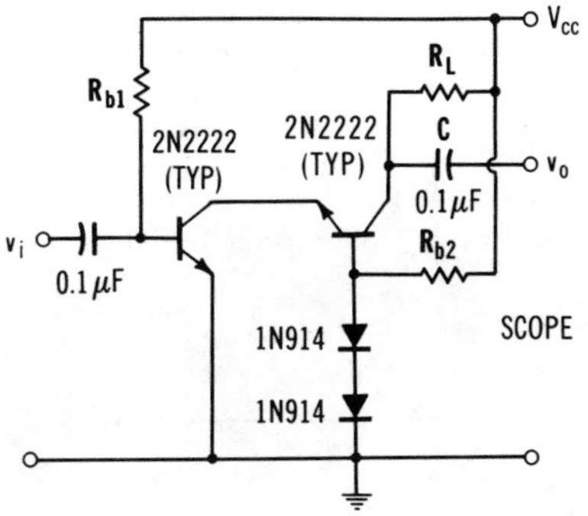

Fig. E-6. Circuit diagram for a cascoded transistor amplifier.

and the emitter-coupled circuits are rather good for this function also.) It has high inherent stability and a somewhat higher overall voltage gain than an emitter-coupled amplifier, as the gain of the first stage is in fact unity when the device currents are equal. (The simplest form of this circuit operates preferentially with equal currents in the two devices.) It has a lower input impedance at the base of the first device because of the common-emitter configuration used and the fact that the second device acts as a collector load for the first.

The cascode amplifier can be used as a mixer as well as an amplifier by introducing a local-oscillator signal into the normally "grounded" base for the second transistor. This transistor must be biased sufficiently above ground so that the input transistor can amplify and introduce a signal into the output transistor but, as has been noted many times, only a volt may be required for this purpose.

The cascode amplifier can also be used as a basic element of an oscillator, but the frequency selection circuit used with it must provide phase inversion. This means that it should be used with either Hartley or Colpitts configurations, where a tapped tuned circuit is used with the input coupled to one end and the output to the other. Normally only a single transistor is used with either of these basic circuits, but there are some potential advantages to the use of a cascode circuit, particularly at high frequencies.

THE DARLINGTON AMPLIFIER

This amplifier, as noted earlier, is built from two transistors, and so connected that the emitter of the first transistor introduces a signal into the base of the second transistor. The collectors of both are often connected together. Both transistors are either pnp or npn types. This configuration can have fantastically high current gains, in a range from 1000 to as much as 25,000 or more, with stability. For that reason, it is ideal for use as a split-load phase inverter. The collector currents for both devices flow through the collector load, as can be seen in Fig. E-7. Also, virtually all of both emitter currents flows through the emitter and the collector loads, leading to a balanced output voltage. Only a very small fraction of 1% of the current can typically contribute to unbalance with this circuit, whereas, from 0.5% to several percent may be obtained with a single transistor.

As long as the input transistor is biased so that it will draw some base current in normal operation, the Darlington configuration operates effectively with no problems. If, however, the first device is biased in an "off" condition, the I_c leakage current flowing through it from the collector to the emitter

469

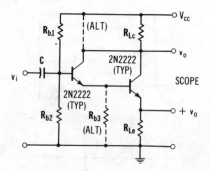

Fig. E-7. A Darlington split-load phase inverter.

allows the second device to draw current, and the control of the output device is lost. This can be readily corrected by placing a resistor from the combined emitter-base connection to the emitter return for the second device. This resistance should be small enough to assure that the base bias on the output device keeps it from drawing significant output current. With a silicon transistor, the voltage on the base of the output transistor should be less than 0.4 volt with respect to its emitter when cutoff is desired.

The Darlington compound has most of the advantages of the straight emitter follower (along with its disadvantages) of extremely high nominal-input impedance, very low output impedance at the output emitter, almost exactly unity-voltage gain, very high current gain, etc. It, of course, can show more than unity gain to the output collector if the load impedance there is selected to introduce this condition and, when so used, may not have as many advantages over the simple amplifier as might be desired. With the collectors of the two sections connected together, the Miller-effect capacitive loading from the output to the input can be severe. Where significant gain with minimum Miller effect is desired, the collector for the first device should be connected directly to supply, and the load should be only in the output stage. When that is done, the Miller capacitance is loaded on the emitter of the first stage instead of the base, and can be less detrimental. (It is possible to bias the input transistor to current cutoff if the input signals are large.)

This self-cutoff property through a storage of charge in the output capacitor is encountered with both emitter followers and Darlington compounds when the emitter impedance is large and capacitive. What happens is that the output capacitor (and/or the interstage capacitor) becomes charged when the device(s) become forward biased. The capacitor does not then discharge when the signal voltage decreases, and the device(s) are cut off. The output signal is thus controlled by the charge in the capacitor instead of the input signal.

Control of this condition requires the use of minimum values of both output emitter resistance (consistent with required operating characteristics), and interstage return resistance (for the emitter of the input transistor). However, this condition can be noncompatible with the magnitude of output signal required. The use of these stages using relatively small signals and followed by "push-pull" amplifiers is clearly the way to avoid the problem.

EXPERIMENT 1
The Inductive-Kick Experiment

The basic principle of the inductive-kick phenomenon has already been explained. It is used in the ignition systems for most gas engines. The phenomenon can be demonstrated by the use of a "switching" magnet in a magneto, and by the interruption of a flow of current in a coil (as in the ignition coil used with an automobile engine). As you will observe in the experiment, the inductive-kick phenomenon cannot be detected when current is interrupted in a resistance. In this experiment, you can use any transformer or filter choke you may have available, a resistor, an spdt switch, a neon bulb (NE-2), and a 1.5-volt flashlight battery (either a "C" cell or a "D" cell). The circuit that is to be wired up on your solderless breadboard is shown in Fig. E-8. Resistor R should have between 2- and 10-ohms resistance.

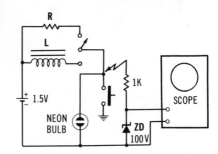

Fig. E-8. Inductive-kick test circuit.

Step 1.

Wire up the circuit as shown in Fig. E-8, and connect the neon bulb (one without an internal series resistance) across the push-button switch. Then, set the spdt switch so that the resistor is in the circuit.

Step 2.

Press the push-button switch and release it. Does the neon bulb flash when the circuit is closed? Does it when it is opened? Explain.

There is no energy stored in a magnetic field here. Energy is being converted into heat. As a result, there is no way that energy can be dumped into the neon bulb for dissipation in a form that would activate the bulb.

Step 3.

Connect the scope across the neon bulb, but be sure that you insert a 1000-ohm resistance in series with its input lead. Again, push and release the push-button switch, and observe the results. Explain your findings.

When the push-button switch is open, there is a voltage of 1.5 volts across the terminals of the neon bulb, which is dark. When the push-button switch is pushed, the voltage disappears. It reappears without any noticeable overshoot when the button is again released.

Step 4.

Disconnect the scope temporarily from the circuit, and change the switch so that the inductor (or transformer winding) instead of the resistor is in series with the switch. Again, press the push-button switch. Hold it down for a few seconds and release it. What do you observe now?

This time, the neon bulb flashed when the button was released. It was dissipating the energy that was stored in the magnetic field of the inductor and, in fact, it attempted to generate a voltage high enough to keep the current flowing. Fig. E-9 shows the voltage pulse generated by the induction coil.

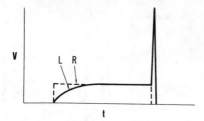

Fig. E-9. Induction-coil voltage pulse.

Step 5.

With the battery disconnected, reconnect the scope across the push-button switch terminals using a series resistor to protect it. The zig-zag connection shown in Fig. E-8 should be made to the top of the push-button switch as shown in Fig. E-8. The zener diode is used to limit the input voltage at the scope. The scope should be set for its minimum-sensitivity condition, 50 or so volts per centimeter. The zener diode should be approximately a 100-volt unit. Set the sweep speed near the slowest rate so that you can release the push-button switch during a sweep. You may still have to try several times before you get the turn-off impulse just where you can see it well. The vertical deflection will be substantial on breaking the circuit, but there will be almost no change on making the circuit. Try this and verify it. Discuss your results.

EXPERIMENT 2
Testing the Tunnel or Esaki Diode

The first thing you will need to discover is the characteristic nature of tunnel diodes, showing in particular the negative-slope region. The second thing you will need to discover is just what the nature of the negative immittance is and how it can interact with typical passive circuit elements. (Remember, the term *immittance* is used to represent either an impedance or an admittance when it is not desirable to specify which is intended.) You should be particularly concerned in finding out whether the tunnel diode is a negative admittance device or a negative impedance device. This can be established by learning whether the immittance becomes negative by going through a region of zero impedance or a region of zero admittance. This determination is easily made by using a scope to plot out the full characteristic curve for the device and by then analyzing the resulting plot. This can be further verified by examining the behavior of the device as you vary the series or parallel load resistance associated with the diode. A circuit that can be used for testing is shown in Fig. E-10.

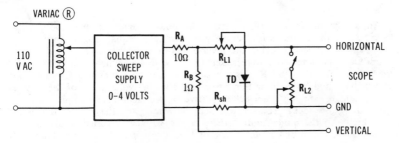

Fig. E-10. Circuit for testing tunnel diodes.

Step 1.

Wire the circuit shown in Fig. E-10 on your solderless breadboard. You will notice that an arrangement for increasing the series resistance in the diode circuit is provided and, also, that a switch and variable resistance are provided for paralleling a resistor (used as a conductance) with the diode. In addition, both current- and voltage-metering points are provided for use with a scope. Once again, you need to minimize the amount of resistive shunt introduced for current-measurement purposes, as you need to be able to carefully control the total series resistance in the circuit. The sweep-voltage supply you used for plotting transistor characteristic curves (set on its lowest voltage range) should be ideal for the power supply. Not more than 2 or 3 volts overall are needed for making the test.

Step 2.

Set the load resistor R_{L1} to a value near zero, and the sweep power-supply voltage near zero. Turn the system on and slowly increase the voltage with the variable transformer. You want to be sure that the total current flowing through R_A and R_B is at least 20 to 50 times the amount of current that you expect will flow through the tunnel diode. This is necessary in order to keep the source impedance low enough to give meaningful results. A good value with many diodes is 200 milliamperes or more. This will keep the source impedance small enough that it will not introduce any problems. If more current sensitivity is needed on the vertical axis, insert an operational amplifier that has a gain between 10 and 100, whatever is required. The current-metering resistance needs to be as small as you can make it.

Slowly increase the supply voltage, and observe the trace that develops on the scope. Describe what happens as the voltage increases to as much as 1 to 2 volts, and what happens as it is increased further.

As the voltage passes 100 millivolts, the characteristic curve will either be invisible or it will be sweeping downward, depending on the total resistance in the circuit. As it increases to 0.5 volt or more, it is once again increasing at a rapid rate. The region of particular interest to us is the region in which the current decreases with increasing voltage.

Step 3.

Adjust the maximum voltage from the power supply so that the positive segments of the characteristic curve are of roughly equal height, and so that the segment between (having the reverse slope) is clearly visible. Under these conditions, the total source resistance for the diode, which is approximately the sum of R_B, R_{L1}, and R_{sh}, is too small to lead to switching. The load line sweeps back and forth along this curve as the supply voltage out of the rectifier varies, and it can only intersect the curve in one spot (the various points or spots along the trace). The scope is actually tracing out the intersection of the load line with the diode characteristic as the voltage is varied. Make a copy of the diode curve and sketch a set of load lines at each of several load resistance values on the graph in Fig. E-11, as we have done in Fig. E-12.

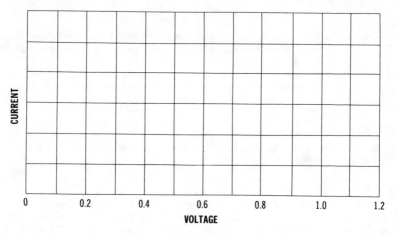

Fig. E-11. Graph for Experiment 2, Step 3.

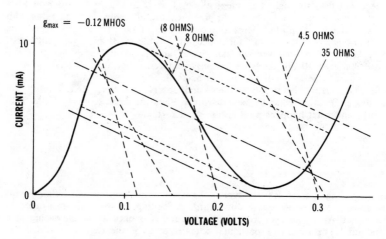

Fig. E-12. Typical load line sets for a tunnel diode.

You should observe that you can have load lines intersecting the curve in one place, or crossing and tangent at one place. The only other alternative is a load line cutting the curve in THREE places. *All of the load lines that intersect in more than one place have a slope corresponding to a resistance HIGHER than the critical resistance which represents the STEEPEST point on the decreasing portion of the curve.* In other words, as long as the load resistance is less than a critical value, there can be ONLY ONE intersection. If the resistance is greater than that minimum value, however, it is possible to find a source voltage for which there will be THREE intersections. This is the normal characteristic of a negative conductance or a negative admittance device. You can also plot out the characteristic curve for the device using a suitable very-low-voltage regulated power supply and your sensitive metering elements.

Step 4.

Increase the load resistance and readjust the supply voltage as required, and monitor the trace on the scope as you do so. At some point, the trace in the middle of the decreasing section (negative admittance section) will start getting just a little dim, and as the resistance is increased (R_{Li} is being increased), the dim segment will lengthen. Turn the resistance back to the point where this action first starts and measure the resistance of the load, including all three elements. Its reciprocal should equal the conductance determined off the curve. You have set the load resistance R_L to the value for which:

$$g_1 R_L + 1 = 0 \qquad \text{(Eq. E-4)}$$

where,

g_1 is the magnitude of the negative conductance of the device,
R_L is approximately the sum of R_B, R_{L1}, and R_{sh}.

In other words, the product of the maximum value of the negative conductance and the load is unity in magnitude here. This is a simple way of finding the value of the negative conductance. (Another way that it can be done using R_{L2} is described in Step 7.) Make this test using your sample device.

Step 5.

Step-by-step, increase the value of the load resistance R_{Li} in the diode circuit of Fig. E-10. In each step, increase the peak-source voltage as required so that the peak current on the higher-voltage positive-slope section exceeds the peak current nearly 100 millivolts. As you do this, you will notice that the trace blanks out more and more in the center of the negative-slope section. It is actually jumping from one segment to the other (switching is taking place). You will also notice that there will be a stretch of the contour from zero volt part way up the section that is brighter than the balance (up to the point at which the line disappears), and that there is also a segment on the upper part of the second visible section that is brighter than the balance of that trace. The bright segments are traced both up and down as the voltage varies, whereas the dimmer segments are traced only one way, either up or down. Switching keeps the segment from being traced both ways.

Make careful tracings of the characteristic curve for the diode, and plot out the switching points as was done in Fig. E-13. Use the graph given in Fig. E-14 for your plots. Calculate the equivalent resistance corresponding to the transition lines, and see if it checks with the resistance value you are using for R_{Li}.

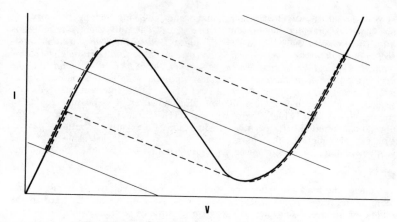

Fig. E-13. Sample switching contours for a tunnel diode.

The switching region can be called an "hysteresis region," and the dim segments mark the slow segments of the switching hysteresis loop. The invisible lines mark the fast segments.

Step 6.

Continue to increase your load resistance, step-by-step, and at the same time increase your peak voltage as required, so that the peak current on the higher-voltage positive-slope segment is always greater than that at the 100-millivolt region. As you do this, you will observe that the dim segments become longer and longer. What do you interpret this to mean?

The tunnel diode is acting more and more like a multivibrator and less and

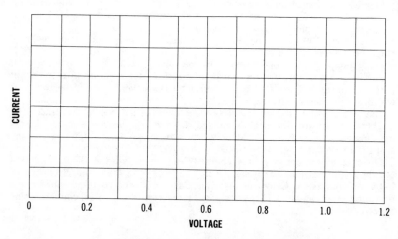

Fig. E-14. Graph for Experiment 2, Step 5.

less like a normal oscillator. This is because of the very high "excess gain" in the center. You should have found that as you reduced the resistance, the curve was traced in full, indicating that the effective "loop gain" became too small for the circuit to oscillate.

Step 7.

Now switch R_{L2} into the circuit, and set the value of the resistor to at least ten times the value obtained for $|g_1|^{-1}$. Switch this resistance into and out of the circuit, and observe any effect it may have. Then reduce the value of R_{L2} to half its previous value and repeat the test. Continue doing this until the contour you plot on the scope always slopes upward. Plot these curves on the graph in Fig. E-15 and explain what you have observed.

In the positive-slope region, as you reduce the resistance, the slope becomes steeper and steeper. In the negative-slope region, it rapidly becomes less negative, and the segment showing the negative slope becomes shorter and shorter until finally there is *no* segment with a negative slope.

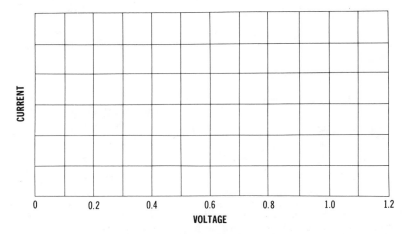

Fig. E-15. Graph for Experiment 2, Step 7.

Oscillation will only occur *if* there is a segment where, with the net impedance involved, the combined characteristic curve has a small segment with a negative slope. Oscillation can only occur over an impedance range where the tuned impedance is high enough so that the negative area occurs, and it will only occur within a frequency range where the impedance is of that high value. A parallel-tuned circuit will *only* have that high a tuned impedance near resonance, and then only if the LC ratio selected is correct. This experiment finishes the conclusive proof that these devices are negative admittance devices.

Step 8.

It would be interesting to repeat this experiment with a trigger diode (or a diac) to find out how it behaves. With this device, you should find precisely the converse of what you found with the tunnel diode; that is, the full trace will be visible with *maximum* series resistance and an appropriate voltage, and the hysteresis effect with the vanishing segments of the curve will appear as the load resistance is reduced. This is typical of a *negative resistance* type of device. If you check the curves, you will find that, in this instance, the curve turns negative through points at which the current increases or decreases with *no change of voltage,* indicating zero resistance. Repeat all steps of the preceding experiment using a trigger diode.

EXPERIMENT 3
The Trigger-Diode Switch

The trigger diode is not normally used for the purpose described in Experiment 2. It is usually used to provide a "jump" in the voltage that is applied to the control gate of a silicon-controlled rectifier (through the reduction of voltage across the diode). Step 8 of the preceding experiment has shown that there is at least a trace of a "negative resistance" characteristic in this device but now it is important to find out just how these devices behave in their normal operating environment.

Step 1.

Connect up a trigger diode using the circuit shown in Fig. E-16. Notice that this circuit consists of a phase-shift network containing a capacitor C and a

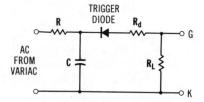

Fig. E-16. A trigger diode phase-shift circuit.

variable resistor R, a trigger diode, and a diode load (R_L). The voltage to be applied to this network must be at least 50 volts and probably should be nearer 100 volts, unless the diode can be triggered by a voltage of less than 20 volts. The load for the diode must be large compared with resistance R, and small compared to the diode resistance before breakdown. A 100,000-ohm resistor should be a good trial value. (In an actual circuit, it is the gate resistance of either an SCR or a triac that is the load for the trigger diode.) A variable transformer will provide a useful way to examine the effects of source voltage on the circuit. Start the circuit operating, and record what you observe as you vary both the source voltage and the variable resistance:

The higher the voltage, the longer the conduction cycle on the trigger connection. The length of the conduction cycle can also be controlled by varying the resistance value.

Step 2.

The phase-shift circuit used in Step 1 does not provide a constant-voltage variable-phase control voltage. For that, a center-tapped transformer is needed, with the RC network connected across the outer terminals, and the phase-shift output taken from the junction between the resistance and the capacitance. The trigger diode connects from the resistor-capacitor junction through the load resistance back to the center tap on the transformer (the usual configuration). The circuit is shown in Fig. E-17. It is

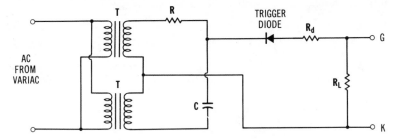

Fig. E-17. An improved version of Fig. E-16.

possible to get a single transformer that has a high enough voltage on either side of the center tap, but you may have to do it this way instead. Test the trigger capabilities of both of the circuits. Explain your results.

The advantage of the circuit in Fig. E-17 over the circuit given in Fig. E-16 is that you have independent controls over the phase and amplitude. The ability to vary both independently means that you can learn more about the control characteristics using it. You should use Lissajous figures for the evaluation of the phase properties.

Step 3.

You should plot some waveforms as you observe them while using the circuits given in Steps 1 and 2. You should use both Lissajous figures and conventional sweep test-measurement methods, as each will disclose important

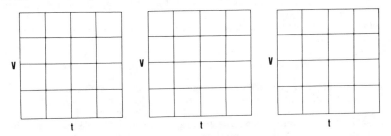

Fig. E-18. Graph for Experiment 3, Step 3.

information about the behavior of the circuits. Sketch the effect of voltage on the waveform, and note any phase effects you may be able to detect. Using the graphs given in Fig. E-18, use a separate graph for several different voltage levels, and note your comments:

You will find that the level of triggering is sensitive to the resistance when using the configuration of Fig. E-16, but it is also sensitive to voltage, particularly at voltages just a little over critical.

Step 4.

In this step, you will use the circuit to trigger either a silicon-controlled rectifier or a triac to turn on a light. Isolating circuits should be used for monitoring, as transistorized scopes may not operate when connected in the "hot" side of the power line. The basic control circuit is shown in Fig. E-19, and

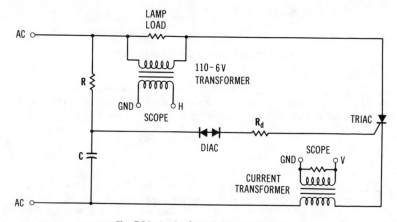

Fig. E-19. A triac lamp load-control circuit.

two possible techniques for isolating the scope input circuits are given in Fig. E-20. (Remember that you will get some waveform distortion with these circuits, particularly if you use an SCR. Transformers do not work very well on direct current.) Discuss your results, and sketch some sample waveforms as viewed when using the isolation devices. (Information on structures for current transformers was given in Experiment 8 of Chapter 6. A conventional low-voltage transformer that has a resistive load will serve for the voltage-isolation transformer.)

You may not find that the triac conducts in a balanced fashion. The transformers will distort the voltage and current waveforms but, if properly loaded, the results should be adequate. As long as the trigger diac is able to provide the switching pulse required, and the triac load is sufficient to keep it switched, you should have excellent control on your load. The intensity with which a lamp load will glow should be easily controlled with this circuit.

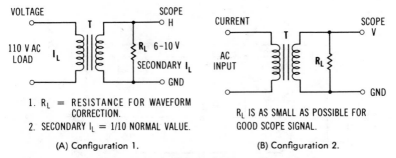

1. R_L = RESISTANCE FOR WAVEFORM
   CORRECTION.
2. SECONDARY I_L = 1/10 NORMAL VALUE.

R_L IS AS SMALL AS POSSIBLE FOR
GOOD SCOPE SIGNAL.

(A) Configuration 1.   (B) Configuration 2.

Fig. E-20. Current and voltage isolation transformer circuits.

EXPERIMENT 4
The Emitter-Coupled Amplifier

The emitter-coupled amplifier is a transistorized version of the cathode-coupled amplifier. It can be built either with two npn transistors, two pnp transistors, or their equivalents using field-effect devices. These experiments will concentrate on the emitter-coupled version, but it is suggested that you try using FETs as well. There are a variety of useful forms for this circuit as an amplifier, ranging from a straight amplifier through a phase-splitting amplifier and the specialty circuits that are considered later in Experiments 5 and 6. You should construct at least the npn and the pnp transistor versions of some of the following circuits and, in the mixer experiment, you probably should build a pair of circuits based on the N-channel FET combination as well.

Step 1.

Set up a simple emitter-coupled amplifier based on npn transistors on your solderless breadboard, using the circuit given in Fig. E-21. You will notice that in the circuit diagram a load has been indicated for each transistor. You will need to test the circuit both with matched loads and with the input-load side shorted to observe the effects of this change. You will need to vary both the emitter resistance and the collector load resistances so that you can determine the effect of these values on the overall gain of the circuit. The minimum value of emitter resistance that you will need to use will be established by the use of the equation:

$$R_e = (2\Lambda L_e)^{-1} \qquad \text{(Eq. E-5)}$$

where it is assumed that the emitter currents for both transistors are the same. If not, the term $2I_e$ may be replaced by the sum of the two emitter currents. They may be adjusted to be equal by balancing the collector voltages (with two equal load resistances in place) by adjusting the base current for one of the transistors to achieve the balance. A differential voltmeter may be used for this adjustment. Load resistances capable of providing voltage gains of 10, 20, 50, and 100 should be chosen, and the characteristics of the circuit tested with both the loads in place and with the load on the input side shorted out.

Step 2.

Once you have gotten the feel for how this circuit amplifies with a minimum of emitter resistance, you will want to repeat the tests using larger emitter resistance values, such as $2R_e$, $5R_e$, etc. You will also find it interesting and useful to replace the emitter resistance with the "transistor resistance" which

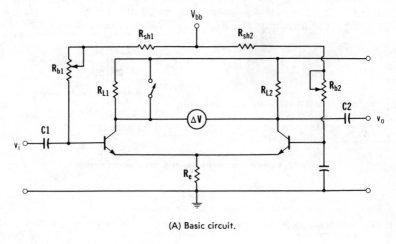

(A) Basic circuit.

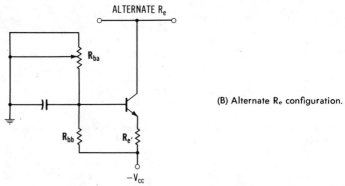

(B) Alternate R_e configuration.

Fig. E-21. Emitter-coupled amplifier test circuit.

can make the circuit appear to have a nearly infinite value of R_e. This is also shown in Fig. E-21. Select the measurements that you will make based on what you have measured in Chapters 3 through 5, and describe your results in the following space. (Record your data in Tables E-1 and E-2.)

With the transistor emitter-current stabilizer, the potentiometer should be adjusted to bring the base voltages on the two transistors to approximately reference-ground potential. The voltages listed in Tables E-1 and E-2 probably represent the minimum set to be measured. You can get the effective collector impedance on the input stage down to zero by bypassing the load resistance in that collector circuit to ground. Then, you can keep track of your dc balance during the test. You will need to compare the frequency response of the ampli-

Table E-1. Data for Experiment 4, Step 2

V_{cc}				
R_L				
V_{c1}				
V_{c2}				
V_e				
v_{o1}				
v_{o2}				
v_i				
v_e				
K_{v1}				
K_{v2}				

fier (with the collector of the first transistor bypassed) to the frequency response measured without the bypass. Using the value of R_e that is defined by Equation E-5 has the effect of reducing the overall gain by a factor of three times.

If a balanced signal is applied to the two inputs for the emitter-coupled amplifier (equal voltages but opposite signs), then the value of the emitter resistance is less critical. It functions to assure rejection of the common-mode

Table E-2. More Data for Step 2

V_{cc}				
R_L				
V_{c1}				
V_{c2}				
V_e				
v_{o1}				
v_{o2}				
v_i				
v_e				
K_{v1}				
K_{v2}				

signal, or a signal of the same polarity applied to both bases. It also linearizes the system. When only one input is used, as is normal with emitter-coupled amplifiers, then the circuit does perform a common-mode rejection function and, in so doing, reduces the effective voltage gain to less than half the balanced value.

Step 3.

A transistor may be substituted for the emitter resistor as noted previously, and its base-to-emitter voltage can be adjusted to provide just the required amount of current for the main emitter-coupled amplifier. Run a further series of tests on the circuit shown in Fig. E-21, using the transistor return circuit instead of R_e. Describe your results.

Your common-mode rejection becomes very high with the transistor current stabilizer in the emitter-return circuit. Signal balance will be very good if the collector currents in the two transistors of the main amplifier are balanced. (Note that a *constant voltage source to both bases* can help to assure this.) The available voltage gain will limit to one-half the theoretical gain for the basic undegenerated circuit as noted above. But, common-mode rejection will keep the common-mode gain to much, much less than unity.

Step 4.

Repeat the above tests using pnp transistors instead of npn transistors. Look carefully for any possible differences in behavior. Describe your results.

It is doubtful if you found any significant differences in the overall behavior of these two circuits beyond the replacement and the inversion of voltages. (All the transistors in the circuit, whether two or three, must be replaced by pnp devices.)

EXPERIMENT 5
The Emitter-Coupled Amplifier as an Oscillator

The emitter-coupled amplifier makes an ideal oscillator because it can provide good amplification with a minimum of losses from such degrading factors as the Miller effect, and it has an effectively zero phase shift from input to output (except for transit-time delay—the time required for the signal to pass through the devices). Since the overall loop feedback for oscillation with an oscillator requires a *voltage gain* that is *just over unity,* with the average value decreasing smoothly to unity as the amplitude builds up, it is not desirable for the circuit to be designed for a voltage gain of more than 5 or 10. The design should plan for the step-down and losses in the feedback-coupling circuit to reduce the effective loop gain to the required value in a stable manner. A variety of coupling circuits can be used, and each of the following Steps will consider a different circuit. For high-frequency operation, the use of npn devices is preferred. For this reason, the use of npn devices will be specified, although where operating frequencies are modest, pnp devices are equally effective.

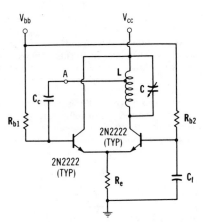

Fig. E-22. Circuit for an emitter-coupled LC oscillator.

Step 1.

The simplest oscillator uses a tuned circuit in the output collector, with a tap feeding back to the input base. The suggested circuit configuration is shown in Fig. E-22. Since it is a little hard to specify the optimum operating conditions without knowing what active devices and tuned circuit are being used, it is suggested that a variable-voltage power supply for controlling the two base currents be used, and that the collector voltages are limited to 5 volts. The base-source voltage can be increased until oscillation starts. The position of the tap on the coil can also be readjusted if necessary. Also, a tuned circuit with a higher LC ratio may be installed to provide a higher tuned impedance. You can also test the circuit with an oscillator by breaking the circuit at point A, and injecting the test signal into the input base, observing the output at the tap on the inductance. The tuned circuit may be checked with a dipmeter to find its resonant frequency, if it is above 2 MHz. Describe what you observed.

This circuit has been suggested for the first experiment because it makes the most effective use of the characteristics of the various components. You need stability, minimum loading, and a properly controlled variation of amplification for an effective oscillator. With this configuration, you minimize the amount of capacitance that will be reflected from the base onto the tuned circuit, and you also minimize the amount of load resistance reflected onto the frequency-determining circuit, thereby minimizing the effect on circuit Q. In addition, the effective source impedance presented to the input base is such that a voltage source is assured, the input transistor functions as a transconductance device, and you will get the kind of reduction of the average voltage amplification around the loop that is required as the oscillation amplitude increases. (The decoupling due to base-spreading resistance further assures this.) Test the circuit and record your comments on its operation.

Step 2.

A modification of this circuit is useful for those oscillators that use frequency selective crystals which have small series-resonant resistances. Under these conditions, a crystal is inserted between the emitters, as shown in Fig. E-23, and a tuned circuit capable of assuring operation in the proper mode is placed in the output collector circuit as is shown. You will need to be concerned with the input admittance of the second stage in this circuit because, if it is too high, it will hamper the development of an adequate coupling between the two emitters, particularly if the series resistance of the crystal is overly high. Wire this circuit on your solderless breadboard using a standard crystal, and arrange

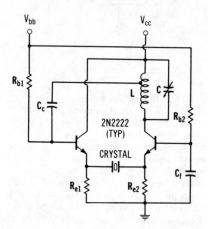

Fig. E-23. An emitter-coupled crystal oscillator.

the circuit so that the input admittance of the second stage can be varied through the variation of the supply voltage on the base of this stage. If your scope has a good enough frequency response, you may observe the circuit behavior as you vary the operating conditions. (The crystal operates in the series mode in this circuit.) Be sure to tune the collector circuit to maximum when you get this circuit to oscillating weakly. Then, vary the operating conditions for both transistors and report your observations.

If the crystal has too high a series resistance, you may not be able to get it to oscillate. In that case, it may be better to select N-channel field-effect transistors rather than bipolar transistors, particularly in the second stage. A higher tuned impedance in the collector circuit may also be helpful. (There can be a substantial coupling loss across the crystal in the emitter interstage circuit.) You can examine the behavior of the circuit by breaking the feedback coupling between the output collector and the input base as shown in Fig. E-22, and applying on the input a signal from a signal generator that is at the crystal frequency. Then, you can observe the signal on the output collector as you vary the oscillator frequency slowly around the suspected resonance. On either the emitter or the collector of the output stage you should be able to detect a response at the correct frequency.

If you find that the output peak voltage is less than the input, you need to change the LC ratio of the tuned circuit to give a higher impedance. Of course, loading from the scope could give you trouble unless you are using a very-low-capacitance probe. If you do not have one, the use of a few hundred ohms of resistance connected in series often will help. The input signal amplitude will drop, but the loading effects will also decrease materially. Once the loop gain becomes greater than unity, the circuit should oscillate unless there is a serious phasing problem. However, a limited amount of phasing problem can be compensated for by a slight retuning of the tuned circuit. Vary the inductance or the capacitance over a very small range and see what happens. Trim very carefully, however!

Step 3.

Magnetic coupling from the tuned circuit is a particularly convenient method of transferring energy to the input, but polarity is of course critical. The input coupling link should have a reactance that is roughly equal to the input impedance of the input base circuit. Once you have established a link of the appropriate inductance (see Experiment 8 in Chapter 4 for design information on inductors), you slide it toward the collector coil, feeding the input base from a signal generator, and observing the results from the link on the scope. The input signal should be used to deflect the horizontal amplifier if the frequency is low enough; otherwise, it is possible to heterodyne both the input and the output to a frequency that is low enough to observe. The experiment on mixers can be useful for developing heterodyne circuits. A diagonal line on the scope with a slope that places it in the first and third quadrants is what is required here. You adjust the coupling until the return voltage is slightly greater than the input voltage and, then, complete the circuit. It should oscillate. Describe what you have observed.

You can check the scope calibration by putting the input signal on both the horizontal and vertical inputs of the scope, and adjusting the amplifications for a 45° slope in the first and third quadrants. Both the horizontal and vertical deflections are then equal for equal signals. The scope is now calibrated. When adjustment of the link gives you more vertical deflection than you observed on calibration, the oscillator is ready to go.

Step 4.

Take one of the earlier tuned-amplifier circuits that you made for the discussion on rf amplifiers, and modify it for use as an oscillator using the emitter-coupled circuit. Set up your test procedures and align the circuit and try it out. Record your notes on how you did this in the following space.

Step 5.

Repeat at least one of the preceding steps using pnp transistors and, then another, based on N-channel FET devices. Make notes of what you found.

EXPERIMENT 6
The Emitter-Coupled Amplifier as a Mixer or Modulator

The emitter-coupled amplifier is ideally suited to function both as a mixer and, to a limited extent, as a modulator as well. For it to perform these functions, however, it is essential that the base of the output transistor be able to modulate the transconductance of the combination so that the voltage gain obtained on the signal that is introduced into the base of the input device will be varied linearly by the modulating signal.

Step 1.

First, it is necessary to find out the size of the resistor which must be used as an emitter return to optimize mixing. If it is too small, the coupling will be too small; on the other hand, if the resistor is too large, the operation of the two transistors will be excessively linearized. In the one case, some mixing will occur but it may be inefficient. In the other, the degeneration introduced can largely eliminate the mixing action.

For this test, you will need to operate at a low frequency (so that circuit strays are not important), but still at a high enough frequency so that the difference frequency can be less than a few percent of the test frequencies (yet still high enough to be useful with the scope). If you have a crystal-controlled source with a divider that can provide around a 100-kHz output, you can use it for one of your sources. An audio oscillator set between 96 and 99 kHz will be fine for the other source. You will need to be able to vary the amplitudes of both of these frequencies.

Step 2.

Set the currents in the oscillator circuits for both transistors at 1 milliampere without signal drive by adjusting the base resistors and/or the base-return voltage. The collectors have been returned to a 5-volt supply to make the variable supply available for that purpose, but any voltage source between 1.5 and 5 volts should be satisfactory (see Fig. E-24). Set up the circuit on your solderless breadboard in accord with the configuration of Fig. E-24. The 1000-ohm resistor (R1) is bypassed by both a ceramic 0.1-μF capacitor and a 10-μF electrolytic capacitor to eliminate all of the output signal on the collector of the input transistor. The output collector is bypassed with a capacitor of adequate size to eliminate primary frequency components. (The circuit shows a tuned configuration, but for this test, it should be replaced by a 1000-ohm resistor and a capacitance of about 0.02 μF.) Then, only the difference-frequency components will appear at the output. Adjust both of the input signals to approximately 100 millivolts for the initial trial, and then, vary the emitter-return resistance. Start at 100 ohms, and decrease the resistance step-by-step, to 47, 22, 10, 4.7 ohms, etc. Select the value of resistance that

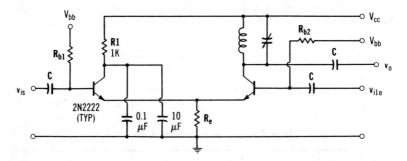

Fig. E-24. A basic modulator test circuit.

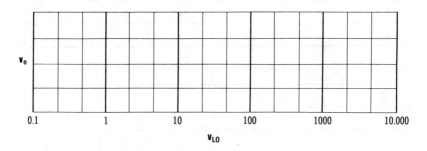

Fig. E-25. First graph for Experiment 6, Step 2.

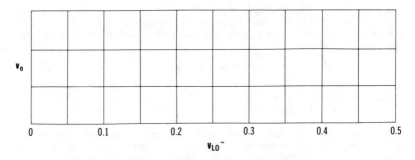

Fig. E-26. Second graph for Experiment 6, Step 2.

appears to you to give the most effective mixing action. When you reach that value, vary the magnitude of the voltage that you have applied to the base of the output transistor, recording the amplitude of the difference-frequency output voltage as a function of this input voltage. Plot your results on the graph diagram shown in Fig. E-25. Then, again vary the value of the emitter resistance to find its optimum value, and plot this on the graph in Fig. E-26. Repeat the test to see if there is any significant interdependence. Record your data as you go in Tables E-3 and E-4. (Value v_o is the voltage output at the difference frequency.) Discuss your results.

Table E-3. Data for Experiment 6, Step 2

R_e				
v_o				
R_e				
v_o				

Table E-4. More Data for Experiment 6, Step 2

v_{LO}				
v_o				
v_{LO}				
v_o				

As you increase the amplitude of the mixer LO voltage on base 2, you are increasing the change in transconductance in your output device. For square-law devices, the conversion transconductance can be defined in terms of the approximate equation.

$$g_c = 0.25(g_{m1} - g_{m2}) \qquad \text{(Eq. E-6)}$$

where a linear variation of transconductance with second-base instantaneous voltage is assumed. This assumes also that the mixing signal is sinusoidal rather than square wave. Up to a point, the conversion conductance will increase, and then it may level or decrease in practical situations.

Since you may be using a square-wave mixing voltage, and your device is likely operating on an exponential relation, you cannot expect to find this equation to be more than an approximation. You may find it interesting to plot a curve of the coefficient of Equation E-6 as a function of the ratio of the peak-to-peak value of v_{LO} to 0.026 millivolt. Remember that the value of g_m is approximately ΛI_c. From these data, you can see if you can get the value to use for the multiplier instead of the 0.25 value the author has given you. (For small values of LO signal, square-law conditions do apply.)

Step 3.

The next test is to evaluate the circuit as a modulator. In this instance, the input to the first base is the modulation, and the input to the second is the carrier, which may be either a sine wave or a square wave.

Why do we use this configuration? With a modulator, it is important that the message signal be applied to the carrier in as linear a manner as possible. In order to gain linearity, emitter degeneration is a must. In fact, it must be sufficient enough that over the full range of amplitude of the desired signal, it must be linear. This may mean that you will need to change the emitter-return resistor. (This is almost a certainty!) As before, you apply a 100-kHz carrier-frequency signal on base 2. You should proceed as you did in Step 2, first varying the emitter resistance to observe the conversion efficiency and linearity. *Do not forget to remove the bypass across the output collector load resistance.* You may replace the RC circuit with an LC circuit tuned to the output frequency. You will need a small coupling capacitor into your scope, possibly about 100 pF, to remove the dc and low-frequency shift components. As you vary the emitter resistance, change your carrier drive on the second base to find out how effective your circuit is overall. You are looking for a perfect modulation envelope on the scope, as anything less is a distorted waveform. (A triangle wave could be used as your signal source, and the envelope then would be that same triangle wave. You might find it easier to recognize imperfections in that wave than with a sine wave.) Describe what you have found.

You probably discovered that you needed to use a much larger emitter resistance than that used with the mixer, and you probably also had difficulties getting a good waveform out of your modulator. If you have a triangle waveform available, you probably found it was helpful in examining the operating characteristics of your modulator. Modulation is not a "small-signal" operation, as the linearization requires increased voltages and increased emitter-return resistance.

EXPERIMENT 7
The Cascode Transistor Amplifier

A very popular form of amplifier for use with high-sensitivity receivers is the cascode amplifier. It is similar in many ways to the emitter-coupled amplifier in its overall properties, but it differs in one important feature. Instead of connecting the two emitters together to achieve parallel injection into the output transistor, the transistors are actually connected in series with the collector of the input transistor connected to the emitter of the output transistor. The emitter of the input transistor is grounded, as shown in Fig. E-27.

The principal advantage of this configuration is that the full transconductance per unit current of the input stage is available to the output stage, helping to minimize the internal noise of the combination. At the same time, the input impedance of the first stage is lower than with the emitter-coupled combination, causing a loss of at least part of the advantage otherwise to be expected. The emitter for the second stage provides the load for the first stage, leading to a voltage gain of unity if both devices are passing the same level of current (the normal situation). Hence, the input stage acts as an impedance transformer with a voltage "gain" of unity, thereby minimizing both the probability of oscillation and, also, any problems from Miller capacitance effects.

This circuit is convenient for the input stage of counters if the following stage is properly coupled to it and the supply voltages are properly chosen.

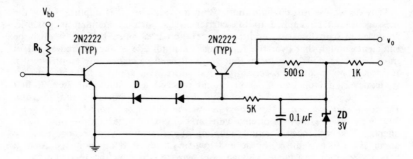

Fig. E-27. Cascode-counter drive circuit.

A few tens of millivolts normally will switch the input transistor on and off and will switch the second transistor in the same manner. In addition, it will create the voltage change required to actuate a Schmitt trigger or a similar circuit. Since the output transistor will be either turned on or turned off, higher values of input signal on the input transistor will have little effect.

Step 1.

Wire the cascode circuit shown in Fig. E-27 on your solderless breadboard. You will need to be able to introduce a variable-frequency, variable-amplitude test signal into the input base, and observe the output across the output collector as you vary the collector supply voltage. The circuit values suggested should lead to approximately 3 milliamperes current in the output transistor when it is turned on, and should produce an overall voltage change of about 1.8 volts. This should be enough of a voltage change to switch a Schmitt trigger unit like the one contained in a 7413 IC (which can then drive a counter chain). Try the circuit, and note your comments in the following space. Vary the collector voltage for the cascode circuit and see if you can find a better voltage value. Also, vary the currents to the two bases to see if you can improve the circuit operation and record your results.

You will find that the collector supply voltage is critical if you want to optimize both the frequency response and sensitivity. The circuit will work with higher collector supply voltages if the cascode load is increased, but this decreases the peak operating frequency. Further, if the level of base current is wrong, then too much input signal will be required to make the circuit switch. When properly adjusted, it works quite well, however. For maximum response, use either a 74LS13 or 74LS14 Schmitt trigger. (Each circuit in these devices functions as a NAND gate or inverter, but because of the Schmitt action, it has different input threshold levels for positive- and negative-going signals.)

Step 2.

It is interesting to see how this amplifier behaves for linear signal processing. For this test, use either an audio frequency or a low radio frequency—one that

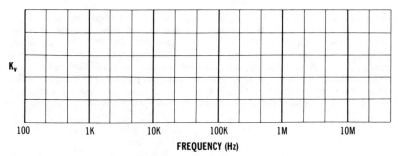

Fig. E-28. Graph for Experiment 7, Step 2.

you can observe on the scope. Use the same basic circuit, but disconnect the "integrated circuit" part of it and couple the cascode output directly into the scope. You may find it helpful to remove the two bias diodes on the base of the second transistor and increase the base-bias resistors so that there is approximately 1 volt at the collector of the output transistor. Then, measure the gain of the circuit as a function of frequency. Record the sample data in Table E-5, and plot a curve of K_v versus frequency on the graph given in Fig. E-28.

Table E-5. Data for Experiment 7, Step 2

f				
K_v				
f				
K_v				

EXPERIMENT 8
The Use of Darlington Compound Circuits

There are several important functions that can be accomplished more effectively with Darlington circuits. Perhaps the best known of these is the transistorized version of the split-load phase inverter. An ordinary transistorized version of this is not suitable because of the unbalance resulting from base current. The amplifications to the emitter and to the collector, as a result, cannot be equal with the single-transistor phase-splitting circuit. Since the Darlington circuit may have a current gain of as much as 5,000 to 20,000, however, the unbalance when this circuit is used becomes negligible. The two outputs may be followed by emitter followers, a pnp one for the collector, and an npn for the emitter. This will provide matched low-impedance sources for a push-pull final amplifier circuit. This circuit is also useful for phase-shift networks, including phase-shift oscillators, and for specialized filter circuits like those used for single-sideband generators. You will probably use analog integrated circuits for many of these functions but, nonetheless, it may be useful to know a little about equivalent discrete circuits.

Step 1.

A phase splitter based on a Darlington amplifier is a very convenient circuit, as you will find. You may use an integrated Darlington amplifier, or you may make one from a pair of 2N2222 transistors or other transistor pairs if you prefer. In either case, it is convenient to test one of these combinations just to learn how they behave. You will become more familiar with their characteristics and their identification by so doing (they are not always marked adequately). To start, therefore, try this combination in your TeeDeeTester, or its equivalent, and also test it in your sweeper. In particular, notice the differences in the input characteristics. Test at all four leads, so that you can recognize the characteristics of four-lead devices. Note what you have learned.

When you have this combination properly connected, you will have two diode drops between the output emitter and the input base, and one between the input base and the collector. If you have a fourth lead, you will have one diode drop between it and the emitter and the collector. Otherwise, Darlington transistors behave much like an ordinary transistor, except that they may appear to have less gain on a current-gain test with a TeeDeeTester, as the forward bias is marginal because of the number of diodes to be overcome.

Step 2.

When you get the leads connected correctly, you will have a very high current gain potentially available, and the input current to the base will be at most a thousandth of the output current (with a properly adjusted good unit). Mount a Darlington transistor circuit on your solderless breadboard, and adjust for a few milliamperes of output current. A collector supply voltage of 5 volts (with a collector load resistance of about 400 ohms) should be satisfactory. Then, measure the small-signal current gain. Next, open the input base lead. How much current now flows in the collector circuit? What is the minimum collector current at which you can get a reasonable current gain? (Make the measurement.) What limitations does this put on your use of these devices?

You will have observed high gain *only* if the output current dropped significantly when you discontinued the bias from the input base. Unless you drain off the leakage current in the input unit to the emitter return through a resistance, you may not be able to turn this combination off. It is very important to recognize this. Once you command more than the minimum current through the device, however, you will be able to get good results.

Step 3.

Once you have verified the general properties of the Darlington componud circuit, you are ready to build your first application circuit. The first circuit is the phase-splitter circuit (Fig. E-29). Starting with a 10-volt power supply, connect a 1000-ohm resistor between the collector and the positive supply (for an npn Darlington), and a 1000-ohm resistor between the output emitter and the supply return. Then, connect a resistor from the input base to the collector supply of such a size as to give a current of about 2 milliamperes in the resistors. (If your Darlington transistor demands more current to operate properly, redesign the circuit to assure the proper level of current for a high current gain.) You will need a 1- to 2-volt voltage drop across each resistor. You now have a phase splitter, and you will find that the signal voltage developed across each resistor is (depending on the matching precision of your resistors) equal in amplitude. A Lissajous figure test will show that they are

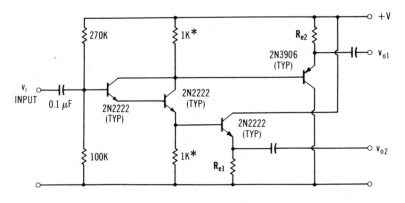

NOTE: *MATCHED RESISTORS

Fig. E-29. A Darlington phase-splitter network.

out-of-phase. Vary the amplitude of your input test signal (a frequency of 1000 Hz will be fine), and observe the amplitude at which the unit loses its linearity. (It may differ on the emitter and the collector, so be careful!) Record your results.

You probably will find that the maximum signal that you can handle without some distortion will have a peak-to-peak value of a little over one-half the static voltage across the resistors. Again, it is easier to detect distortion if you use a triangular shaped waveform.

Step 4.

This circuit may require a slight readjustment so that you can add the emitter followers after the respective outputs. For this reason, increase your device overall current so that not more than one-third of the supply voltage is across the transistor. You do this by increasing the input base current. (You may want to increase V_{cc} if you have been using 5 volts.) Then, you can connect the pnp emitter follower to the collector circuit as shown in Fig. E-29 (again assuming that Q1 is an npn Darlington), and connect an npn emitter follower to the emitter circuit. If you wish, you can now connect the output to the input of a push-pull power amplifier. You will need to use capacitive coupling into the power amplifier so that you can control the bias on the power stages independently. When you get the circuit working, vary the parameters at your command, the input current to the Darlington, the signal level, etc., to see what you get, and record your results.

Step 5.

You may want to use your Darlington configuration differently. It makes an ideal universal phase-shifter. Take the circuit for either Step 3 or Step 4, and connect a 0.01 μF capacitor to the collector-side output, and use a 500,000-ohm audio potentiometer as a variable resistance, as shown in Fig. E-30. Connect your sine-wave oscillator to the input of the Darlington, and connect a scope to the emitter output. The output tap between the capacitor and the variable resistance is then connected to the vertical input of the scope. To start, set the resistance value at about 100,000 ohms, and adjust the frequency of the oscillator so that you get a circular pattern on the scope (adjustment of the horizontal and/or vertical gains may be required). Read the frequency. Then, rotate the shaft of the potentiometer to change the resistance. You will notice that the peak vertical deflection remains essentially constant, but that the circle pattern changes into an ellipse, in one direction for lower frequencies and, in the other direction for higher frequencies. In the limit, it will become a

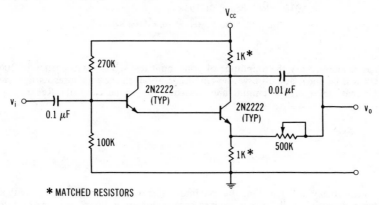

Fig. E-30. A Darlington phase-shifter configuration.

diagonal line at 45° in one direction or the other. Vary your frequency, and measure the resistance value that reestablishes the circle. Also, change capacitors and repeat the test. Record your data in Table E-6. Do the values fit the following equation?

$$2\pi fCR = 1 \qquad \text{(Eq. E-7)}$$

The values should satisfy this equation within component tolerances. Can you use this arrangement as a means of measuring capacitance? For measuring frequency? Explain.

As long as you know two of the three parameters in the equation adequately, you can easily measure the third. Precision resistors should be used in place of the 500,000-ohm potentiometer for this kind of measurement.

Table E-6. Data for Experiment 8, Step 5

f				
C				
R				
f				
C				
R				

Step 6.

The Darlington configuration can be used to make an excellent "ladder RC," or phase-shift, oscillator. (It is best with this configuration to have separate collector leads on the transistors.) For this application, you can decrease the common-emitter resistance to about 30 ohms—just enough to give a reasonable amount of degeneration. You also need phase inversion with these oscillators, a current gain of at least 1000, and a voltage gain of a magnitude of 35 or more, all of which you can get easily.

Two ladder configurations are commonly used with these circuits, like the configurations shown in Fig. E-31. One is based on a "high-pass" RC configuration using series capacitors and shunt resistors, and the other on a low-pass configuration using series resistors and shunt capacitors. This latter form has significant advantages in that it tends to be freer of harmonics, and it can be used with a three-gang variable tuning capacitor. You will find it interesting to try both configurations. (Needless to say, the Darlington device can be replaced by an op amp.) Since the operating frequency of the two configurations is different, you will want to use the same capacitors and resistors in first one configuration and then the other. Your solderless breadboard is ideal for making up the circuit.

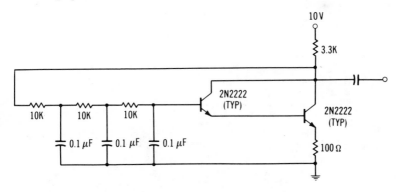

Fig. E-31. A Darlington ladder oscillator.

Each RC section is supposed to give you 60° of phase shift with this circuit. This is not strictly true since each successive stage loads the previous one. Nonetheless, the basic frequency equation takes the form:

$$2\pi fRC = \gamma \qquad\qquad \text{(Eq. E-8)}$$

where,

γ (or Gamma) is the value to be determined.

You should attempt to estimate what the value is after you get the circuit to working. You can break the circuit at the Darlington input and use a scope to check the voltage gain. You can check the phase shift using a Lissajous pattern. Introduce the oscillator signal into the input base of the active configuration, and put that signal also into the horizontal input of your scope. The output of the frequency-selection ladder network is applied on the vertical plates. Assume that Gamma will lie within the range of 0.25 and 4.0, and select your test frequency in that way. Adjust the test oscillator until the scope presentation is a diagonal line in Quadrants 1 and 3, and be sure the gain exceeds unity. Then, close the circuit and see how close the output frequency is to the test frequency! Describe and explain what you find.

You will find that with the low-pass ladder configuration, the value of Gamma is approximately 2.45, whereas, with the high-pass ladder network, it is 0.408. The value of amplification required of the amplifier for oscillation to start is −29 or more. The *effective current amplification,* however, should exceed 1000, as the current gain *must* be very large compared to 29. The input to the amplifier for that reason must not load the output of the ladder network.

The Characterization of Active Devices

You are no doubt aware that many active devices are not characterized in a way that is helpful to us as the users. As you have also seen, there are ways to get around at least some of the listed limitations but this is not possible with all of them. This discussion is not designed to be either extensive or detailed, but will point up some of the deficiencies that exist in order to help you to avoid the problems that they possibly can cause.

BIPOLAR TRANSISTORS

The discussion in this book has perhaps shown you rather clearly that the main aspect of importance with respect to the beta parameter is the minimum value that it is likely to have. Since the transconductance efficiency for bipolar devices under low injection conditions is approximately unity (but not under high injection), some data on values and turning points is of importance to users. Some indication of the range of values for the spreading resistances, emitter, base, collector, is also vital to the user, but apparently only values of base-spreading resistance are quoted in data sheets, and then only for special devices. This is why it is important to find out what these parameters do in a circuit.

As noted previously, the two limit frequencies are important to users. The first of these is the frequency below which that noise in excess of thermal noise is encountered. The point at which this actually becomes significant for a given device can indicate something about the probability of the device being prone to an early failure. (For example, surface barrier transistors had very high f_{n1} values.) The second limit frequency can be estimated from existing data. This frequency is probably the more stable of the noise parameters. It is caused by the granularity of the current flow, and it is a function of the actual transit time of individual minority carriers across the base region. Above this frequency, there is a degradation of the effective noise figure for the device. The probability of a multivibrator functioning at a rate above this frequency is rather small also. It is one of the fundamental parameters for a device. Its value can be estimated by dividing the maximum operating frequency by the square root of the beta. However, because of the poor information that is available on the beta and maximum frequency, the resulting estimate may be off by up to one-half an order-of-magnitude when determined this way.

FIELD-EFFECT TRANSISTORS

FET devices are transconductance control devices, but they are rated as having a range of possible transconductance values *and* a range of drain current values. The fact that these values are interdependent is neglected. As with bipolar devices, the transconductance-per-unit drain current is relatively stable, and can be used to advantage both by circuit designers and circuit repair craftsmen. (Remember that it is not unity with FETs, however.) Under normal conditions, the transconductance-per-unit-current efficiency of these devices is more nearly 0.01, with the result that they have significantly more inherent power-handling capability than a bipolar device that is operated in the same way. Either a plot of typical transconductance as a function of ouput current or a plot of the efficiency factor as a function of this current would be of immeasurable help to the user of the device.

While the low transconductance efficiency in FET devices has been attributed to source-spreading resistance, it appears that the phenomenon is more like the phenomenon encountered in electron tubes, with the limitation being primarily a consequence of the uncontrolled flow of carriers in the center of the channel and, secondarily, a consequence of a space-charge region behind the gate. This combination will act as an equivalent resistance in many respects, but not as a *fixed* resistance. Presence of a space-charge type of phenomenon like that encountered with vacuum tubes is the only explanation that can account for the excellent noise-figure properties in these devices. The noise figures that are encountered in the very-high-frequency part of the radio spectrum are much lower than might be expected otherwise. The spreading resistance explanation is also rendered doubtful by the properties of VMOS devices, since the internal structure of this configuration tends to maximize the effectiveness of the diffusion mode of operation and leads to improved power FET devices.

For many years the application of the insulated-gate field-effect device was held back by charge-trapping problems in the insulating layer between the channel and the gate electrode. This trapping led to slow drifts in the channel conductivity and, as a result, drifts in the operating characteristics. An associated problem must still be guarded against, however, That is the "zapping" of the insulator layer with static electricity. Protective clamping diodes are often built into the device to help protect against this static electricity problem, but the problem can still be encountered. In short, *these devices are very delicate* and must be handled with care.

The peak operating frequency of a FET device is a very important parameter for users. It is a function of the channel length between a space-charge region behind the gate and the drain region. For practical purposes, this region ends where the gate ends, or where a relatively high conductivity drain region starts. The best high-frequency FET devices today are built as "vertical" devices, and can look something like planar transistors with grooves cut in them. With these devices, there will be an npn or a pnp structure, and the channel will be established either on the edge of the center region by the gate field, or it will be established by growing a lightly doped oxide of the same polarity as the source and the drain. The channel then develops against a "base" region, and it can be exceedingly short. The result will be an extremely high-frequency device. The vertical device has much better high-frequency characteristics than the lateral device, and it is actually easier to make.

OTHER DEVICES

With other kinds of active devices, including tunnel diodes and all kinds of electron tubes, it is critical that small-signal parameters that are independent of small differences be chosen, and that they be characterized in terms that represent a minimum dependence on the small differences. With electron tubes, for example, these parameters are transconductances. They should be presented in terms of the output current, just as with the bipolar and the field-effect transistors. It is vital that this relation be known in terms of the proper voltages as well as the currents. With triode tubes, as well as with transistors, the proper voltage is the output voltage. With tetrode and pentode tubes, the proper voltage is the screen voltage. In this last case, the plate voltage should exceed the screen voltage at the point of maximum current through the tube.

It can be anticipated that all kinds of solid-state two-port devices will follow this same basic set of rules like transistors and tubes, and you can safely use them in configuring a circuit for their use. In addition, the transconductance-per-unit-current efficiency defines either the output voltage, or in the case of tetrode and pentode tubes, the auxiliary control voltage applied to the screen. In each case, the critical voltage should be selected as small as possible as long as it complies with both the required operating conditions and the moderate power dissipation. If it cannot comply, then a different device is required. The reciprocal of the transconductance efficiency gives a good starting point for either establishing the required control-signal voltage or the required output-supply voltage.

PARAMETERS FOR SWITCHING DEVICES

With these devices, the switching points and the voltage and current maintenance points are the points that are critical. Typical curves are sometimes adequate with SCR devices, particularly if they are supplemented with detail information that indicates their ability to withstand high switching rates. With a device like the SCS, which has two control leads, there will be a switching condition for use with one gate with the other gate disabled and, also, switching conditions for use with both gates. Often the data are rather skimpy for many of these kinds of applications, beyond the fact that they can be used for specialty applications. With any of these devices, if you are planning on making a critical circuit, it is probably wise to run tests using a rather large sample of the chosen device to be sure that proper operation is possible. In addition, the manufacturer of a critical device should be consulted to ensure that you are not exceeding some characteristic that the manufacturer does not plan to maintain. The specifications data sheets frequently are insufficient to assure the desired operation otherwise.

Index

VMOS FET, 251, 257
Voltage
  common-mode, 51
  doubler rectifiers, 47-48
  gain and dissipation, 159-162
  gain, transistor, 86-87
  peak inverse, 386
  references, 327-330
  regulator, uses of, 402-404
Voltmeter, 374
  digital, 451

Volt-ohm-milliammeter, 373, 456

W

Waveform sources, 381
White noise, 149

Z

Zener diode, 24
Zener/trigger diode rectifier test set,
  449-451

READER SERVICE CARD

To better serve you, the reader, please take a moment to fill out this card, or a copy of it, for us. Not only will you be kept up to date on the Blacksburg Series books, but as an extra bonus, **we will randomly select five cards every month, from all of the cards sent to us during the previous month. The names that are drawn will win, absolutely free, a book from the Blacksburg Continuing Education Series.** Therefore, make sure to indicate your choice in the space provided below. For a complete listing of all the books to choose from, refer to the inside front cover of this book. Please, one card per person. Give everyone a chance.

In order to find out who has won a book in your area, call (703) 953-1861 anytime during the night or weekend. When you do call, an answering machine will let you know the monthly winners. Too good to be true? Just give us a call. Good luck.

If I win, please send me a copy of:

_____

I understand that this book will be sent to me absolutely free, if my card is selected.

For our information, how about telling us a little about yourself. We are interested in your occupation, how and where you normally purchase books and the books that you would like to see in the Blacksburg Series. We are also interested in finding authors for the series, so if you have a book idea, write to The Blacksburg Group, Inc., P.O. Box 242, Blacksburg, VA 24060 and ask for an Author Packet. We are also interested in TRS-80, APPLE, OSI and PET BASIC programs.

My occupation is _____
I buy books through/from _____
Would you buy books through the mail? _____
I'd like to see a book about _____
Name _____
Address _____
City _____
State _____ Zip _____

MAIL TO: BOOKS, BOX 715, BLACKSBURG, VA 24060
!!!!!PLEASE PRINT!!!!!